A History of Women in Psychology and Neuroscience

Titles currently available in the
Trailblazers of STEM series by Dale DeBakcsy:

A History of Women in Medicine and Medical Research

A History of Women in Astronomy and Space Exploration

A History of Women in Mathematics

A History of Women in Psychology and Neuroscience

A History of Women in Psychology and Neuroscience

Exploring the Trailblazers of STEM

Dale DeBakcsy

First published in Great Britain in 2024 by
Pen & Sword History
An imprint of Pen & Sword Books Limited
Yorkshire – Philadelphia

Copyright © Dale DeBakcsy 2024

ISBN 978 1 39903 235 3

The right of Dale DeBakcsy to be identified as
Author of this Work has been asserted by him in accordance
with the Copyright, Designs and Patents Act 1988.

A CIP catalogue record for this book is
available from the British Library

All rights reserved. No part of this book may be reproduced or
transmitted in any form or by any means, electronic or mechanical
including photocopying, recording or by any information storage and
retrieval system, without permission from the Publisher in writing.

Typeset by Mac Style
Printed in the UK by CPI Group (UK) Ltd, Croydon, CR0 4YY.

Pen & Sword Books Limited incorporates the imprints of After
the Battle, Atlas, Archaeology, Aviation, Discovery, Family History,
Fiction, History, Maritime, Military, Military Classics, Politics,
Select, Transport, True Crime, Air World, Frontline Publishing, Leo
Cooper, Remember When, Seaforth Publishing, The Praetorian Press,
Wharncliffe Local History, Wharncliffe Transport, Wharncliffe True
Crime and White Owl.

For a complete list of Pen & Sword titles please contact:

PEN & SWORD BOOKS LIMITED
47 Church Street, Barnsley, South Yorkshire, S70 2AS, England
E-mail: enquiries@pen-and-sword.co.uk
Website: www.pen-and-sword.co.uk
or
PEN AND SWORD BOOKS
1950 Lawrence Road, Havertown, PA 19083, USA
E-mail: uspen-and-sword@casematepublishers.com
Website: www.penandswordbooks.com

For Uncle Paul,
Who inspired everyone around
him with a passion to write.

Contents

Introduction		x
Chapter 1	Christine Ladd-Franklin and the Colour Wars of the Late Nineteenth Century	1
Chapter 2	The Woman of a Thousand Brains: Augusta Déjerine-Klumpke, Pioneer Neuroscientist	5
Chapter 3	Self-Remembrance: Mary Whiton Calkins's Adventures Among the Atomists	10
Chapter 4	The Two Montessoris: The Rise, Fall and Rebirth of an Educational Revolution	15
Chapter 5	Margaret Floy Washburn and the Motion of Thought	19
Chapter 6	Generations: The Neuroscience Dynasty of Cécile and Marthe Vogt	22
Chapter 7	Brief Portraits: The Nineteenth Century	28
Chapter 8	Workers as Humans: Lillian Moller Gilbreth and the Founding of Industrial Psychology	49
Chapter 9	Filled With People: The Teeming Mental Spaces of Melanie Klein	53
Chapter 10	Helene Deutsch, As-If Personalities, Adolescent Friendship and the Art of the Quiet Revolution	58
Chapter 11	Speaking Culture to Psychoanalysis: Karen Horney's Gender Revolution	64
Chapter 12	Bringing Science to Psychoanalysis: The Many Survivals of Sabina Spielrein	69

Chapter 13	Tsuruko Haraguchi: The Strenuous Road to Becoming Japan's First Woman Doctor of Psychology	73
Chapter 14	Of Gifted Children and the Banality of Menstruation: The Educational Psychology of Leta Hollingworth	77
Chapter 15	Come Together? Inez Prosser and the Psychological Impact of Mixed Schooling Systems	81
Chapter 16	Children are People: The Life and Science of Anna Freud	84
Chapter 17	Neuroembryology in Wartime: Rita Levi-Montalcini and the Discovery of Nerve Growth Factor	88
Chapter 18	Mary Ainsworth, Infant Anxiety and the Case of the 'Strange Situation'	91
Chapter 19	Separate: Mamie Phipps Clark and the Psychology of American Segregation	95
Chapter 20	Virginia Satir and the Art of Family Communication	99
Chapter 21	More than the Sum of their Parts: Eleanor Maccoby's Studies of Children in Groups and the Adults they Become	103
Chapter 22	Building a Kingdom in the Brain's Unfashionable District: Brenda Milner's Century of Neuropsychology	108
Chapter 23	Edith Graef McGeer, the Great Neurotransmitter Race and a Glimpse towards the End of Alzheimer's	114
Chapter 24	Virginia Johnson and the Development of Effective Sex Therapy	118
Chapter 25	Beyond Nature vs. Nurture: The Enriched Heredity of Marian Cleeves Diamond and the Evolutionary Psychology of Leda Cosmides	123
Chapter 26	Brief Portraits: The Early Twentieth Century	128
Chapter 27	Signs: Ursula Bellugi and the Neuroscience of Language	231
Chapter 28	When Memory Has Gone: Suzanne Corkin's Journeys through the Hippocampus	235

Chapter 29	Filling in the Gaps: Naomi Weisstein's Active Brains and Activist Life	238
Chapter 30	Making Working Memory Work: The Multidisciplinary Neuroscience of Patricia Goldman-Rakic	242
Chapter 31	Life on the Grid: Nobel Laureate May-Britt Moser and the Fine Art of Knowing Where You Are	246
Chapter 32	Dr. Tania Singer and the Neuroscience of Empathy	249
Chapter 33	Dealing: Dr. Iris Mauss and the Science of Emotion Regulation	252
Chapter 34	Brains In Love and Brains Alone: The Social Neuroscience of Stephanie Cacioppo	255
Chapter 35	Achtung, Brainy: Grace Lindsay and the Mathematical Modeling of the Human Brain	258
Chapter 36	Brief Portraits: The Modern Age	261

So... What Does That Mean?: A Glossary of Frequently Used But Kind of Weird Terms   294
Selected Bibliography   300
Index   304

Introduction

In her classic 1933 book, *Seven Psychologies*, Edna Heidbreder expressed the excitement and tension in the burgeoning field of psychology by noting, 'One can hardly cross the threshold of the lively young science without suspecting that all is not peace and harmony under its roof-tree; that the bands of workers one encounters there represent not only a division of labor but a state of internal strife.' And so it was – at that moment in the history of psychology, experimentalists, functionalists, behaviourists, structuralists, Freudians and Gestalt psychologists represented so many warring camps, attempting to establish themselves not as specialists each uncovering complementary pieces of a much larger puzzle, but as the true monarchs of the psychological kingdom.

The newness of the science, the relative paucity of the existent data as against the ambitions of the field to explain something as complicated as human behaviour and cognition, bred an early tribalism born of defensiveness in the face of the mockery from more established branches of scientific endeavour, resulting in a proliferation of approaches to the brain and its workings arising from struggle, and carried out with often evangelical fervor. One of the early challenges in writing this book was making the decision whether to include *all* women who contributed to the development of psychology and neuroscience, even those who belonged to branches of study that eventually petered out or are regarded as non-scientific today, or to focus only on 'the winners', those who either by luck or instinct associated themselves with approaches to psychology that we, a century-and-a-half later, approve of.

I have elected for the former – part of the thrill of scientific history for me is the complicated tale of how we got from there to here, including all of the initially promising but ultimately futile lines of investigation that gave us useful information even in failure. This, then, is the story of women in psychology in all of its complex messiness, containing both the tales of overwhelming triumph – the discovery of the physical underpinnings of working memory by Patricia Goldman-Rakic, the mapping of the brain's functional regions by Augusta Déjerine-Klumpke, the discovery of the grid cells responsible for our sense of where we are in a room by May-Britt Moser, the mapping of the human sexual response by Virginia Johnson – as well as its dead ends, controversial outgrowths, and moments of overzealous faddishness – the 'decade of testing' in the 1920s, the vogue for Psi research in the 1930s, the flirtations of psychologists with eugenics in the 1910s, and, always and forever, the controversy surrounding the theories of psychoanalysis (which I encourage you to approach with an open but critical mind in the pages to come).

Psychology is one of the younger sciences we'll be talking about in this series, dating informally back to the Enlightenment, when thinkers like Julien Offray de La Mettrie (1709–1751), Luigi Galvani (1737–1798), and Alexander von Humboldt (1769–1859) began positing purely mechanical and anatomical explanations for behaviour, reflex, memory, and development. While these explanations had predecessors in antiquity, including Epicurus and Lucretius, and had noble if isolated successors in the persons of Pietro Pomponazzi (1462–1525), who proposed getting rid of the soul altogether as a useful explanation for why we act as we do, Pierre Gassendi (1592–1655), who held that our perceptions and ideas are the result of the motion of atoms within the brain, and Giovanni Borelli (1608–1679) who performed a series of brilliant if stomach-churning experiments that established that nerves and muscles must communicate through some form of non-gaseous, perhaps liquid chemical medium, it took the more widespread eighteenth century decline of classically-filtered religion as our monolithic source for ideas about the nature of humanity to bring such ideas into their own by removing the intellectual restrictions and spiritual assumptions that had hemmed in the efforts of even the most determined neural researches of the seventeenth century, such as Thomas Willis (1621–1675), François Pourfour du Petit (1644–1741), Georg Ernst Stahl (1659–1734), and Herman Boerhaave (1668–1738), who by and large limited themselves to the study of the physical mechanisms underlying the coordination of muscle movement, blood flow, tissue responsiveness, and organ function, without venturing much further than classical philosophical notions in expressing the religiously controversial connections between physiology, chemistry, and personality or behaviour.

Psychology as a professional, university-based area of study formally separate from philosophy had to wait until the mid to late nineteenth century for its formal beginning, which is usually located either with the arrival of Jean-Martin Charcot and Alfred Vulpian at the Salpêtrière in 1862 and the subsequent drive to relate unusual behaviours and physiologies to conditions of the nervous system, or with the founding by Wilhelm Wundt (whom we shall encounter many times over the course of this book) of the Institute for Experimental Psychology in Leipzig in 1879. Wundt's introspective methods, though more scientific than the philosophy-infused attempts of his predecessors, would hardly pass muster today, but represented an important change of orientation for psychology, away from theory-down philosophical analysis, and towards a methodology that broke the mind and sensation down into their smallest components to be analysed from the ground up.

Women were early entrants into the study of psychology and neuroscience, with the 'official' start date of women in psychology being often given as Christine Ladd-Franklin's 1887 publication of a paper on horopters (which we'll talk more about later), though with less of an Anglophone bias I think we should probably push that date back either to 1885 and Augusta Déjerine-Klumpke's published description of the condition that now bears her name, Klumpke's Palsy, or Susanna Rubinstein's 1878 publication of a book of essays on the relation between psychology and aesthetics. Wherever you put its beginnings, by the 1890s women were studying psychology in significant numbers. Mary Whiton Calkins, after studying for some time with psychological pioneer William James, began teaching psychology at Wellesley in 1891. In 1893, Ladd-Franklin became the first woman inducted into the American Psychological Association. Milicent Shinn released her groundbreaking dissertation in child psychology in 1898, while two years earlier a promising doctor by the name of Maria Montessori began her observations of phrenasthenic children. The decade closed on another milestone as, in 1899, Lillien Jane Martin started teaching psychology at Stanford University.

In spite of this prominent First Wave of women psychology researchers and lecturers, and in particular the prominent gender inclusivity of early psychoanalysis, by the 1960s women were still earning psychology PhDs with something on the order of half the frequency of men. Though these women included some titans in the field of psychological research, it was not until the 1980s that parity began to be achieved, with a vengeance. Not only did women achieve equal representation with men in the late twentieth century, but by the early twenty-first century they came to dominate the academic scene. In 2021, the American Psychological Association reported women made up 76 per cent of psychologists under 30, 81 per cent of those between 31 and 35, and 69 per cent of psychology workers overall.

The doctoral and early stage career numbers betoken a future flooding of the psychological field with women researchers and practitioners in a trend that is the full inverse of what we have seen in, for example, physics, for the past century-and-a-half. The gulf between the two has not gone unnoticed, with much resultant thought given to why one field has been able to turn under-representation into robust over-representation, while the other putters along at roughly the percentages it has had since dirigibles were the hot travel ticket.

Perhaps, in an age that has grown frustrated with the perceived stagnation of headline-grabbing new discoveries in physics research, the constant new insights into the functioning of the human mind, and new technologies available to probe those insights, are simply a more appealing option for those confident enough to pay no heed to the stigma of psychology as a 'softer' science. Perhaps psychology as a profession that actively welcomed women in force into its halls from an early date as an extension of its keen interest in infant and child development, in which field it believed that women offered particularly useful insights and approaches, benefitted from that pioneer effect, allowing it to establish an academic culture that had the presence and value of women researchers baked into its foundational experiences and structures, and thereby to attract more scientifically engaged young women into its practice than other fields that were still struggling to shake off their ossified paternalist practices of an earlier era. Regardless of how it began, the movement is now in full swing, and if women were able to achieve so much in an age when they were consistently outnumbered by two to one in classrooms and by even more than that in psychology and neuroscience departments, the probable extent of their future contributions boggles the imagination, and I am both envious of, and a bit worried for, the historian in 2123 who has to sit down and attempt to encapsulate the vast empires of the mind that will be illuminated by women researchers in the century to come.

Chapter One

Christine Ladd-Franklin and the Colour Wars of the Late Nineteenth Century

Colour is among the most familiar of our sensations, and at the same time also one of the most foreign. Whereas any child can tell you that Big Bird is yellow, any philosophy major can just as easily put you in a state of profound existential unease by informing you that, in actuality, nothing is yellow, that yellowness is a consistent illusion woven by brains doing their best to provide us with a map of our surroundings based on a very limited sampling of the electromagnetic radiation around us.

Just like the image of yourself in a mirror, yellow exists only in your head.

'It's all lies!' however, doesn't *quite* do as an explanation of our experience of colour, and in the early nineteenth century scientists started developing our first scientifically robust theories of how colour works, both on the physical level of the biology of the eye, and the mental level, in the interpretation of the light taken in by the retina. Within a century, two rival schools of colour theory had risen, positing seemingly mutually contradictory explanations for our sensation of colour, which one individual, who also happened to be one of her era's most significant logicians and feminists, managed to reconcile through the daring use of evolutionary theory. She was Christine Ladd-Franklin (1847–1930), America's first woman psychologist, and an individual who never did anything by halves.

Part of the explanation for Ladd-Franklin's unique sense of intellectual self-determination can be found in her upbringing. Born in Connecticut to a merchant father and an avidly feminist mother, her youth corresponded to the blossoming of American feminism as a movement. The great Seneca Falls Convention that established property rights, education, and suffrage as major goals for the women's movement met the year after Ladd-Franklin's birth, in 1848, and at the age of 5 young Christine was taken by her mother to hear a speech by the women's right activist (and attendee at the first Women's Rights National Convention in 1850) Elizabeth Oakes Smith (1806-1893).

Ladd-Franklin was brought up to expect great things of herself, and brook no restriction on the exercise of her mind or capacities, and when it was announced in 1861 that a college for women was being established by Matthew Vassar, she set herself fully to the task of convincing her family that she be allowed to attend. Her mother had passed away in 1860, and shortly thereafter she was shipped off to live with relatives, of whom the most important to convince were her aunt, who had money, her grandmother, who had moral authority, and her father, who had the ultimate say in her life's course. The aunt was all in, and offered to finance Ladd-Franklin's way. The grandmother was reluctant, feeling that Christine would be too old upon graduating to find a good marriage, but Ladd-Franklin assured her that, as there was nobody of their immediate circle that she wanted to marry, and that her ordinary looks weren't likely to win many suitors anyway, she wouldn't miss out on any marriage prospects by leaving home. Somehow 'I'm not good-looking enough to marry' convinced her grandmother, and after some initial profound resistance from her father, he was also swayed by her will to attend, and in 1866 she joined Vassar's second class, graduating in 1869 in mathematics.

While at Vassar, she had the opportunity to study with Maria Mitchell (for more about whom, see the Astronomy volume of this series), the most famous woman scientist in America at the time, and was inspired by the idea of pursuing a life in scientific research, but upon graduation she came up against the realities of American life for an educated young woman. Teaching was still the expected career for a college graduate, and Ladd-Franklin, for nearly a full dismal decade, fell into the profession. She despised the work, stating with her characteristic directness, 'Teaching I hate with a perfect hatred.' Fortunately, the galloping pace of late nineteenth century American higher education came to her rescue again, and in 1876 Johns Hopkins University admitted its first graduate students to pursue their PhDs. She wrote to a professor there, James J. Sylvester, for permission to attend the university. Fortunately, he was already well acquainted with her, through her publication of a series of solutions to mathematical problems posed by the *Educational Times*, and prodded Johns Hopkins into allowing her to attend his classes.

The exceptional nature of her work as a student, and her clear gift for both mathematics generally and logic specifically, compelled the university to allow her to take classes from other professors besides Sylvester, and by 1882 she had completed all of the work required for the awarding of a PhD, but was ultimately denied the degree she had earned on account of her gender (in fact, Johns Hopkins would not get around to awarding her that degree until forty-four long years later, on the occasion of its fiftieth anniversary).

The same year she completed her work at Johns Hopkins, Ladd-Franklin married mathematician Fabian Franklin, and in the next two years gave birth to two children, only one of whom survived past infancy. Sometime during these years, her attention swung from her purely logical exercises to psychology, and particularly to the study of vision. In 1887 she wrote a paper on the subject of horopters, which are the sets of all points in a plane that make the same angle with your eyes as a point you are actively focusing on. This was a topic that had been worked on previously by two figures whose research would play a large role in Ladd-Franklin's life, Hermann von Helmholtz and Ewald Hering, and which Ladd-Franklin proposed an experimental method for determining in her 1887 work.

That paper represented the first time an American woman had published a piece of psychological research, and is the source of her claim to being America's first woman psychologist. As interesting as horopters are, however, the real drama in the late nineteenth century study of vision came with the nature of colour, which Ladd-Franklin would study in Germany from 1891 to 1892 at Göttingen and Berlin. At the time, two warring clans of colour theorists had arrayed themselves behind two (on the surface at least) quite different theories. The Young-Helmholtz theory, first proposed by Thomas Young in 1802 and then elaborated upon by Helmholtz in 1850, held that the eye distinguishes colour through the use of three different types of cones, one that is receptive to long wavelengths of light (like red), one responsive to short wavelengths (like violet or blue), and one responsive to intermediate wavelengths (like green), and that the brain tells us what colour we are seeing at a location by interpreting the different degrees of activation of each of those three cone types, with white being the result of equal stimulation of all three.

It was a solid theory, but its detractors believed that it did not explain yellow well (which they believed was a fundamental colour and not a mix of red and green), and more convincingly that it did not explain why colour blind people, supposedly deficient in one of their cone types, still managed to see white perfectly well. These people turned towards the rival theory of Ewald Hering, who hypothesised that colour works through sets of opposed colours, red-green, yellow-blue, and white-black. We are only capable of processing one or the other of each set of opposed colours, Hering hypothesised, a limitation which he used to explain why we can imagine a 'reddish-blue' colour or a 'yellowish green' colour, but not a 'reddish green' or 'yellowish blue' one.

This 'tetrachromatic' theory put yellow back in an elemental position, which fans of yellow were super excited about, but had problems of its own in terms of what shades of red and green he had to pick as the fundamental ones in order to make the combinational math come out right. From our perspective today, we know that both theories were in fact, in their essences, correct, and that Young-Helmholtz explains our experiencing of light at the retinal level, while Hering explains how that information is processed by the brain. At the time, however, there seemed to be no common ground between them, until Ladd-Franklin published her seminal paper, 'A New Theory of Light Sensation', in 1892.

Ladd-Franklin sought to synthesise these two views, which she saw the clear merits of, and used the still-controversial (particularly in America) theory of evolution (which had only been published in 1859) to provide the explanatory mechanism for her theory. She noticed that certain types of colour-blindness were more common than others, and held that the explanation for that fact lay in the evolution of the physical structures underpinning our sense of sight. She theorised that animals first developed a general black-white sense of colour that allowed them to binarily detect light as either Present or Not Present. Eventually, we evolved the ability to detect two different bands of radiation, with one receptor for detecting long wavelengths (yellow), and another for short (blue). Later still, another refinement of our colour senses evolved, with our yellow receptors splitting into long-wave specialists (red) and mid-wave specialists (green).

Further, Ladd-Franklin theorised, the eye still carries with it vestiges of its evolutionary past, with the highly developed central regions of the retina carrying the recently-evolved tri-colour receptors, while as you work out towards the edges you travel back in time, arriving at the purely light-dark sensing rods towards the periphery of the retina. It was an elegant theory that allowed the retention of the best elements of Young-Helmholtz and Hering, and explained why red-green colour blindness is more common than blue-yellow colour blindness, the former being a more recent, and therefore less stable, addition to our optical repertoire.

The evolutionary theory of colour was a major result which put her in the front rank of American scientists, earning her election to the American Psychological Association in 1893, and invitations from around the world to talk on the subject of colour. Though a worldwide scientific figure, however, Ladd-Franklin had perpetual difficulty finding a professional position equal to her accomplishments. First at Johns Hopkins, then at Columbia, she had to fight to obtain guest-lecturing positions that amounted to something on the order of one class a semester, often without pay. Part of the explanation for her troubles might have lain with her outspoken feminist beliefs. Unlike the crop of women psychologists who would follow in her wake, and who grew up in a more Victorian atmosphere, Ladd-Franklin's early life was full of the fire of the early women's movement and the crucible of the Civil War, and she was not afraid to confront important members of the psychological establishment when she believed them to be in error, particularly castigating Edward Titchener for his decision to exclude women from his important Society of Experimental Psychologists on the grounds that the presence of women would prevent the men from being allowed to smoke, and would restrain them from using the full spectrum of vigorous and emphatic language they wanted to employ.

Though lacking in official position, however, Ladd-Franklin was a well-known presence in American life, publishing hundreds of articles, letters, and reviews on logic, colour, politics, and feminism, as well as her book, *Colour and Colour Theories*, which presented all of her mature thought on the subject of colour, and served as a classic in the field for decades thereafter. That book was published in 1929, the year before Ladd-Franklin's death in New York at the age of 82, and it is to be hoped that the almost universal praise for the volume was some solace at her life's close for the professional career that never became what it ought to have been, the decade spent teaching when all she wanted to do was learn, and the years of frustration trying to make American science live up to its best ideals of open discourse and equal opportunity.

FURTHER READING: There is, almost inexplicably for an individual of her high level of achievement in so many areas, no single volume devoted to the life of Christine Ladd-Franklin. Both *Untold Lives: The First Generation of American Women Psychologists* (1987) by Scarborough and Furumoto and *Women in Psychology: A Bio-Bibliographic Sourcebook* (1990) have nice chapters devoted to her, with *Untold Lives* focusing more on her struggles against Titchener, and the *Sourcebook* more on her scientific output. Her book, *Colour and Colour Theories* is available in reprint editions, and is also available to peruse online for free, which is more than can be said for a number of her key articles in psychology, which *Science* and the *American Psychological Association* have locked away behind hefty paywalls. If you're really good at deciphering early twentieth century handwriting, Ladd-Franklin's diaries from 1866 to 1873 have been scanned and made available online as well, which covers her Vassar years and early mathematical work rather than her psychological studies.

Chapter Two

The Woman of a Thousand Brains: Augusta Déjerine-Klumpke, Pioneer Neuroscientist

The brain is ready. It luxuriated in a solution of potassium bicarbonate for a year, and then in a thick mixture of nitrocellulose for another month and a half, and now it is finally ripe for the slicer. A woman places it in a cylinder, as she has countless times before, and fixes it in place on her microtome, advancing it upwards ever so slightly for each new pass of the blades, which will change this once whole brain, which held a lifetime of memories, triumphs, and frustrations, into some 2,000 individual sections to be stained and analysed for the presence of lesions or abnormalities.

The woman is Augusta Déjerine-Klumpke (1859–1927), the place is an almost comically small back room at Bicêtre Hospital soon to become famous as the birthplace of the century's most comprehensive examination of neuroanatomy, and the year is 1890, solidly in the centre of France's golden age of neuroscientific research. Though known today primarily as 'the wife of eminent neuroscientist Jules Déjerine,' she was at this time, at age 31, already a scientific celebrity, an award-winning researcher and trailblazer with a reputation for procedural rigor and deep familiarity with the international neuroanatomical literature. Born in San Francisco, she was the second oldest of six siblings, the majority of whom went on to international fame in widely different fields of accomplishment.

Her father had come to California in 1849, originally for the prospect of striking it rich in the gold fields, but ultimately, and wisely, deciding to make his money off of the sale of California real estate instead of taking his chances as a prospector. When she was 2 years old, her older sister Anna (later a famous painter) had a fall that led to osteomyelitis, a bacterial bone infection, which frontier doctors were understandably ill-equipped to treat. So, in 1866, her mother packed her and her three sisters up, and headed for Europe, where better results were hoped for from experts in France and Germany. For nearly two years, they remained in Europe while Anna underwent different treatments, and Augusta began to learn the German language, which would prove invaluable later in life.

Ultimately, Anna's condition did not appreciably improve in Europe, and she would have trouble walking for the rest of her life, and the family returned to California, where two more siblings were produced, in 1868 and 1870. In spite of the new additions, Augusta's parents' marriage was in trouble, and her mother requested a separation, taking her now six children with her to Europe in 1871. Augusta went to a German boarding school, and in 1873 was reunited with the rest of the family, who were living at Clarens, in Switzerland. Here, Augusta continued her studies of French while her mother wondered what to do with her prodigiously gifted daughters.

An obvious solution would be to send them to the University of Zürich, which had a fine medicine school that allowed women to enroll. A significant number of women who took up that opportunity, however, were from Russia, and had a reputation for espousing at-the-time fashionable nihilistic philosophies (a reputation not entirely unfounded – remember from the mathematics volume of this series that Sofia Kovalevskaya, who left Russia to study mathematics in France around this time, was also the author of the 1890 novel *Nihilist Girl*). As far as Augusta's mother was concerned, that disqualified Zürich as a place to send her daughter. Eventually, a friend pointed out the obvious – that if she wanted a place that could offer musical training for Julia, artistic training for Anna, scientific training for Dorothea (who we met in the Astronomy volume of this series), and medical training for Augusta, she could not do better than Paris, where Madeleine Brès had just, in 1875, become the first woman to earn a French medical degree.

Paris it was, then. Here Augusta began her studies in 1877, and for neuroscientific research, she could hardly have been in a better spot. The legendary Jean-Martin Charcot and Alfred Vulpian were both there, and both at the height of their powers. Guillaume Duchenne, the great pioneer in the use of electricity to study neuro-muscular afflictions, had only passed away two years previously. Here the great work was being done, relating maladies of the body to problems of the brain, centred at hospitals that poured out new research at dizzying paces – Bicêtre, the Hôtel-Dieu, the Charité, and most of all, Charcot's Salpêtrière. Augusta began by working as assistant to Professor J.A. Fort while studying anatomy, winning the *Médaille de Vermeil* in 1878 for her work. Fort encouraged her to apply for an externship position, but for several years her requests were turned down on account of her gender.

In 1880, she joined the clinic run by Professor Hardy as a trainee. The previous year, a promising researcher by the name of Jules Déjerine had taken up a post there as chief of the clinic. Jules had been in a disappointing romance before which convinced him that he would never find somebody who would share his serious and deep enthusiasm for science, and that the best course in life would be to remain a bachelor, married to his work. Then he met Augusta, and very soon letters singing her praises were wending their way back to his mother, praising the sharpness of her mind, and the depth of her devotion to research: 'The more I see this young woman, the more I like her, and you know, my dear mother, that this is not a galloping love like the last one, it is a rational love which has arrived slowly, and which could not be more serious. It is a love of the minds, as we say between us. And you know, my dear mother, that is true love.'

Besides a blossoming romance, the early 1880s also saw Augusta join forces with the unstoppable Blanche Edwards to give women access to French externships and internships. Edwards wanted nothing so much as to become a surgeon, a virtual impossibility without access to those opportunities. In 1882, Edwards's lobbying opened the doors of the externship track, allowing Augusta to extern at the Hôtel-Dieu in the mornings and then attend demonstrations and lessons from Charcot, Sée,

Fournier, and Besnier during the rest of the day. In 1883, while observing a patient at L'Hôpital de Lourcine, she noticed that a patient with an injury to the brachial plexus (a bunching of nerves that branch outwards from the neck) causing an expected arm paralysis also displayed unusual pupil constriction not described in the literature. Using facilities placed at her disposal at the laboratory of Alfred Vulpian (who was the primary mentor of Jules), she studied the effects of different brachial plexus lesions on dogs. Other doctors had assumed that the patient simply had Erb's palsy, the result of a lesion to the C5-C6 root nerves, but which didn't explain the pupil constriction. Digging deeper, Augusta found that damage to the C8 and T1 nerve roots produced both phenomena.

She published the results of her research in 1885, won the Godard Prize for the discovery in 1886, and today the condition is known as Klumpke's Palsy or Déjerine-Klumpke's Palsy. Also in 1885 she and Edwards broke through the internship barrier, thanks again to Edwards's gift for lobbying. At the time, internships were granted by participation in a contest which generally required preparation via a series of courses. Though they were given permission to compete for the internships, the medical establishment conspired to close the doors of any preparation courses to them, and so they had to make do as best they could between private teachers and their own studies. In 1885, Augusta won a temporary intern position, and in 1886, a full internship.

For the next three years, she interned at some of France's most important hospitals, working with some of medicine's leading lights. In 1888, however, she and Jules finally married, and Augusta cut short her internship program to work with Jules at Bicêtre, where the pair would remain until 1895, with Augusta earning her PhD in 1889 for a dissertation which improved upon the existing classifications system for polyneuritis. Here at Bicêtre, Augusta worked at preserving, sectioning, staining, sketching, and analysing the brains of patients possessing different maladies to determine what lesions and abnormalities might underlie those conditions, compiling the data that the pair would eventually publish in the 1,200 page *Sémiologie du Système Nerveux* (1900) and the two volume *Anatomie des Centres Nerveux* (1895, 1901), which presented not only Augusta's sectioning and staining methods, but also the couple's novel projection method for obtaining more detail out of microscopic observations. These were definitive texts, creating mappings of neurological conditions to the underlying brain areas involved with heretofore unknown levels of clarity, thanks to the illustrations drawn by Augusta and the careful analytic work she and Jules performed to make minute distinctions in the subcortical structures involved in different diseases.

Nowhere was Augusta's analytic skill more on display than in the great Aphasia Duel of 1908, with in one corner the great student of Charcot, Pierre Marie, and in the other, that of Vulpian, Jules Déjerine. The debate between the greatest students of France's most legendary neurological researchers was the Wrestlemania main event of its time, though today the topic of that debate, a disagreement over how many types of unique aphasia (language processing problems caused by brain damage) there are, couldn't seem less likely to be the stuff of high drama. But high drama it was, and Augusta's analysis of the anatomical differences in the lesioning behind two different types of aphasia dealt Marie the knockout blow, a defeat Marie would not be quick to forget, or forgive.

From 1895 to the First World War, the Déjerines, after a round of succession drama occasioned by the death of Charcot in 1893, worked at the Salpêtrière that Charcot had ruled for over three decades, continuing their research not only into distinct new neurological syndromes, but into new approaches for the treatment of patients who had been diagnosed with incurable neural diseases by over-eager neurologists, when in fact all they needed was a sympathetic doctor and rest to find their balance again. With the coming of the First World War, Jules volunteered his medical and administrative services while Augusta set up hospital care for the war's paraplegics and hemiplegics, developing new procedures to help them re-train their bodies, researching the impact of different war injuries (and discovering heterotopic ossification, or the growth of bones fragments within soft tissues, in the process), and creating work training programs so that even those most severely injured could find self-respect and occupation after the war. For these services, which she was actively involved in until 1920, she was made an officer of the Legion of Honor in 1921.

In 1917, Jules's health, which had been on the decline for some time, finally snapped, and with his passing, his old adversaries saw a chance to strike at his legacy through his wife. Pierre Marie, elevated to the Salpêtrière chair, and Jules's great rival, gave Augusta fifteen days to clear out their working space and leave. She did, taking the treasure trove of specimens and data they had collected over the last nearly three decades with her, and donating 10,000 francs to the establishment of a foundation to oversee their care. Meanwhile, the French scientific establishment that had done so much to hinder her rise made up for its previous behaviour in some part by honoring her in her final years, electing her a member of the *Société de Biologie* in 1924, while the *Académie* was planning to elect her to the first available vacancy, but did not get the chance to do so before her death in 1927. She left behind a daughter, Yvonne, who would go on to be a neuroscientist herself, as well as a determined guardian of her parents' legacy, a host of wounded veterans who would honor her memory for the rest of their lives for her personal and innovative care, a treasure trove of neurological specimens and records denoting in exquisite detail the structure and symptoms of damaged brains, some fifty-six scientific articles, three absolutely foundational works in the history of neuroanatomy, and a nation that remembered her as a beloved adopted daughter who took the baton of Charcot and Vulpian and carried it securely into a new century.

FURTHER READING: There is actually quite a bit about Augusta Déjerine-Klumpke out there, particularly if you can read French. She wrote an autobiographical fragment which covers the years up to her marriage, and which is a useful source to resolve many of the contradictory dates you see concerning these early years in different online accounts. *Passion Neurologie: Jules et Augusta Déjerine* (2017) by Michel Fardeau is an interesting source which gives some nice context about the French neurological scene at the time of the Déjerines arrival, though the book concentrates more on Jules than Augusta. Harder to find is Gauckler's 1922 biography of Jules Déjerine,

Le Professeur J. Déjerine, which benefits from knowing personally all the main players but which correspondingly also needs to be taken with a pinch of salt. The Déjerines main neuroanatomical volumes, meanwhile, are readily available in reprint editions, and are worth it for the hundreds of illustrations alone.

Chapter Three

Self-Remembrance: Mary Whiton Calkins's Adventures Among the Atomists

By 1910, the woman whose brilliance had forced the doors of Harvard University open to women (if only in an unofficial capacity) and who had overseen the development of the nation's first psychological laboratory for women was striving resolutely against a psychological establishment which held that virtually every one of her ideas was wrong. The Freudians and experimentalists, who usually couldn't find a polite word to say about each other, were nonetheless united around the one opinion, that Mary Whiton Calkins (1863–1930) was a pioneer, with a mountain of respectable work behind her, who was throwing her life away on philosophical irrelevancies.

By 1960, some three decades after her death, those very same philosophical irrelevancies were roaring to life again, part of a vanguard of 'new' ideas propelling psychology past its various atomist, behaviourist, and psychoanalytic tribalisms into a new psychological landscape dominated by thoughts of the unitary self, the role of exterior feedback in the generation of personal psychology, and a growing skepticism about deep abstraction in explaining the motivations of individuals.

Whereas time is the great enemy for most psychologists, who live to see their ideas grow a bit dimmer in the public imagination with every passing year, victims of new trends and fresh data, Calkins is, if anything, more popular and esteemed now than she was a century ago, leaving us to wonder how it was that a mind so ahead of its time ever found its way through the stifling orthodoxies of her era.

Calkins was born in Connecticut on 30 March 1863, at a time when the armies of Robert E. Lee stood at the high water mark of their northward advance, threatening a full-scale invasion that was not staved off until the Gettysburg conflict of early July. It was a frightening world to be born into, but the great conflict raging at the time little touched Calkins's family. Her father Wolcott was a Presbyterian pastor, and a graduate of Yale who was away studying in Germany at the war's start, only returning in 1862. Mary was the first child of eight, all born safely outside the possibility of military service. Protected by profession and status from war service, Wolcott was able to devote himself to organising the education of his children, teaching Mary German alongside English, a bi-lingualism that would serve her well as a psychologist during a time when many of the landmark texts in the field originated in Germany. Throughout her life, he was a strong advocate of her achieving whatever level of higher education she desired, and had no fear of approaching the nation's greatest educational authorities directly with his pleas to win her access to lectures and institutions historically barred to women.

In tension with this broadening of horizons during Calkins's early years, however, were two events that narrowed them precipitously. When she was around the age of 10, her mother suffered a profound breakdown after seeing her children through a wave of illnesses that left her own health ever afterwards in a precarious state. That alone would have not been enough to determine the confines of Calkins's future, however, because of the existence of her sister, Maud, who could be counted on to take up the traditional youngest sister role of parental caretaker. But in 1880 Maud died suddenly of rheumatic fever, leaving Mary the sole daughter of the family, and therefore as the presumptive familial caretaker, it being out of the question in that era for any of the sons to take up that role.

The caretaker role would eventually place strict limits on Calkins's radius of movement, a limitation which would have been more impactful had the family not moved, just prior to Maud's death, to Newton, a town near Boston and therefore to one of the beating hearts of early American psychology, Harvard University. As Calkins graduated high school, however, psychology and Harvard lay well on the horizon. At the time of her matriculation at Smith College in 1882, her eyes were fixed upon philosophy and classical literature, particularly the Greek tradition. She studied Greek privately in Greece during the year off from college she took following the death of Maud, and again while on a European tour with her family following her Smith graduation in 1884.

Almost immediately upon returning from Europe, Calkins was offered a position as a Greek instructor at Wellesley to fill a last minute vacancy. She proved so competent and popular as a teacher that the administration selected her for a new position they were creating, that of professor of psychology. Up to that point, Calkins's training in psychology consisted of precisely one class taken at Smith, and she understandably felt herself under-prepared to take on the responsibility for teaching classes in the topic, insisting that, if she were to accept, she could not begin until after taking advanced courses from whatever institutions might accept her.

Were she a man, the obvious course, given her facility with German, would have been to take a year and study in Germany with Wilhelm Wundt, but most of the people she approached for advice warned her off Germany as an environment not yet accepting of women students. That left her with the United States, where she considered the University of Michigan (where John Dewey was located) or Yale (where G.T. Ladd was located) as institutions where women could gain experience in psychology of a variously official nature. The best of all worlds, however, would have been to study at Harvard, which would allow her to stay with her family, and to study with one of America's (and the world's) reigning psychological and philosophical superstars, William James.

The problem, however, was that Harvard was one of the most resistant American universities to the idea of women's education, going so far as to create a separate program and location, referred to as the Annex, where women could learn material from Harvard professors, segregated from the actual university, and without the ability to earn university credit for their studies and work. While talking with noted Harvard philosopher Josiah Royce (1855–1916) about what classes she might take from the

Annex, he recognised that her situation was not like that of most undergraduates. She was a university instructor seeking higher level courses for the purpose of augmenting her own lectures, not some prospective freshman off the street. He felt she should be granted special access to study at Harvard itself, and William James agreed with him.

Her initial application was turned down by the university, but upon her father's renewed pleas to Harvard's president to reconsider, she was ultimately given permission to attend Harvard seminars, as long as it was understood that she was doing so as a guest and not a member of the Harvard student body. Perhaps predictably, all the male undergraduates who were signed up for the William James seminar she would be attending dropped out of the class, which was both discouraging, as a none-too-subtle sign about how her presence on campus was regarded by other students, and a boon, as it gave her one-on-one time working directly with James at precisely the moment when his influential two volume *The Principles of Psychology* was emerging into print.

Simultaneous with the incredible opportunity to study individually with one of the titans of early psychology, Calkins also studied with Edmund C. Sanford at Clark University, who taught her about experimental psychology, and was instrumental in suggesting the equipment she would need to outfit the psychology lab at Wellesley she was responsible for creating. Together, they performed an early experiment on cataloguing the content of dreams, each keeping a record of the dreams that they had, the contents of those dreams, their vividness, the locations and topics involved, the times of night they occurred, and so on, generating thereby one of the first data-based, rather than theory-based, accounts of what humans dream. Calkins would later publish their results in 'Statistics of Dreams' (1893), which laid out their conclusion that dreams are, for the most part, either manifestations of events happening directly to the sleeper (as when a person dreams that they are being buried alive when a pillow happens to fall over their face) or continuations of events, places, and themes experienced throughout the course of regular waking life, and generally the most mundane of those events, rather than the big dramatic moments of life. We dream about reading books or riding subways or doing homework, not about the time that our father died in our arms, or our wedding nights.

It was an important result which would go on to be almost entirely submerged in the wave of deep dream interpretation brought about by Sigmund Freud's publication of *The Interpretation of Dreams* in 1899, but which would rise to the surface again after the passage of many decades brought with it some skepticism about the omnipotence of psychoanalytic theory to explain the underlying structure of all dream content in terms of ego-mediated expressions of the id.

On the strength of her studies with some of America's most prominent psychological minds, Calkins felt confident in setting up Wellesley's own psychological laboratory and teaching its first psychological classes in the fall of 1891, bringing women students for the first time into regular, credit-granting contact with the methods and findings of modern experimental psychology. For the better part of a decade, she and her students working in her lab produced a paper a year detailing phenomena including synaesthesia, dreams, memory, aesthetics, and children's intellectual development, including a pair of 1896 papers on association that developed the paired-association technique that has become a mainstay of psychological testing since.

In 'Association: An Essay Analytic and Experimental', Calkins outlined a series of experiments seeking to determine whether frequency, vividness, recency, or primacy is the most important in causing humans to associate two objects together. In this experiment, subjects were seated before a white screen, and shown colours followed by numbers, and were tested later on their recall of which colours were associated with which numbers. She found that subjects were able to associate 'green' with 'fourteen' better if those two were displayed several times together during the observation part of the experiment (frequency) than if that pairing was the first thing they saw (primacy) or the most recent thing they saw (recency), and further that varying the physical properties of the numbers (their physical size, or the number of digits) instead of their colours did not compete for memorability next to frequent association. The result was interesting, but the technique was what the psychological community truly latched onto, and is found in university psychological experiments to this day.

While carrying on this work at Wellesley, Calkins continued her studies at Harvard, now under Hugo Münsterberg, until she had reached the point where she had done enough work to earn a PhD. Both James and Royce once again approached the university, asking them in 1895 that she be given the degree that her studies had merited. The university refused, offering her a PhD only in 1902 under the provision that it be considered a 'Radcliffe' PhD and not a 'Harvard' PhD (Radcliffe College had evolved from the Harvard Annex in the mid-1890s and was primarily an undergraduate institution at the time). Calkins turned the degree down, noting correctly that this was simply Harvard's way of trying to get out of offering Harvard PhDs to women who deserved them, by recasting them as degrees from an institution that didn't even have a graduate school, a deception of which she wanted no part, even if it meant not receiving the doctorate that she deserved.

In the 1900s, Calkins's focus shifted from experimental psychology towards what she would call 'self psychology', an approach to what psychology was fundamentally about that was also fundamentally out of step with its time. Instead of the atomist approach of Edward Titchener (1867–1927) who sought to reduce each mental event to its basic constituent elements, and saw no reason to talk about something as large, ungainly, subjective, and unmeasurable as a 'self', and unlike the later behaviourists who only dealt with outward phenomena they could quantitatively measure, Calkins proposed that all psychology should flow from the self, a unitary being not to be confused with the soul, and its interactions with other objects and individuals. When reviewing studies in recollection, for example, in her classic 1915 paper 'The Self in Scientific Psychology', she pointed out that, basically, recollection and recognition are about the self. When I see a copy of *Amazing Spider-Man* #349 I recognise it instantly as being a copy of the first comic book I ever bought – the recognition that it is not a foreign and new object is mediated by my previous engagement with it as a child, whole in my identity. We are caught in a dance with the objects around us, which we are constantly taking into ourselves in different degrees, and that whole nexus of self and object and the lines between them is something that can't be, Calkins asserted, meaningfully reduced to atomised general mental events. Sometimes, we are more than the sum of our mental processes, and making sense of how we approach the world simply breaks down when we leave out our concept of ourselves as a single and continuous being.

Arguing for the virtue of thinking of a self that other psychologists were keen to throw away as subjective and unmeasurable dominated Calkins's work during her last decades. In fact, when asked to provide an autobiographical sketch for the collection *History of Psychology in Autobiography* (1930), Calkins spent the bare minimum amount of space possible talking about the development of her ideas from 1890 to 1900, and the vast majority of the 'autobiography' in a long and detailed argument for the validity of self-psychology, so sure was she of the basic correctness of her insight and the need to use every conceivable platform to present its virtues.

Calkins retired from her Wellesley post in 1929 after forty-two years of service to the university, years which saw her create a place where women could, unfettered by the need to scrape and beg for academic recognition, carry out modern experiments, and see their results in print, and which saw her personally author four books and more than a hundred articles on psychology, both in the experimental and self varieties. After her death in 1930, her ideas about our experience and mental models of ourselves as single indivisible units, and particularly her ideas about the importance of our interactions with external objects in the formation of our mental world would only grow more popular the more we learned about the brain and how it models itself and adapts to cues from the outside world. Even the psychoanalysts, who had looked down upon her early dream work as important but primitive, came around to aspects of her self psychology, with Melanie Klein in particular focusing on the importance of internalisation in the formation of character and behaviour. Over the course of her long and prestigious career, Calkins had achieved many firsts – the first woman to attend regular Harvard seminars, the first person to establish a psychological laboratory at a women's university, the first woman president of the American Psychological Association (1905) and the American Philosophical Association (1918), but when it comes to legacy, that of Mary Whiton Calkins will, I think, always be this: she was the individual who kept psychology aware of the importance and indivisible features of the vast and seemingly unmeasurable self until technology could catch up and allow us to begin capturing that self at last, in all of its dream-having, colour-associating, *Spider-Man* recognising glory.

FURTHER READING: Calkins's autobiographical sketch is a good source about her education and career during the 1890s, and is a great statement of her thoughts about self psychology going into her final year of life, but isn't as great in providing details about her early life, and errs on the side of diplomacy in detailing her struggles with Harvard. To fill in those gaps, Laurel Furumoto's essays about Calkins, found in *Untold Lives* and *Women in Psychology* are useful, with the former focusing more on the details of her attempt to gain admission to Harvard, and the latter containing more about the content of her work. Most of her articles are available for free on the internet, including those mentioned above and her paper 'Synaesthesia' (1895) detailing the stability of synaesthetic properties over time in some 200 surveyed individuals.

Chapter Four

The Two Montessoris: The Rise, Fall and Rebirth of an Educational Revolution

Had Maria Montessori died in 1913 at age 43, at the height of her fame and insight, this would be a pretty straightforward article about somebody who saw a problem and used her own profound scientific instinct to make the world incalculably better. The first half of her life she moved from triumph to triumph, expanding the horizons of what education might do for a child with a system that not only achieved core competencies earlier than anyone thought possible, but engaged children as agents of their own learning progress. It is, however, difficult to both found a movement and to see it through its mature years, and for all of Montessori's early innovation, her later career featured a marked resistance to new ideas and a tight control of administration and marketing from which it would take years for the method to recover.

That decline is a harrowing and cautionary story, but preceding it was a series of triumphs unprecedented in the history of public education. Montessori was born in 1870 in an Italy only recently united by the singular diplomatic genius of Cavour. Her native country was, and I'm being charitable here, a sloppy sloppy mess at the time. A conjured amalgam of former papal states, Habsburg possessions, and dirt poor southern territories, there was little uniting these regions except for the vague feeling that once, a while ago, all of them had something to do with the Renaissance. What was needed was a universal education system to raise the shockingly low standards that abounded throughout Italy – a new, united and educated generation to steer the ship of state into the future.

Maria Montessori was a product of that intense sense of the future and its possibilities. Her mother actively encouraged her in every bold and unorthodox step of her early career, and her father, while not always happy with her startling life choices, nevertheless refrained from getting in her way. At the time, students out of elementary school had a choice between taking either a classical or a practical track for high school. Most girls, if they continued their education at all, went for the classical track, with its training in ancient languages and literature. Maria, however, opted for the practical, with its modern languages, science, and math. Initially, she wanted to be an engineer, but once she submerged herself in the sciences, she found herself beckoned by medicine.

This was, to all Right Thinking Italians, madness. No woman had ever been accepted at the University of Rome to study medicine. Maria's father was concerned lest their family become the laughing stock of Rome, but somehow (possibly through the intercession of the Pope) Maria was accepted to study, and proved herself one of the greatest students in the history of the college. Granted, this wasn't too hard, as Italian universities of the late nineteenth century were famously amongst the most slovenly run and ill respected institutions of Europe. Students showed up, or more often didn't, heard a couple of lectures, took some tests, and got their degrees. Most were in it for the social standing a degree conferred, and so made the absolute minimum of effort in attendance and study. Expectations were crushingly low, but Maria, to the shock of everybody, seemed to actually *want* to learn, showed up for every lecture, and filled every moment with books and questions. She was easily made a doctor with the overwhelming recommendation of the faculty and thereupon began her practice.

Initially, she had no thought of specialising in the science of education. Her field was a biological anthropology which sought to use scientific measurements to determine psychological types. Her early writings focus on such matters as the relation of nose ratios to secretiveness or madness, a line of inquiry which was to have disturbing consequences in the early twentieth century in the hands of eugenics-leaning governments. Fortunately, an experience at the university's psychiatric clinic deflected her attention onto her true path.

Common practice at the time dictated that the 'mentally challenged' all be lumped together in barren rooms to prevent overstimulation of their imbalanced minds. Maria noticed that, after meals, the children would fling themselves down on the floor looking for crumbs and food scraps. The other doctors looked with disgust on the practice as an example of their mental deficiency, but Montessori saw it differently. What she saw were children so starved for mental stimulation that they were turning to scraps and crumbs to get it. These children didn't need less sense training, they needed more.

She was soon given the chance to put her ideas into practice at the Orthophrenic School, where she developed new methods to teach and develop the senses of such children, and then tethered that primary sense development to intellectual learning. In doing this, she was working in the tradition of Itard and Seguin, whose research in the early nineteenth century had demonstrated the potential of using a sense-based approach to help foster learning in the developmentally disabled. By working with blocks and feeling the shape of cut out letters, they were able to eventually teach abstract concepts to children who were given up as lost by the rest of the medical world. Montessori felt she could extend and systematise their work, and was soon pulling off minor miracles at the Orthophrenic School, teaching the children to first distinguish the crude sensory differences of objects, and then through a process of refinement, bringing them to more abstract understanding of the world and their function in it.

It was a culminating moment in the history of special education, but she soon realised, with a clear instinct for the psychology of children, that the methods she was using with the patients at the Orthophrenic School could also be used to improve education for all children. But before she could apply that knowledge, Montessori had a personal struggle to overcome. She had fallen in love with another doctor at the school, and had a child by him. Of course, it would destroy her fragile reputation to publicly acknowledge having born a child out of wedlock, and so she was faced with keeping the child but losing her career or continuing her work but remaining a stranger to her own son.

She chose the latter. For the first fifteen years of his life, Mario Montessori's mother was a passing acquaintance in his life, and until her death she continued to refer to him publicly as her nephew, a role he understood and came to accept. She left the child behind to be raised by her family and returned to her work.

That work led to the establishment in 1907 of the revolutionary *Casa dei Bambini*, an experimental school that was the original idea of some low-rent landlords seeking a way to keep the children of their buildings from running wild and defacing property during the day. They decided to create a small school in the building, and called upon the world-famous Dr. Montessori to design the program and oversee its implementation. Given free rein, she developed the system that continues to be used in Montessori schools the world over.

Traditionally, children were held to be incapable of learning to read before the age of 6, and were expected to sit still and be lectured at over the course of a day by way of education. Montessori, by observing children at play, discerned a thirst for understanding their environment and mastering new skills. So, she organised her school around that sense of independent mastery. The children would have a choice of activities in a large cupboard that they could take out and play with for as long as they wanted. The teacher would show them how each activity worked, and then leave them to figure the rest out on their own. The children naturally worked their way from the simple challenges (placing cylinders in the right shaped holes) to the more fine-tuned motor applications, and demanded more.

So, Montessori decided to try teaching them to write and read through a senses-first approach, crafting letters for their hands to trace and letting them hear the noise of the letter as they felt its contours. And, very soon, those children began putting their letters together to make words, writing everything they could think of anywhere that they could find (a task made easier in Italian by the fact that things are actually written as they sound, unlike the 'knight' and 'through' bestrewn wrecks of English spelling). Once they had that down, reading was a comparative snap. While the national schools had children just starting to struggle with their first copybooks at age 6, Montessori's children were writing full sentences at age 4.

Not only that, but visitors to the *Casa* noted how orderly and attentive the children were, how they took turns serving each other at lunch, and how engaged they were in their own learning processes. Reports of the school's miraculous results flew over Europe and across the sea to the United States, while Montessori found herself besieged with letters from teachers curious to learn the method. On a tour through the United States in 1913, she was treated like an A-list celebrity, her lectures instant sell-outs wherever she went.

And that's where a movie version of her story would likely stop. Challenges faced and overcome, fame attained, all is well. However, the following decades saw Montessori settling into a long and often dictatorial struggle to preserve the purity of her method. She resigned her official positions, making herself financially dependent on sales of her learning apparatus and teacher training course fees. She steadfastly refused to let anybody but herself train teachers in the Montessori Method. Worse, she insisted that her system was absolutely complete, that any of her disciples who spoke of merging it with other educational theories or even altering the order of the apparatus was a traitor to the movement. As a result, she cut the Montessori technique completely off from other developments in the field of education, and particularly from the important ideas of Dewey and Kilpatrick in the social education of children.

She did important work in her later years, especially in overseeing the development of Montessori schools in India, but her refusal to update her methods, to scientifically test her assertions, or to allow the training of teachers outside of her immediate control all hindered the evolution of her educational philosophy and practice. When she died, the Montessori Method was a phantom of an idea in the United States, where it once seemed poised to take over the educational system entirely. It would take a new generation with fresh concerns to revive her concepts and restart the Montessori movement we know today.

However lamentable the end, there is no doubt about the ultimate impact. Take a walk down the toddler aisle at your local Target, and what you'll find is device after device aimed at the sensory training that Montessori made famous. Those techniques, and the underlying idea of the importance of agency in education, have, when combined with Dewey's principles of school as a social and creative space, formed the core of our modern educational system. And, in an age when more testing is the answer to every educational problem, perhaps it's time to step back and consider Montessori's fundamental wisdom again, about how children, through learning, become themselves.

Chapter Five

Margaret Floy Washburn and the Motion of Thought

Margaret Floy Washburn (1871–1939) was the first American woman to receive a PhD in psychology (though not, as we learned from our time with Ladd-Franklin, the first to *earn* one), and was among the most significant psychologists of the early twentieth century of any gender. Born in New York, a lack of friends during childhood left her much time to study and read by herself, placing her well in advance of her peers in school, and allowing her to graduate high school and enter college at the age of 15. Like many women of her era, her school of choice was Vassar, where she graduated in 1891. As she explained in her 1930 autobiographical essay,

> At the end of my senior year I had two dominant intellectual interests, science and philosophy. They seemed to be combined in what I heard of the wonderful new science of experimental psychology. Learning of the psychological laboratory just established at Columbia by Dr. Cattell, who had come a year before from the fountain-head, the Leipzig laboratory, I determined to be his pupil, and my parents took a house in New York for the year.

James McKeen Cattell (1860–1944) had only just become the head of the psychology department in 1891. Cattell would grow to become a key figure in making psychology a respectable branch of science in the United States, particularly in the experimental form he had learned in Germany from Hermann Lotze at Göttingen and Wilhelm Wundt in Leipzig, but upon Washburn's entry to Columbia he was just setting out on his life's course, without any institutional preconceptions about restricting women's entry into the university system.

Washburn studied with Cattell for one year as an auditor, since Columbia did not yet accept women as graduate students, before moving onto Cornell where America's other great Wundt disciple, E.B. Titchener, was just starting out on his mission to bring introspective analysis to American psychological practice. Titchener took on Washburn as his graduate student (his first), as he would later do for Cecilia Parrish and Eleanor Gamble, and set her to work studying how visual imagery affects the mind's perception of tactual distances, which was an extension of the work she had already begun with Cattell. In 1894, she received her PhD, and Titchener valued her findings so highly he sent her dissertation on to his mentor Wilhelm Wundt, who had it translated and published in 1895 in his prestigious journal *Philosophische Studien*.

For the next six years, Washburn taught at Wells College, where she was offered the position of psychology department chair for the reason that the university president desperately wanted to teach Greek and desperately did *not* want to teach psychology, and was therefore searching for somebody to take the responsibility from his shoulders. While at Wells, she began to find her own path as a psychologist, becoming interested in finding a new way between the hyper-refined analysis of Titchener and the stream of consciousness approach of William James. By 1900, however, Washburn was growing restless, and after brief stints at Cornell and the University of Cincinnati, she found herself in 1903 back at Vassar as a professor of philosophy, where she would remain until 1937.

It was at Vassar that she would perform the research that placed her squarely at the centre of American psychological life. Her most lasting contribution originated in an experiment she performed in 1905 on colour perception in a brook fish. Its ability to distinguish between colours without the possession of a cortex fascinated her, and she set out to collect as much as she could in the literature on animal cognition and psychology. The result was her 1908 book *The Animal Mind: A Textbook of Comparative Psychology*, which was an exhaustive collection of all the work done to that point on animal psychology, boasting a bibliography of some 476 titles. The book ran into multiple editions (in which the bibliography would ultimately expand to 1,683 titles), and was for decades the standard text in the field of comparative psychology, which took as its central thesis the idea that strict behaviourism, with its focus only on what could be observed externally (though that's a characterisation that would only be solidified in 1924 with John Watson's book *Behaviorism*), was too limiting a concept when dealing with animal psychology. She argued that animals exhibited not qualitative, but rather simply quantitative, differences in their mental faculties, and must be judged as existing on a scale of reasoning and consciousness with humans, rather than alone in the separate sphere to which behaviourism had necessarily consigned them.

Washburn's other great activity was the development of her motor theory of thought. As expressed in her paper 'The Function of Incipient Motor Processes' (1914) and her subsequent book *Movement and Mental Imagery* (1916), Washburn conceived of consciousness as arising from a tension between conflicting movement impulses, and thought as originating in the body's reaction to objects. For Washburn, thought doesn't happen without motion, of some sort. Over the course of *Movement*, she gives several compelling examples of how this might be so – when you think of the sound of a word, it feels like your lips are on the verge of saying it, or when you pull to mind a visual image, it feels like your eyes are straining to track it. Washburn states that the appropriate motions associated with a sound or an image in fact *are* being made, just on a level that is not consciously noticeable.

So far, so plausible. I can't think of an image of Superman zooming past my field of vision without also feeling my eyes trying to move in that direction, or my neck muscles twinging in anticipation of swiveling my head to catch another few final glimpses. When we head into the field of thoughts and reasoning patterns, Washburn heads into thornier terrain, maintaining still that even in its most abstract form, thought is movement. Everything we conceive comes from a stimulus we once experienced. Each stimulus, when received, is registered in unique and characteristic ways by an arsenal of motor responses throughout our body, from the obvious motion of the eyes, lips, and eardrums, to the subtle and unmeasurable movements of internal organs. Thinking, then, is a complicated symphony of movement resulting from our body combining the characteristic stored patterns of motion associated with each idea that forms the chain of thought.

Reading *Movement* today is an interesting exercise in personal befuddlement. There is a difficult line here between the 'maybeness' that accompanies reading a lot of introspection-based psychology, and the fact that, with a few substitutions of terms, a great number of these ideas actually turned out to be... kinda true. The discovery of mirror neurons, and with them of our neural tendency to fire pathways associated with an action even when we are just watching that action, has, the more you think about it, something of a family resemblance to a number of the points Washburn makes throughout *Movement* about the unconscious encoding and replaying of experiences when bringing them to mind.

Though her book did not draw much attention upon first publication, by the time she wrote her memoirs in 1930 she was gratified to note that what she considered to be her most important contribution to psychology seemed to be gaining a second wind in the estimation of the psychological community.

In 1937, Washburn suffered a stroke that ended her career at Vassar, and just two years later she was dead. The first woman to receive a PhD in psychology was also the second woman to assume the presidency of the American Psychological Association (in 1921), the first woman elected to the Society of Experimental Psychologists, the guiding force behind the publication of sixty-eight undergraduate psychological papers, and the author herself of nearly two hundred articles, Margaret Floy Washburn not only paved the way for women to enter psychology, but played a key role in bridging the gap between the warring philosophical and experimental branches of American academic psychology, an example we would do well to keep in focus still.

FURTHER READING: *Movement and Mental Imagery* is, I'm happy to say, readily orderable in reprint editions, and reading it makes for a fun game of, 'Is this ridiculous, or is it brilliant?' which I still can't satisfactorily answer. Washburn's autobiographical notes are available free online from the University of Toronto's psychology classics site, and is a great document not only for her life but for the state of psychological practice of her time. Finally, as expected, you can find brief accounts of her both in *Women in Psychology: A Bio-Bibliographic Sourcebook* and Scarborough and Furumoto's *Untold Lives: The First Generation of American Women Psychologists*.

Chapter Six

Generations: The Neuroscience Dynasty of Cécile and Marthe Vogt

'There was a tunnel, and at the end a beautiful light, and I heard the voices of my family calling to me.' For centuries this, the standard refrain of those brought back from the brink of death, assured us of a warm and comfortable Something waiting for us beyond the boundaries of this life. Then, in the early twentieth century, along came pathoclisis, the brain child of the greatest neuroscientific power couple this side of Jules and Augusta Déjerine, Cécile and Oskar Vogt. After years of studying brains in distress, they observed that certain affronts to the brain, like a lack of oxygen or the presence of toxins, instead of affecting all cerebral tissue equally, tended to hit particular regions of the brain more severely, creating a staggered response to extreme conditions whereby some parts (such as the occipital and temporal lobes responsible for visual and auditory processing) start blinking out before others, presenting us with comforting if misleading lights, tunnels, and familiar sounds on our way off this mortal coil.

It was not the first nail that the Vogts drove into the coffin of humanity's spiritual sense of self, and it would certainly not be the last, for if there was one thing that drove Oskar and Cécile through all the decades of struggle that would mark their years together, it was a multidisciplinary devotion to studying the connections between the structure of the brain and our experience of the world and ourselves as mediated by that brain. Using brain sectioning methods ten times finer than those of the Déjerines, a glistening battery of distinct staining techniques, and new technologies that allowed the non-invasive measurement of brain activity, they were able to map some two hundred cortical areas of the brain associated with distinct functions.

For both Oskar and Cécile, the rise to the pinnacle of the international neuroscientific community originated in unconventional soil. Oskar Vogt (1870–1959) was from a Danish family that only became German with Prussia's absorption of Schleswig-Holstein in 1864, and was inspired from a young age by the implications of Darwinism for the study of psychology, and the relation of genetic and structural factors to human behaviour, perception, and disease, in spite of the fact that these interests landed him into repeated conflict with his religiously-inclined elders and teachers. Cécile Mugnier (1875–1962), meanwhile, was born in Annecy, located in the mountainous Haute Savoie region of eastern France. Her father, who had never married her mother, died when she was 2 years old, and her subsequent life's course and education was directed by a wealthy aunt who had in mind for her a career as a nun. Cécile, however, had other plans in mind, and was distinctly skeptical of the church's description of the Judaeo-Christian God as both all-powerful and benevolent.

Having lost the patronage of her aunt, Cécile proceeded forward with her education through private teachers, and by 1891 she had received her high school's highest award for academic distinction, and in 1893 took the baccalaureate examinations that earned for her a *Baccalauréat ès sciences* in September and a *Baccalauréat ès lettres* in October. With her interest in the connections between psychology and the body, it was all but given that she would make her way to Paris, where the warring camps of the Déjerines at the Salpêtrière and Pierre Marie at Bicêtre were daily probing the mysteries of the brain's structure and function. By 1896, she was installed with Marie at Bicêtre, where she was distinguished by the quality of her work and her passion for brain research. It almost all came crashing down, however, in 1897, when she discovered she was pregnant. Though we do not know conclusively who the father was, Vogt's biographer Birgit Kofler-Bettschart has a more than plausible theory that it was oral surgery pioneer Hippolyte Morestin, who delivered the child discreetly in 1898. Having a child out of wedlock could well have ended both Cécile's career and her prospect of future marriage, but fortunately in 1898 she met a man entirely indifferent to societal expectations for women, and who saw Cécile solely in terms of the quality of her mind and the depth of her passion for scientific research – Oskar Vogt.

Besides their love of brain research, the two had much in common – both born in smaller towns to elderly fathers who passed away early in their lives, both religious non-conformists, and both committed to the idea that the key to psychology lay not in the formulation of abstract theories, but rather in the multidisciplinary investigation of the brain's structure and function. Vogt was at the time studying at the lab of the Déjerines, while Cécile was working with their arch-nemesis Pierre Marie, lending the romance a certain intoxicating Montague and Capulet vibe, with love winning over academic rivalry in the end. In her diary, Cécile wrote, 'Our relationship did not begin with great passion, but step by step formed itself into an ever deeper connection. And he loved Claire [her illegitimate daughter], which was an additional positive point.'

While Cécile remained in Paris to finish her doctoral degree (awarded in 1900 for her paper, '*Etude sur la myélinisation des hémisphères cérébraux*,' which employed Weigert staining methods to investigate the placement of myelinated neurons in the cerebrum), Oskar was in Berlin, laying the groundwork for a private brain research institute to be funded by donations from the Krupp family, whose wealth from ammunition manufacturing was virtually without limit at the time, and whose good-will Oskar had earned by treating several members of the family with hypnosis while working at Alexandersbad. The pair married in 1899, and soon founded the *Neurologische Zentralstation*, the first in a series of research institutes the pair would erect over the course of the next half-century.

Having a private research centre at such a comparably young age was both a blessing and curse, however, as it focused the concentrated ire of the Berlin neurological establishment firmly upon their persons. How dare these young upstarts circumvent the holy towers of academia and secure independent financing to pursue questions of interest to them without the direction and control of their elders? The Vogts' success, combined with Oskar's combative personality, would earn them enemies at this stage of their careers that would dog their steps for decades to come.

For the moment, however, all was well, and the Vogts set out on a joint research program that encompassed cytoarchitectonics (how neural cells are arranged into distinct regions in the brain), genetics, psychology, and psychotherapy. Cécile made strides in studying the architecture of the thalamus that laid out several thalamic centres and their connections (1909), and diseases of the extrapyramidal system, which resulted in the re-discovery of a malformation of the corpus striatum that produced slow involuntary facial and hand movements (1911).

Through a variety of staining methods that highlighted different cell communities within the brain, new sectioning methods that allowed the slicing of brains into some 20,000 sections instead of the 2,000 or so characteristic of the Déjerines' early work, and new techniques in clearly photographing brain sections for reference, the Vogts were rapidly building up an international reputation for the quality of their analysis of the brain's structure. In 1911, when the Kaiser Wilhelm Society for the Promotion of the Sciences was founded with the intent of pouring resources into scientific study, it was apparent to many that a brain research division should be created, but the question lingered, who was to run it? It was a battle between the reputation the Vogts had built up and the power of the German academic machine, which was decided ultimately in the Vogts' favor after the Krupps swooped in with a million mark donation that was more or less dependent upon Oskar Vogt being named director of the new brain research foundation, which earned the Vogts even more enmity from a scientific establishment ill-disposed towards them, but which also ensured them of a place to work, where they could determine their own research program, for the next two decades.

Though appointed head of the Kaiser Wilhelm Brain Institute in 1914, the combined effects of the First World War and the post-war economic depression pushed the construction of a new building for the institute to 1931, at which point the Vogts' daughter Marthe was brought on as director of the chemical division, about whom more in a bit. The Vogts survived the 1920s through the continued support of the Krupps, the awarding of a Rockefeller grant, and support from the Soviet Union somewhat grimly tied to the affair of Lenin's brain, which Oskar was hired by the Soviets to analyse in order to find biological evidence for Lenin's mental superiority. During this time, they pressed forward with their research of the function of different parts of the brain through direct electrical stimulation of primate brains, and later through some of the first uses of electroencephalogram technology to measure human brain activity.

Returning to the subject of Lenin's brain we are treading on tricky ground, for part of the Vogt research program was related to the comparison of brains of exceptional individuals, including those exceptionally brilliant, who often donated their brains to the Vogts for study, and those exceptionally troubled, including criminals whose brains were sent to the Vogts after their owners' execution. As part of their larger study of what brain areas are associated with what physiological functions, they endeavoured to find out what parts of the brain might be more well developed in a great poet or mathematician versus a murderer or paranoiac. To the Vogts, this was just a matter of figuring out What Does What in the brain, but to a certain party rising to prominence in Germany in the late 1920s, it was research that rang eerily familiar to their own eugenicist leadings, and it would not be long until the Vogts had to figure out how to approach the Nazis.

Cécile and Oskar both despised the Nazis, but recognised their power, as leaders of a major scientific research institute in Germany that was shielded by the Krupps, whose weapons were a critical component of Hitler's plans, to potentially protect individuals working for them who would otherwise face persecution. So, while Cécile continued analysing and cataloguing the growing inventory of the Vogt brain library (her centrality being attested to by the fact that the abbreviations used throughout the system are of the French terms for the items in question, rather than the German ones), Oskar played up to Nazi high officials how well his study of superior brain types matched their goals, while protecting the Jews and women on staff who, by Nazi law, ought to have been relieved of their positions.

Though able to charm his way into the good graces of many higher Nazi officials, his cachet with the foot soldiers of the Nazi regime, as represented in the local SA squad, was less well developed, and on two occasions he had to sit by while his institute was raided by SA men, and his staff taken away for rough questioning. By the mid-1930s, it was clear that even his association with the Krupps, and theoretically *Uebermensch*-leaning research program, couldn't long protect him from those elements in the Nazi Party who had grown resentful of his independence. In 1935, Marthe left the institute for London, where she began a career every bit as successful as that of her parents, making this as good a time as any to check in on Marthe's story. She was born in 1903 in Berlin, some ten years before her sister, the future developmental geneticist Marguerite Vogt. She was from a young age her father's companion on trips to gather the insects he needed for his genetics research, and was a student at the Auguste Viktoria-Schule from 1909 to 1922, where her favorite subjects were mathematics and physics. In 1922 she matriculated at Berlin University, where she divided her time between the study of medicine and chemistry without the need to worry about supporting herself that was such a constant among her classmates during those lean and uncertain years.

Following her graduation in 1927, she did her graduate work at the Kaiser Wilhelm Institute for Biochemistry, earning her PhD in 1929, by which time she was working under Paul Trendelenburg at the University of Berlin, where she learned the endocrinological techniques that would define the most fruitful part of her career. Her experience in endocrinology, medicine, and organic chemistry made her the perfect person, in spite of her relative youth, to take up the leadership of the chemistry division of her parents' Brain Research Institute in 1931, where she carried out research, in line with her parents' theory of pathoclisis, about how drugs differentially interact with different parts of the brain.

By 1935, she recognised the writing on the wall, however, and took up Sir Henry Dale's invitation to work with him at the National Institute for Medical Research. Here, in London, her real work would begin, analysing the neurochemicals that seemed to play a role in coordinating neural action. She worked with Feldberg and Dale on the role that acetylcholine plays in motor nerve signaling, and launched herself from that foundation to a study of what role acetylcholine might play in the brain itself, which was distinctly new territory at the time. With the arrival of the Second World War, however, she was classified as an enemy alien in London, and was due for incarceration when her colleagues rallied around her with letters of support and got her reclassified as a 'friendly alien', allowing her to continue her work. After receiving a Cambridge PhD in pharmacology in 1938, she began working at the Pharmacological Laboratories at Bloomsbury Square in 1941, where she initiated her work (to be continued after the war at Edinburgh) investigating a plethora of neurochemicals, including acetylcholine, adrenaline, noradrenaline, dopamine, and serotonin. In a seminal 1948 paper, co-authored with William Feldberg, she demonstrated that acetylcholine plays a role in brain signaling similar to that which she had observed it playing in motor nerve-muscle communication. Through dozens of papers minutely reporting the results of her painstaking bioassaying of positively miniscule quantities of neurochemicals present in different parts of the brain during different activities, she convincingly argued for connections between increased concentration and activity that elucidated the role that neurotransmitters and neuroglandular secretions played in the brain.

In 1952, Marthe Vogt would be honored for her work with induction into the Royal Society. Her parents, meanwhile, were navigating with their own idiosyncratic aplomb the political landscape of post-war Germany. In 1936, they had left behind the Kaiser Wilhelm Institute at Berlin-Buch to set up yet a new brain research institute at Neustadt, in Southern Germany, where they continued their work with a small but dedicated group of researchers through the war years, working independently as was their wont while other neuroscientists latched onto the Nazi engine as their best means of advancement. After the war, the Vogts moved somewhat seamlessly back into scientific life, having been for so long so far from the centres of power of the Nazi establishment, a fact resented by their Berlin colleagues, whose de-Nazification path was less clear, and who bristled at Oskar's ongoing accusations of collaboration with the regime. For Cécile, the war years were lean but largely happy ones, if punctuated by Oskar's infidelity (which some acquaintances of the time claim she not only knew about, but helped facilitate, as a matter of keeping the family together), with the pared-down Neustadt institute pushing forward its research agenda with new techniques, including the Caspersson method for using UV light absorption to track the activity of DNA and RNA during neuron genesis and growth, which led them to new theories about a future day when genetic manipulation could be used to combat neural diseases, a day we are just now approaching.

Oskar passed away in 1959, and after a year of preparations to ensure a worthy successor to their nearly sixty years of joint work, Cécile left Germany in 1960, to join Marthe in England, where she remained until her death in 1962. Marthe, for her part, would continue her research for another two decades, including investigations into the release and re-uptake of serotonin, and into the development of new drugs that treated neural conditions caused by neurotransmitter imbalances, a line of research which has spawned billion dollar industries today but was at the time in its tentative infancy. She continued publishing sporadically throughout her retirement in the 1970s and into the 1980s. In 1988, she moved to La Jolla, California, where her sister Marguerite had worked as a geneticist at the Salk Institute since 1963. With the onset of Marthe's Alzheimer's disease, Marguerite increasingly took up the role of caretaker to her older sister, who passed away in 2003, one day after her 100th birthday.

From the publication of Cécile Mugnier's dissertation in 1900 to the release of Marthe Vogt's last paper in 1988, mother and daughter had between them charted the regions of the brain, catalogued its variable response to different chemicals and threats, outlined connections between its different processing centres, elucidated which brain areas corresponded with which physiological functions, discovered the first neurotransmitters at work within the brain, and developed novel approaches to treating brain diseases harnessing new ideas emerging from genetics and pharmacology that echo into modern practice. In the process, they might have taken away some of the comfort at the end of the tunnel, but for all of the knowledge they gave us about the universe we carry with us within the thin confines of our skulls, it seems a small price to pay.

FURTHER READING: The best book for Cécile Vogt is the aforementioned *Cécile Vogt: Pionierin der Hirnforschung* (2022) by Birgit Kofler-Bettschart, but if you want something in English, a close second is Igor Klatzo's *Cécile and Oskar Vogt: The Visionaries of Modern Neuroscience* (2002). Klatzo was a neuroscientist who worked with the Vogts during their Neustadt era, so his insights into their daily lives during this era are interesting, though I would say the focus of that book is much more on Oskar than Cécile, and of the daughters, on Marguerite more than Marthe. For Marthe, there isn't a standalone book on her life, but Alan Cuthbert's celebration of her work for the Royal Society upon her death in 2003 is a nice overview of the full scope of her neuroglandular and neurotransmitter researches of the 1940s through 1960s.

Chapter Seven

Brief Portraits: The Nineteenth Century

Susanna Rubinstein (1847–1914):

The first woman to earn a doctorate from the University of Bern, Susanna Rubinstein was born in Czernowitz on 20 September 1847, at a time when that city was a part of the Austro-Hungarian Empire (today it is the city of Chernivtsi, Ukraine). After the early death of her mother, Rubinstein was encouraged by her father to pursue her education, and was provided with a series of private tutors, as it would be another half-century until girls in Czernowitz had a high school they could attend. After passing a college-entrance examination, she was able to attend the University of Prague in 1870, followed by Leipzig University. In 1874, she received her historical doctorate from the University of Bern in the field of psychology, for a paper on the nature of the human senses, a topic she would return to throughout her career, including a series of essays published in 1878 on the connection between psychology and aesthetics, a subject she returned to in 1902.

Lillien Jane Martin (1851–1943):

Martin made up for a late start in psychology with a spectacular autumnal intellectual renaissance that achieved a number of firsts in the field. Born in Olean, New York, as the eldest of four children it fell to her to earn extra income as soon as possible after her father left the family. A precocious child who had started her schooling at the age of 4, by 16 she was herself a school teacher, making money for the family while also saving up for her own college fund. By 1876, she had amassed enough funding to attend Vassar College, which had only been established the previous decade. Receiving her BA in 1880, she returned to teaching, working as a high school physics and chemistry teacher in San Francisco, California, for the next fourteen years until she made the sudden decision to resume her studies at the University of Göttingen in 1894, receiving her PhD in Psychology in 1898.

The year after receiving her doctorate, she was back in the Bay Area, and teaching at the recently established Stanford University, where she would remain until 1916. Her Stanford years saw her develop new experimental methods for measuring non-image-based thought, and for determining the psychophysical processes by which we turn the physical act of lifting weights into mental perceptions of that act. During her last year at Stanford, she was raised to the position of psychology department chair, which made her the first woman to ever run a department in Stanford's history. Upon her retirement in 1916 at the expected age of 65, she would have been well within her rights to sink into a respectable retirement, but instead decided to fire up a third stage of her career, this time as a consulting psychologist, beginning with studies of preschool children (culminating in her books *Mental Hygiene* (1920) and *Mental Training for the Pre-School Age Child* (1923)). By the late 1920s, her interest had turned again, to the at the time almost entirely neglected field of gerontological psychology. In 1929, she established a consultation centre for the elderly, which was arguably the first of its type in the United States. Throughout the 1930s, she wrote books on the subject of elderly psychology (she ultimately wrote twelve books over the course of her career), including *Salvaging Old Age* (1930) and *Sweeping the Cobwebs* (1933).

Celestia Susannah Parrish (1853–1918):
Upon her death in 1918, Celestia Parrish was eulogised by Georgia's State Superintendent of Schools as 'Georgia's Greatest Woman' in recognition of her decades of devotion to the cause of Southern education, and of providing students in the South with exposure to experimental methods in the emerging field of psychology. It was the culmination of a life that had begun with anything but promise. The daughter of a Virginia plantation owner, Parrish spent her adolescence at the churning centre of the American slavery system, and at the age of 10 she lost both of her parents to the Civil War, the battles of which swept regularly through her state. She was placed in the care of an uncle, who died in turn in 1868, leaving Parrish at the age of 15 once again without protection or a clear future in a state still trying to regain its footing from the depredations of war and its own deep economic and social wounds. Like Lillien Martin, she turned to the power of her brain to support her siblings, and became a teacher while herself still a teenager.

Parrish found teaching difficult at first, but had little choice but to persevere, and soon she developed a local reputation for the excellence of her craft, taking up a post in Danville, Virginia in 1874 that also gave her access to extending her own education at the nearby Roanoke Female College. By 1892, her star as an educator was definitively on the rise, and she was offered the chair of mathematics at Randolph-Macon Woman's College, which had only been established the previous year. As with many newly established colleges in the late nineteenth century, staff were expected to cover fields outside of their department, and Parrish was given charge of not only mathematics, but philosophy, pedagogy, and psychology. She felt herself woefully underprepared to teach in this last field particularly, and enrolled in the summer of 1892 as a graduate student at Cornell University, working with structuralism founder Edward B. Titchener.

Titchener's approach to psychology was as a grand analogue to chemical analysis – each event could be broken down into its structural elements, as a compound could be broken into its atoms, and analysed in terms of the relative contributions of each of those elements, a method pioneered by one of psychology's founding figures, Wilhelm Wundt (1832–1920). Though not a particular supporter of women in education (and, in some respects, a downright opponent of the idea), Titchener did take on women graduate students, and Parrish in particular took his focus on experimentation and analysis to heart in much of her later work. She published two papers during her time with Titchener, both on the subject of the sensory capacities of skin.

Inspired by Titchener's methods, Parrish established a psychology lab at Randolph-Macon, which was likely the first of its kind in the recently Reconstructed South, and when she resigned from Randolph-Macon in 1902, and took up a position at the State Normal School of Athens, she established an even larger experimental psychology lab. As an educational reformer, and experimental psychologist, Parrish was gaining fame in the scientific community, and when the first edition of *American Men of Science* was published in 1906, she was one of twenty-two women psychologists selected for inclusion. The last chapter of her life, from 1911 to 1918, saw her raised to the position of State Supervisor of Schools for Georgia, work that required her to regularly tour the state's 48 counties and 2,400 schools, attempting to find means of supporting a network of nearly 4,000 teachers, many of whom had little more than a fourth grade education. To this end, she founded Muscogee Elementary School, an experimental school funded by a $10,000 Peabody grant for the training of new teachers in modern pedagogical methods.

Emma Sophia Baker (1856–1943):

Like Christine Ladd-Franklin, Emma Sophia Baker chose colour as her primary (some pun intended) field of psychological study. Whereas Ladd-Franklin, however, focused on the development of the mechanical structures that allowed for the distinguishing of colours, Baker's work centred on how colour combinations are evaluated as more or less pleasing by humans, i.e. on the aesthetics of colour perception. Born in Ontario, she received her PhD from the University of Toronto (the first ever granted in psychology from that institution), for work she carried out in the extensive colour lab of August Kirschmann (1860–1932). Kirschmann had taken over the psychology department at University of Toronto in 1893, just four years before Baker arrived, and by 1900 he had expanded the four room psych lab into a sixteen room juggernaut that carried out world-class experiments in colour perception which were extensions of work Kirschmann carried out as a doctoral student under the omnipresent Wilhelm Wundt. Baker's papers centred around experiments that challenged the reigning conception that complementary colours make the most naturally pleasing combinations to the human eye. She found, contrary to the perceived wisdom, that under experimentally controlled conditions people gravitated to combinations that featured similar colours, rather than complementary ones.

In 1903, Baker received her PhD in psychology, and pursued a career in school administration, as a vice-principal and principal, until she received a professorship in psychology, ethics and economics from Maryland College for Women in 1914, a post she held until her retirement in 1928.

Margaret Keiver Smith (1856–1934):

Margaret Smith is today a faint shadow in the memory of the psychological community, but during her time she worked successfully in a number of intriguing areas of research. Born in Nova Scotia, she carried out most of her studies in Europe, wending her way through the universities of Jena, Thüringen, and Göttingen before receiving her PhD at the University of Zürich in 1900. Her dissertation, *Rhythmus und Arbeit*, focused on experimentally measuring the impact that rhythm had on the quality and quantity of work performed by labourers, differentiating between the psychological impact of rhythms that workers pick for themselves, as when they sing while working, and that of rhythms dictated to them by the machinery they happen to be working on. It was an early experiment in the field of industrial psychology, which would come into its own with the work of Lillian Gilbreth several decades later, and a fascinating attempt to get to the bottom of just *what precisely* it is we respond to when we fall into a rhythm, and what parts of our mind and bodies do the responding.

Returning to New York, Smith spent most of her career at the State Normal School in New Paltz, New York, where she turned her interests towards pedagogy research, culminating in her 1908 paper, 'The Training of a Backward Boy', which tells the story of Willie, a New Paltz boy whose head underwent repetitive rotational motion, but who otherwise seemed entirely unresponsive to the world around him, both emotionally and physically. According to his mother, he would not speak, laugh, run, or play. In her 1908 paper, Smith documents the efforts of both herself and Willie's teacher to give Willie the abilities needed to interact with his peers, including methods of repetition, suggestion, and self-direction that within the space of a year saw a substantial improvement in Willie's ability to select and carry out goals, speak with people around him, understand the concept of play and engage in it, and acquire basic competencies in reading and mathematics.

Julia Barlow Platt (1857–1935):
Here where I live, in California, Julia Platt is known primarily for her role in preserving a stretch of the California coastline that would later become the Monterey Bay Aquarium, famous the world over for the quality of its research programs, educational outreach, and very excellent otters. But long before she became the patron saint of California marine conservation, Julia Platt had a different life entirely, as one of the first women neuroscientists in the world, a career cut short by American academic inertia. She was born in San Francisco, but most of her youth was spent in Vermont, where her father, who died in her early youth, was a state's attorney. Between 1887 and 1898 she studied at a multitude of universities, seeking out the greatest minds in the fields she was interested in, and learning from them directly. Her intellectual travels took her from Harvard to Naples, and Chicago to Munich, before earning her PhD at the University of Freiburg in 1898.

During this time, she was a prolific researcher and writer, producing paper after paper on the growth, segmentation, and differentiation of the nervous system in different species, including work on amphioxus (a type of lancelet), chickens, spiny dogfish, and mudpuppies. This was controversial work that challenged dominant germ layer theories about nervous development, and had she remained in Europe she would likely have been able to continue it to one degree or another. Unfortunately, in 1899 she returned to the United States, and an academic community that simply didn't know what to do with her. Unlike colour theory or child development, which featured a number of women doing research at various institutions up and down the East Coast, embryological segmentation studies were so specialised, and so universally the realm of male researchers, that Platt could not find any position that would allow her to continue her work.

And so it was that Julia Platt settled down to live in Pacific Grove, where in 1931 a disagreement with a local landowner over his attempt to bar the population from access to public land compelled her to run for mayor, a race that she won. Concerned for the preservation of Monterey Bay's unique marine life, including the critically threatened sea otter population there, she petitioned the state to allow Pacific Grove to control its own coastlines, and upon being granted that permission, set up a protected marine region, safe from the heedless industrial fishing practices that would ultimately crash the local economy of Cannery Row. Platt would die four years after winning her historic run for mayor, but the marine preserve that she had created served as the nucleus for a reborn Monterey, centred at the aquarium, which opened in 1984 to display to the world the species that continued to exist thanks to the efforts of a 71-year-old woman who had been frustrated in realising her scientific ambitions, but who never gave up doing what was right by nature.

Milicent Washburn Shinn (1858–1940):

In 1877, Charles Darwin published 'A Biographical Sketch of an Infant', which set out to record the physical, emotional, and intellectual development of a baby during its first two years of life. Whereas the habits and behaviour of older children had been documented in variously scientific ways prior to Darwin, his sketch was arguably the event that brought the detailed study of infants to the broader attention of the psychological community. In the 1880s, American psychologist G. Stanley Hall founded the child study movement, which grew through a series of Hall clubs that dotted the United States. This movement gave many educated women who lacked access to professional scientific careers a means of conducting original research in a manner consistent with the often Victorian expectations of well-to-do American society. Foremost among this group of women was Milicent Washburn Shinn, who was born in the sleepy town of Niles, California (known today as 'America's First Hollywood' for its part in the silent film industry during the 1910s).

One of seven children, Shinn showed early intellectual promise, and was sent to study at the University of California, Berkeley in 1874 at the age of 16, the year after the university opened its admissions to women. After graduation, Shinn took up a position as editor of the *Overland Monthly*, a California periodical that plugged into the post Gold Rush national interest in the history, geography, nature, and development of California. She served as editor from 1883 to 1894, when she returned to Berkeley to begin her doctoral studies, culminating in her dissertation, *Notes on the Development of a Child*, which was published in 1899.

During this time, Shinn was acting as the primary caretaker for her aging parents, which limited her ability to expand her academic career after receiving her PhD. That role of family caregiver also extended to the care of her brother's daughter, Ruth, which gave Shinn the opportunity to closely observe the development of an infant, and record her observations using a novel diary method which became standard among later infant development researchers. The product of her studies of Ruth's development, *The Biography of a Baby*, was published in 1900 and represents a foundational text in the field, originating ideas of the development of practical intelligence, imitation, expectation, sound recognition, and emotional interdependence that became standard in the field.

The Biography of the Baby provided women throughout the country with a model to guide their observations of their own children. Shinn received many of these observations, and used them as the basis for her second volume, *Notes on the Development of a Child*, published in 1907, and focusing on sense development during the first three years of a child's life. With that publication, however, Shinn's role as a psychological researcher largely came to an end. She continued writing, primarily about women's suffrage, and was a member of multiple professional, conservational, and reform societies. Today her home in Niles is preserved as the Shinn Historical Park and Arboretum, with tours available the first Wednesday and third Sunday every month.

Katharine Bement Davis (1860–1935):

Katharine Davis ranks as one of the most intellectually complicated figures we shall spend time with in this volume. At times, her mind dazzles with ideas decades ahead of their time, while at others, it drags with the weight of Victorian encrustations that set her hard against the progressive momentum of her age. She was a reformer whose contact with the day-to-day grind of overseeing the care of society's cast off individuals brought her simultaneously to ever darker conclusions about what might be needed to save society from criminality and to ever more enlightened ideas about the variety and diversity of women's sexual lives. She is, in all ways, a profound puzzle that defies ready categorisation and demonstrates the intellectual complexity of 'the long nineteenth century'.

Davis belonged to a middle class family with a long tradition of reform sentiments, including abolitionism, temperance, and women's suffrage. Though respectable, the family was by no means opulent, and young Katharine had to work as a teacher for several years after high school to earn enough money to attend Vassar College, where she matriculated in 1890. This was some fourteen years after the establishment by Ellen Swallow Richards (1842–1911) of the MIT Woman's Laboratory, devoted to the scientific investigation of the food, air, water, and clothing that formed such a critical part of most women's daily work. Richards believed strongly that better scientific information about the content of food and water, and the composition of different cloth types and household materials, would allow women, and institutions, to make better decisions about what food and material to buy, and thereby provide their families, and thus the nation, with healthier options at lower prices. Davis wanted to be part of this movement, and specialised in the chemistry of food. After Vassar, she continued to study at Columbia's Barnard College, work which resulted in her being appointed in 1893 as the director of the Chicago World's Fair exhibit on ways that labourers could maximise the health impact of the choices available to them on limited budgets.

Her success with that exhibition led to a job offer to become head resident at a college settlement in Philadelphia. The settlement movement of the late nineteenth century was motivated by the idea that, through organisation, and the devoted efforts of educated women, underserviced communities of immigrants and poor workers could be given more opportunities to integrate into society and improve their lots. Davis worked at the Philadelphia settlement until 1897, working in a community of primarily Russian immigrants and Black residents whose condition she collaborated with a young W.E.B. Du Bois to document.

At this point, Davis realised that to have a more than local impact on society, she would need training in political organisation and administration, and in 1900 she received her doctorate in political economy from the University of Chicago. This training, and the contacts she gathered while receiving it, opened the door to the job that would change her life, for better and worse, as superintendent of the New York State Reformatory for Women at Bedford Hills, in New York. Here the intricate web of influences that formed Davis's intellectual background all came into play, tugging her in different directions as she experienced the realities of the American correctional system.

On the one hand, she used her education and influence to attempt to improve the day-to-day lot of the inmates of Bedford Hills, seeking to recast the institution as more of a boarding school with vocational training opportunities than the prison-like edifice she inherited. On the other, her contact with the inmates of the facility increasingly convinced her that criminality was due to deep underlying and inherited mental deficiencies, which led her to identify herself with the eugenics movement that was increasingly popular in late nineteenth- and early twentieth-century America.

One of the cornerstones of eugenics was the idea that sexual promiscuity was a genetically inherited aberration that society needed to be protected from. Davis decided to study women's sexuality further, and used her position as the head of the largely eugenicist Bureau of Social Hygiene (a position she received in 1918) funded by John Rockefeller to carry out that research. What began as an attempt to document the irregular sexual activity of the nation's criminal population, ironically, grew over the course of the 1920s into one of the era's most progressive documents. When Davis began her surveying of 5,000 women and their sexual habits and preferences, it was with a relatively Victorian set of starting assumptions that 'normal' women possessed muted to no sexual appetites. The ground-breaking document she produced in 1929, *Factors in the Sex Life of Twenty-Two Hundred Women*, showed the prevalence of masturbation, homoeroticism, and extra-marital sexual activity even among the most educated and socially refined classes, and sparked a re-evaluation of women's sexuality that would see its most famous fruits in Alfred Kinsey's research of the late 1940s.

In 1927, Davis either retired from the Bureau of Social Hygiene on account of her health, or was let go from it on account of Rockefeller's discomfort with the sexual nature of her research, depending on whom you ask, and moved to Pacific Grove for her retirement, perhaps rubbing elbows from time to time and having intense discussions on the nature of psychology with fellow resident Julia Barlow Platt (I'd like to think so, at least). She died in 1935 of cerebral arteriosclerosis.

Kate Brousseau (1862–1938):
Brousseau had a rich career that encompassed service in the First World War as director of a soldiers' home, teacher of French literature, and advocate for increased women's access to higher education, but it was her work as a psychologist during her twenty-one years at Mills College in Oakland, California for which she has been most remembered. Born in Michigan, she received her doctorate from the Sorbonne in 1904 for her treatise on the psychological necessity of better educational opportunities for the Black population of the American Southern states. She began her career at Mills in 1907, a position she retained until her retirement in 1928, with a break from 1917 to 1919 to serve in the French army as a specialist in the rehabilitation of injured soldiers. It was while in this position that she began her exhaustive studies of the abilities and limitations of over a thousand children housed at the Sonoma State Home for the Feebleminded, studies which contributed to her landmark work on Down's syndrome, *Mongolism, a Study of the Physical and Mental Characteristics of Mongolian Imbeciles* (it was not until 1961 that Mongolism began to be replaced by Down's syndrome as the official designation for the disorder). In that book, which combined previous research from the historical records with her own observations and studies, she concluded that none of the treatments available at the time were actually effective in combating the disorder, and none of the explanations for its cause were satisfactory. *Mongolism* thereby helped clear the ground of decades of accumulated detritus, which opened space for the powerful genetic explanations of the disease that were made available during the 1950s.

After her retirement from Mills in 1928, Brousseau returned to Los Angeles, where she acted as director of psychological services at the Institute of Family Relations, and developed programs to offer education and advice to couples prior to marriage, and to offer psychological services to see married couples through times of hardship. The year before her death, Brousseau published *Psychological Service at the Los Angeles Institute of Family Relations*, an account of the work being done by the institution.

Mary Lawson Neff (1862–1945):
As noted above in the case of Katharine Bement Davis, eugenics was the great looming presence over the medical field in the early twentieth century, and psychology was not spared its influence. While Davis came to her conclusions about the desirability of eugenics through her work in the American correctional system, Mary Lawson Neff saw it as a central element of her work with infants and their nervous development. The author of the book *Mental Hygiene* (1904), Neff was an active proponent of the early identification of potentially nervous children by means of 'Better Baby' contests, of which she was a regular judge. As she explained in her 1916 Arizona Board of Health Bulletin article, 'Better Babies: How Scientific Health Contests Have Made

Possible Much Advancement', having baby contests at local fairs wherein babies are measured, given scores, and referred to doctors to correct any nascent deficiencies was considered not only a good way of catching problems of malnourishment and delayed development, but also could serve as an encouragement to eugenics, by giving recognition to families of desirable genetic stock who produce children within the accepted norms of mental and physical measurement.

Neff's work with the Better Baby contests lies like an ableist elephant across the body of her work, all but obliterating any possibility to see the things of real and lasting worth that she did, like her studies of fatigue and boredom and their impact on the mental well being of children, studies which just grow more important with each passing year of increasingly hyper-scheduled and terminally bored children we regularly produce and hand off to the tender cares of standardised testing and social media, only to wonder why they don't do more free time reading.

Emma Eckstein (1865–1924):

Whenever people are of a mind to take a shot at Sigmund Freud and his legacy, the case of Emma Eckstein is usually among the first bullets they choose to fire. It's hard to blame them – the Eckstein case reveals Freud at his worst on multiple fronts, during a time when he was too much under the influence of Wilhelm Fliess (1858–1928), and too youthfully ready to dictate monolithic reasons and cures for the problems of others. Eckstein came to Freud in 1892 when she was 27 years old and he was 36. She complained of stomach problems and depression, which Freud was all too willing to ascribe to hysteria brought on by excessive masturbation, in full accord with the usual Victorian concerns about the mental impact of any form of self-pleasure. So much would have been par for the course, but where the story goes off the rails is in Freud's subsequent recommendation that Eckstein undergo surgery at the hands of his friend, Wilhelm Fliess. Fliess was an ear, throat and nose specialist who located the source of all human sexual deviations to a naso-genital connection that could be severed through cauterisation or surgery. Freud believed in the procedure, and even had it performed on himself, and felt that it was just the thing to cure Eckstein of her masturbatory excesses.

The procedure did not go well, hindered perhaps by the fact that Fliess left a half-metre of gauze lodged in her nasal cavity, which required a subsequent operation to remove which caused permanent disfigurement. Eckstein suffered the rest of her life with nose bleeds from the failed surgery, a condition which Freud, unwilling to believe in the incompetence of his friend, said was in fact caused by Eckstein herself, and her ongoing hysterical condition.

In spite of these experiences, in 1897 Eckstein began practicing psychoanalysis herself, thereby beginning a long line of women psychoanalytic practitioners, many of whom we shall meet in the pages to come. In 1904, she published *Die Sexualfrage in der Erziehung des Kindes*, a book about the importance of sexual literacy in the upbringing of children which Freud lauded as an important contribution and support to his own theories, thereby beginning another pattern whereby the merit of women psychoanalysts for decades was bound to how much their theories agreed with those of Freud himself.

Yekaterina Konstantinovna Gracheva (1866–1934):

Yekaterina Gracheva's four decades of constant and selfless devotion to the improvement of childcare for the mentally challenged in Russia stands as one of the greatest testaments in this book to the power of one determined individual to affect nationwide systemic change. Before Gracheva, specialised care for mentally non-normative children was simply not a thing. Teachers and nurses were not trained in how to care for their particular needs, institutions did not exist to see to their raising and education, and the idea of developing programs to give such children an eventual vocation was deemed utterly impracticable. Gracheva challenged her nation's indifference towards these children, and traveled through France, Sweden and Germany to study how other countries provided care, psychological support, and educational enrichment for them. Returning to Russia, she established in 1894 an asylum for what were then called '*глубоко отсталого*' or 'deeply retarded' individuals.

Her techniques proved effective, and four years later she was given leave to set up a school to teach 'mentally deficient' children, which she parlayed into a 1900 welfare organisation benefiting the care of the mentally challenged, and in the same year a training program for nurses in best practices for their care. She campaigned for vocational training to allow *отсталого* children to support themselves in later life, and wrote five books on the subject of their needs, abilities, and ways that the state could respond to and support them. Her work was of such value that, while other highly placed intellectuals of the old regime faced various tragedies during the communist assumption power, Gracheva's efforts continued to be promoted, and she continued publishing new works into the 1930s.

Eleanor Acheson Gamble (1868–1933):

Wellesley has, over the course of its rough century and a half of existence, been fortunate to possess among its faculty a string of multi-talented individuals capable of acting the roles of administrator, lecturer, experimentalist, and mentor with equal facility and felicity. Mary Whiton Calkins was one such, as was her successor in the role of head of the department of philosophy and psychology, Eleanor Acheson Gamble. Gamble was born with amblyopia (lazy eye) in her left eye, a condition which compelled her to rely increasingly upon her right eye, until it developed glaucoma in later life. In spite of her vision difficulties, however, Gamble was successful from the start in her aspirations towards experimental psychology. At Cornell University, she studied with Edward Titchener, and her PhD thesis, *Applicability of Weber's Law to Smell* (1898) bore the distinct stamp of his interest in sensory components. (Weber's Law, by the by, is a hypothesis which states that, in order to notice a change in stimulus, the size of the change must be roughly of the same order as the original stimulus – i.e. you notice there are 20 more chickens in a room when the room started off with 20 chickens, but not so much when it started off with 5,000.)

The year she completed her PhD work, Gamble joined the faculty at Wellesley, where she would remain for the rest of her career, becoming a full professor in 1910, and taking over for Calkins as head of the department upon the latter's death in 1930. While at Wellesley, she expanded the operations of its psychology lab, seeing over a dozen graduate students through to their doctorates, and helping some fifty undergraduates to publish papers detailing the research they conducted there. She also carried out her own studies, centred around the topics of memory and senses, the former honed by her time in 1910 studying at the University of Göttingen with Georg Elias Müller, whose studies of memories resulted in the key concepts of consolidation (the brain work required to transfer a piece of learning into long term memory) and retroactive interference (whereby presentation of new material interferes with the ability to memorise any preceding material).

Gamble included her studies on memory and sense in the two volumes of the *Wellesley College Studies in Psychology*, which she edited (published in 1909 and 1916), as well as in her 1909 book *A Study in Memorizing Various Materials by the Reconstruction Method*. She was also actively engaged in mediating between two of her era's conflicting programs for describing the base material upon which psychology works – Titchener's structuralism, and Calkins's self psychology. Her 1933 book *Outline Studies in the Essentials of Psychology* attempted to bridge the gap between Titchener, who thought of psychology as a project of using introspection to break down sensory and mental experiences into their component parts, and Calkins, who as we saw above thought the basic unit of psychology was the self, a being and formulation more fundamental than the body it is attached to. This was a continuation of Calkins's position that her self psychology was not in fact adversarial to her era's reigning structuralist and functionalist approaches. The attempt was later described by the author of her obituary in *The American Journal of Psychology* as 'valiant and fairly plausible.'

Gamble died in August of 1933. Today, in Wellesley College Chapel, there still stands the memorial window to her memory, placed by a university in honor and recognition of her thirty-five years of leadership, in the classroom, in the laboratory, and in the psychological community.

Amy Eliza Tanner (1870–1956):

Amy Tanner had a multi-faceted career in psychology spanning a number of different research interests, achieved through a variety of inventive investigatory techniques. For her 1907 *American Journal of Sociology* paper, 'Glimpses at the Mind of a Waitress,' she went undercover, working as a waitress herself to gather data. She wrote surveys and questionnaires to probe the religious belief structures of children, the moral beliefs of college students, and notions of ideal femininity. She is most famous, however, for her work on the psychology of belief, as laid out in her 1907 paper, 'An Illustration of the Psychology of Belief' and most famously in her 1910 book, *Studies in Spiritism*, which continues to this day to act as a model study for how one scientifically investigates both claims of paranormal abilities, and studies the belief structures and psychological states of those who participate in paranormal rituals.

Studies in Spiritism centred upon the case of Leonora Piper, a trance seance medium popular in late nineteenth and early twentieth century America. Tanner was continuing the work of G. Stanley Hall, who we met previously as the originator of the child studies movement, and who was the president of Clark University, where Tanner was head of the experimental pedagogy department. *Studies in Spiritism* represents the culmination of the efforts of not only Tanner and Hall, but also the larger Society for Psychical Research. It's a fascinating book in many senses, not the least of which being the methodology employed to investigate Piper's claims. My favorite method for getting around a subject's propensity to unwittingly give away information, and a medium's ability to use that unspoken information to guide their cold reading, involved elderly members of the society writing letters and then sealing them before anybody else could see the contents. The society would then wait until that member had died and ask the medium, who was claiming to channel the spirit of the recently deceased society member, to relay the contents of the sealed message. Every time this method was tested, the mediums under scrutiny, in spite of giving very confident dictations of the contents of the letters, failed spectacularly in providing responses even tangentially related to the actual letters in question.

Tanner retired from psychological research in 1919, at which point she bought and operated the Worcester Theater for a number of years to add yet another patch of experience onto her already sumptuous life's quilt. She died in February of 1956.

Oh, and quick tip: If you want to pick up some of her books (which are mostly available in print-on-demand editions), make sure you look her up as Amy Eliza Tanner, because if you just look up Amy Tanner you'll get rather ... different ... book options.

Margaret Drummond (1871–?):
Margaret Drummond was, over the course of the 1920s, a prolific author specialising in the subject of child psychology and its application to education, who has since been almost entirely forgotten to the extent that even ascertaining with certainty when she died has proven an elusive endeavour. She was born in Edinburgh, received her degree from the University of Edinburgh in 1897, and taught at that same institution, beginning in 1929. Her publications include *Dawn of the Mind: An Introduction to Child Psychology* (1918), *Five Years Old or Thereabouts: Some Chapters on the Psychology and Training of Little Children* (1920), *The Psychology and the Teaching of Number* (1922), *Some Contributions to Child Psychology* (1923), *Psychology of the Preschool Child* (1929), and *Gateways of Learning: An Educational Psychology Having Special Reference to the First Years of School Life* (1931).

Beatrice Edgell (1871–1948):
In many ways, Beatrice Edgell was to Britain what Cecilia Parrish and Lillien Martin were to the United States a generation previously – a pioneer not only on account of her gender, but in her adoption of an experimental approach to psychology inspired by German methods. Born in Gloucestershire, Edgell earned her BA in mental and moral sciences from the University of London in 1894 before travelling to Würzburg to

study under Oswald Külpe (1862–1915), an experimentalist who had trained with, you guessed it, Wilhelm Wundt, at the University of Leipzig. Külpe took the experimental methods he learned from Wundt and employed them when he established his own laboratory at the University of Würzburg in 1896, which soon developed a reputation for precise introspection-based psychological testing second only to Leipzig itself. Külpe was interested in proving that some aspects of thought did not rely on or refer to specific mental images, and in ascertaining how the mind goes about deciding what aspects of an object to keep and which to discard when it generates abstractions. Edgell received her PhD from Würzburg in 1901 for her work on the application and limits of experimentation to psychological investigations, becoming thereby the first British woman to earn a PhD in psychology in any country.

Returning to England, Edgell put her experimental approach into practice, studying problems in psychometrics to determine how the mind judges time, with particular interest in the Weberish question of how we judge the temporal midpoint of an event. She also undertook large scale studies of memory in children, discovering differences in the effectiveness of various learning techniques across gender lines after experiments conducted with over 1,200 children. On the instrumental side of psychology, she played an important role in the refinement and calibration of the Wheatstone-Hipp chronoscope, which allowed for a more precise timing of mental events, and thereby opened the number of questions that psychology could hope to answer experimentally. Finally, Edgell was recognised in her time for her application of the insights of psychology to the improvement of societal institutions and other fields of study, particularly with regard to the study of human societies, and the best practices for people engaged in health care.

Edgell taught at Bedford College from 1897 to 1933, a time that saw her produce dozens of articles and three books, including *Theories of Memory* (1924), *Mental Life* (1926), and *Ethical Problems* (1929). In 1927, she became the first British woman professor of psychology, and in 1930 the first woman president of the British Psychological Society. Her full story is told in Elizabeth Valentine's 2006 book *Beatrice Edgell: Pioneer Woman Psychologist*, which is among the hardest to find of the volumes I'll be mentioning in this book.

Gina Ferrero-Lombroso (1872–1944):

Gina Lombroso was the daughter of influential criminologist Cesare Lombroso, who was responsible both for originating some of the greatest criminal reform notions of the nineteenth century, including juvenile court and parole, and for perpetuating some of the most pseudoscientific ideas of nineteenth century theory, including a phrenological and positivistic approach to identifying people possessing criminal traits. Gina was born in Pavia, Italy, and attended the University of Torino, where she received both her doctorate (1896) and her medical degree (1901). Her subsequent career included the popularising of her father's work, as in the introduction she wrote to the 1911 reissue of Cesare's *Criminal Man*, and investigations into the psychology and roles of women in modern society, including *La Donna Nella Vita: Riflessioni e Deduzioni* (1923) and *La Donna Nella Società Attuale* (1927).

Ethel Dench Puffer Howes (1872–1950):

During the late nineteenth century, the one nearly sure-fire way to end your career as an academic woman was marriage. Though one can point to a handful of women such as Christine Ladd-Franklin or Ellen Swallow Richards who, by dint of a series of very unusual circumstances, managed to maintain their professional positions after getting married, the general law of the land (sometimes quite literal, as many states had laws on the books making it illegal to higher a married woman in the teaching professions) was that upon marriage, a woman's intellectual career was done. This was certainly the case for the promising career of Ethel Puffer, who was born in Massachusetts to a family with a strong tradition of women's education.

All four of the Puffer sisters attended and graduated from Smith College, which first opened to students in 1875. Ethel Puffer received her AB in 1891, and after some time at Smith as an instructor in mathematics, traveled to Germany in 1895 to pursue her interest in psychology. Meeting resistance at the University of Berlin on account of her gender, she transferred to the far more welcoming environment of the University of Freiburg, where former Wundt student Hugo Münsterberg took her on as a quasi-adopted member of his family. Münsterberg was an important figure in the history of women in psychology, who not only provided key early encouragement to Puffer at Freiburg, but later to Mary Whiton Calkins, while his ideas were important to Margaret Floy Washburn's long process of disentangling herself from the strict structuralism of Titchener. He was an applied psychologist, with a particular interest in how psychology could be put to use to improve the lot of industrial workers and the accuracy of the legal system (he was, for instance, a central figure behind the scientific study of the unreliability of witness testimony that continues to impact our legal system). He was also a firm believer in the need to communicate the insights of psychology to a non-technical audience, a belief that Puffer would put into practice upon her return to America.

In 1897, Münsterberg was offered a position at Harvard as the successor to William James, who had given up his psychology chair to join the philosophy department, and Puffer followed him there. By 1898, Puffer had wrapped up her doctoral research, but although Münsterberg and several other professors at Harvard wrote testimonials attesting to the quality and importance of her work, Harvard remained adamant that women could not receive PhDs, compelling Puffer to seek her degree from Radcliffe, which granted it at last in 1902.

While her PhD was being sorted out, Puffer taught psychology at Radcliffe, Wellesley, and Simmons, while preparing her research for publication in book form. The resulting volume, *The Psychology of Beauty* (1905), which is comprised of a series of essays seeking to pull aesthetics from the abstract categories of truth and form through which it has been traditionally analysed by the philosophical tradition, and ground it instead in experimental practice to determine what the component parts are to a brain's aesthetic judgments. The book ought to have been the launching point for a multi-decade career, but in 1908 Puffer made the fateful decision to marry civil engineer Benjamin Howes, ending her ability to work full-time at any institution of higher learning. She continued writing papers on aesthetics until the birth of her children in 1915 and 1917 effectively ended her career as a psychologist.

She devoted the rest of her life to women's causes, particularly suffrage and, with the formation of the Institute for the Coordination of Women's Interest at Smith in 1925, to a larger consideration of the systemic roadblocks facing women seeking to live full and engaging lives. The institute existed from 1925 to 1931, and researched the importance of access to childcare, the value of home economics as a means to give women better information for saving their money and time, and issues of women's ongoing fight to obtain equal access to education, and to long-term academic careers. One of the institute's solutions to the perennial issues facing women was a cooperative structure that, by combining efforts of cooking and homecare, would give women more time to pursue other interests. Unfortunately, with the rise of Communism in the 1920s and particularly with its jagged morphing into Stalinism in the 1930s, any solutions that smacked of anything less than pure capitalist pedigree were unlikely to meet with widespread approval, and the efforts and findings of the institute were allowed to go fallow in the subsequent decades, waiting for a new generation, and a new wave of feminism, to take them up again.

Marie Goldsmith (1873–1933):
How do ideas about evolution interface with ideas about psychology? Today, armed with a fleet of biochemical, brain imaging, and genetic tools, we can answer questions of evolutionary psychology with increasing precision (for more on which, see the profile of Leda Cosmides below), but in the early twentieth century, the question about how psychological attributes have evolved over long stretches of time was caught in a web of social and political beliefs that subtly (and often not so subtly) twisted the available experimental data to its own ends. The biologist and anarchist Marie Goldsmith, and even more so her colleague Peter Kropotkin, are outstanding exemplars of this tangle of beliefs and practices.

Goldsmith was born in St Petersburg but spent most of her life in Paris, where she was educated at the University of Paris, where she received her degree in biology in 1894, and her doctorate in 1915. In 1897, she began corresponding with Peter Kropotkin, whose 1902 book *Mutual Aid* sought to 'save' Darwin from the idea of natural selection. As an anarchist, Kropotkin believed that the source of human evolution could be located not in an individualistic and violent Malthusian struggle for limited resources, but rather in a more neo-Lamarckian drive centred around cooperation. Kropotkin leaned significantly on Goldsmith for the biological underpinning of his theory, and she in turn took from him the inspiration to start a Russian language journal, хлеб и воля (Bread and Will), which lasted from 1903 to 1905 and expressed her views on the desirability for anarcho-syndicalism as a positive alternative to the destructive and terroristic anarchism of the late nineteenth century.

Goldsmith published her ideas on evolution in *Les theories de l'evolution* (1909), and throughout the 1910s carried out research in animal psychology, particularly with regard to their reactions to sensory data, which she used as the basis for her ideas about comparative psychology and the evolution of psychological attributes, as contained in her essays '*Réactions physiologiques et psychiques des poissons*' (1915), '*Quelques réactions sensorielles chez le Pouplé*' (1917), and her book *La psychologie comparée* (1927).

Marcelle Lapicque (née de Heredia) (1873–1960):
As the wife of famed neurophysiologist Louis Lapicque (1866–1952), whom she met while studying at the University of Paris, Marcelle Lapicque enjoyed both the benefits and drawbacks of an early twentieth century scientific partnership. Louis was uncharacteristically scrupulous about publicly declaring the independence and import of Marcelle's individual work, but of course that did not stop critics from claiming that the majority of the work they published jointly was primarily the product of Louis's mind. Louis was a specialist on nerve impulses, and was particularly interested in what he called 'chronaxie', or the changes in strength an impulse wave exhibits over time. Marcelle, among the eighty or so papers she wrote, did much work with Louis on chronaxie, but also struck out on her own, particularly in the study of how different poisons impacted nerve function, as contained in her papers '*Action de la strychinine sur l'excitabilité du nerf moteur*' (1907) and '*Influence du suc d'Amanita muscaris sur l'excitabilité du muscle et son inhibition*' (1927). (Fun fact: Amanita muscaria is the mushroom featured most prominently in the Super Mario Bros videogames, and is distinctly poisonous!)

After Louis's death in 1952, Marcelle continued to oversee the Laboratory of General Physiology at the *École Pratique des Hautes Études* until her own death in 1960.

Clara Harrison Town (1874–1967):
In 1905, French psychologists Alfred Binet and Théodore Simon published the first version of their famous intelligence tests, which would play a decisive and fateful role in the United States when combined with native eugenicist policies to identify and institutionalise 'feeble-minded' individuals in the population. By 1910, the psychologist, segregationist, and eugenicist Henry Goddard had taken the Binet-Simon scale and used it to formalise a division of mental undesirables into three categories – idiots (those with the equivalent intelligence of a 2 year old or less), imbeciles (those mentally aged 3 to 7 years), and morons (8 to 12 years), with each category further broken down into three sub-categories based on severity. One of the first institutions to put this system into practice, and to attempt to use Binet-Simon to probe the boundaries between groups, and particularly to find ways of discovering morons who were passing in society as cognitively normal, was the Lincoln State School and Colony, which in 1911 brought Clara Harrison Town on board to run its testing endeavours.

At that point, Town was just two years out from completing her doctorate at the University of Pennsylvania under Lightner Witmer, a founder of clinical psychology, founding member of the American Psychological Association, and inevitably a former student of Wilhelm Wundt. Witmer's focus was a generally positive one, of using psychology to distinguish different types of issues impacting children's ability to perform cognitively at the level of their peers, and to develop ways that schools could aid them to perform at the best of their abilities, going out of his way to work with teachers so that they could recognise students with special needs and offer them extra support.

By the time she had earned her PhD, however, Town had already moved away from the positive approach of Witmer, and embraced a eugenicist policy in line with that of Luther Burbank's 1907 *Training the Human Plant*, Henry Goddard's 1911 piece 'The Menace of the Feeble-Minded,' and Charles Davenport's central eugenicist text, 1911's *Heredity in Relation to Eugenics*. Town believed that the work of evaluating children for mental deficiency should not fall to doctors or school administrators, but rather to trained psychologists such as herself, who would be empowered with the ability to testify in court as to a child's mental ability, and the advisability of his or her transferal to a specialised institution. In 1913 she published her influential article in the *Journal of Criminal Law and Criminology*, 'Mental Types of Juvenile Delinquents Considered in Relation to Treatment,' which broke delinquency into four categories which required variously severe approaches. The last of her four categories, 'the feeble-minded', she describes as follows:

> The simple and appalling fact is that there are in the neighborhood of 280,000 absolutely irresponsible individuals at large in our country. Can we doubt they furnish a large quota not only of our juvenile delinquents but of all criminals? ...This vast host of feeble minded individuals have proven themselves incapable of protecting or supporting themselves and have become a burden to society... The one measure which would bring protection both to the feeble minded and to society, and save the coming generation also from the burden of an increasing host of feeble-minded, is complete segregation of the whole class.

Well, folks – they can't all be heroes.

Town believed that, though the cost of permanently incarcerating the feeble-minded would be high, the savings by reducing the criminal population would more than make up for it, and the value from a eugenics perspective by preventing them from reproducing, was more valuable still. In 1915, her testimony was central to the passing of Illinois House Bill 655, which approved the permanent institutionalisation of those deemed feeble-minded by a qualified psychologist. After her work at the Lincoln State School and Colony, Town went on to a series of faculty and director positions, including a stint as chairman of the clinical section of the American Psychological Association.

Helen Thompson Woolley (1874–1947):
If Clara Town serves as our prime example of psychology wielded in the service of evil, there is a strong case to be made for Helen Thompson Woolley being among the prime examples of it being wielded for good. Through a life encumbered by cascading reversals of fortune, Thompson somehow managed for three decades to hold onto her vision of psychology as a tool to provide people with greater freedom, self-knowledge, and opportunities for actualisation. The daughter of a shoe manufacturer and inventor of middling means, Thompson was valedictorian of her high school, which gave her access to a University of Chicago scholarship that allowed her to go to college when many of her contemporaries, stung by the financial crunch accompanying the Panic of 1893, were compelled to stay at home.

At the time, the University of Chicago was the stomping ground of a number of philosophical and psychological luminaries, but the example of John Dewey (1859–1952) in particular seemed to leave an indelible impression upon the young Thompson. His project of reforming the American schooling system by paying attention to the education of the whole child, rather than the rote learning of a standard set of facts, was realised in much of Thompson's work of the early twentieth century. After receiving her undergraduate degree, Thompson worked under James Rowland Angell for her doctorate. Angell was a founder of the functionalist school of psychology, which held that the job of the psychologist was to ascertain how the mind engages with the environment, and improve our functioning by developing behaviour patterns that take environmental factors into consideration. This outward looking psychology was the opposite of the introspective experimentalism of Titchener's structuralism, and was another component of Thompson's more socially active approach to psychology.

Thompson's work with Angell centred around a rigorous set of experiments to determine the extent of gender differences across a variety of different cognitive capacities. The results, published in her book *The Mental Traits of Sex: An Experimental Investigation of the Normal Mind in Men and Woman* (1903) found that, while women performed better in tests of memory, association, and sensory discrimination, men performed better on motor tasks and tests of ingenuity. The differences between the two groups, however, though significant, were slim, and not nearly of the towering order expected by the popular conceptions of the era. Critics who refused to believe the good performance of the women during the trials accused Thompson of using an unfair sample, because all of her subjects were college students, and college women of the era were held to be more 'manly' in their attributes than women who did not seek a higher education. Generally, however, Thompson's results were received as important contributions to the psychology of gender, and the spirit of them would be continued through the studies of Leta Hollingworth a decade later.

In 1905, Thompson married Paul Woolley, a pathologist whose work, and general restlessness, kept the couple on the move in the subsequent decades, disrupting Helen's ability to find lasting academic positions. Somehow, however, she found a way, no matter where Paul's peripateticism landed them, of finding significant work to do in her field. Two of her major contributions during this era were her work for the

Bureau for the Investigation of Working Children, and her time as a psychologist at the Merrill-Palmer School. Helen joined the bureau in 1911 after Paul moved the family to Cincinnati. At the time, a law was on the books stating that children could remove themselves from school at the age of 15 if they had proof of employment, and Helen's work for the bureau included compiling data on over 5,000 adolescents to assess the impact of early entry into the work force, results she published in her 1914 book *Mental and Physical Measurements of Working Children*, and which played an important role in the establishment of compulsory education laws and child labour bans in the United States.

After Paul uprooted the family once more in 1920, Helen found work at the Merrill-Palmer School as a staff psychologist and assistant director, where she helped establish one of the country's first nursery schools, which served as a testing ground for new approaches to early education, and a training ground for nursery school teachers. Throughout both her time at the bureau and at Merrill-Palmer, she began to doubt the utility of the IQ testing that was growing increasingly popular in some psychological circles as a means of evaluating the intelligence and capacities of young children.

By 1923, Paul was on the move again, this time as a result of health concerns, and he moved to California for tuberculosis treatment while Helen remained in Detroit to earn money for the family, which she accomplished by picking up a second job at the Teachers College in New York. Her marriage to Paul at this point was very much on the rocks, and her mental health was not helped by the stress of working two different jobs in two different cities, while attempting to raise children. The final one-two punch seems to have occurred in May of 1926, when doctors found an abdominal tumor which required a hysterectomy and appendectomy to remove, followed by Paul's request for a divorce in 1927. Helen plunged into a deep depression, culminating in an attempted suicide, and in 1930 she was let go from her position at the Teachers College. Subsequently, she would be unable to find steady work, and grew increasingly dependent on her children for support until her death in 1947 from an aortic aneurysm.

June Etta Downey (1875–1932):

Wyoming born and raised June Downey was, for the first quarter-century of her life, devoted to a life path of classics and philosophy, until the summer of 1901 when she happened to take a course offered by E.B. Titchener, which sparked her interest in psychology. She enrolled for graduate studies at the University of Chicago, where she was supervised by James R. Angell, whom we just met as a positive force in the story of Helen Thompson Woolley. Her PhD work centred around handwriting, which was to be an interest for the rest of her career. Downey's program employed handwriting as a tell-tale indicator of one's fine motor control, and used differences in handwriting when under emotional distress, or during episodes of distraction, to measure the mechanisms of the mind's fine-tuned muscle control.

During a time which, as we have seen, was obsessed with different schemes for the measurement of an individual's intelligence, Downey was an innovator in seeking methods by which something more elusive, one's temperament traits, might be tested, measured, and compared. She called this her Will-Temperament system, which she developed across a series of publications, including *The Will Profile: A Tentative Scale for Measurement of the Volitional Pattern* (1919), and *Will-Temperament and its Testing* (1924), and which employed handwriting analysis as a means of classifying individuals according to a few basic temperamental types: the hair-trigger, the accurate, and the willful. Though today this has the whiff of state fair hucksterism, at the time it was an important step forward in developing criteria to measure non-cognitive aspects of an individual's psychological makeup.

From 1915, Downey served as head of the department of psychology and philosophy of her alma-mater, the University of Wyoming, becoming thereby the first woman to head a psych department at a state school in the United States. She wrote seven books, seventy articles, and a hefty collection of poems, plays, and even the official school song of the University of Wyoming. She died of stomach cancer at the age of 57.

Helga Eng (1875–1966):
The first woman to receive a doctorate in psychology in Norway, Helga Eng is known today as a key figure in the field of experimental pedagogy. In 1909/1910 she studied with Ernst Meumann, the founder of the field, whose *Vorlesung zur Einführung in die experimentelle Pädagogik* went through three editions (1907, 1912, 1920). Employing Meumann's methods, she completed her PhD in 1912 on children's use of abstract terms, and successfully defended it in 1913. She returned to Germany shortly thereafter to study further with Meumann and, inevitably, with Wilhelm Wundt.

In 1918, she published *Kunstpaedagogik*, her book on the importance of art pedagogy in the education of children. This was an important work in helping shift the focus of education away from an omnipresent interest in maximising the results of intelligence testing, and towards a structured approach that supported children's imaginative expression in a manner that did not fully neglect the development of a self-discipline that would allow the child to eventually achieve a harmonious life in society. This was an expression of what she called her 'realistic humanism' – a drive to support the full spectrum of children's emotional and creative lives while still aiming them towards the skills they would need to function as adolescents and adults.

Published in 1926, her most famous, enduring, and endearing work, *Children's Drawing*, was translated into five languages and followed the evolution of her niece Margrete's drawings from the age of 10 months to 8 years. This represented the first ongoing case history and analysis of a child's creative artistic output. She continued the story in 1944's *Margrete's Drawing: From the 9th to the 24th Year*.

In 1938, Helga Eng was appointed as professor of psychology at the University of Oslo, and proceeded to found there the Pedagogical Research Institute. In 1940, she had reached the mandatory retirement age, but so crucial was she to the functioning of her department that the university asked her to remain in her post for a further eight years, finally allowing her to retire in 1948 at the age of 73.

For the curious, as of this writing, Helga Eng and I still haven't walked in the glow of each other's majestic presence.

Grace Helen Kent (1875–1973):

Grace Kent was one of her era's most noted experts in psychometrics, or the development and evaluation of tests used in the gathering of psychological and clinical data. At the University of Iowa, she studied for her Master's degree with Carl Seashore, who was known for the quality of his musical ability metrics, and then proceeded in 1905 to Harvard, where she worked with Hugo Münsterberg, whom we have already met on several occasions for his willingness to mentor women psychologists at the beginning of their careers. In her subsequent career, she developed the Kent-Rosanoff Free Association Test with Aaron Rosanoff. The test, published in the *American Journal of Insanity* in 1910, is still in use, and features a hundred emotionally neutral terms which subjects are asked to produce free associated responses to, the results then compared to a list of associations provided by a control group of a thousand 'normal' individuals. She also performed research into the utility of repetitive tasks as a therapy for dementia patients, work for which she received her PhD in 1911, and developed a famous series of geometric puzzle tests to measure the mental ability of children without reference to linguistic capacities. She spent the last decades of her life evaluating the utility of different children's mental tests, publishing her results in *Mental Tests in Clinics for Children* (1950).

Chapter Eight

Workers as Humans: Lillian Moller Gilbreth and the Founding of Industrial Psychology

Humans have been building structures out of ceramic brick for 5,000 years, and for 4,900 of those years the method had hardly changed: throw a bunch of bricks at the foot of a bricklayer and then watch as he bends over, picks one up, places it, and then proceeds to do that another 600 to 1,000 times over the course of his 11 hour workday. For millennia it continued to be done that way, for no other reason than that was How It Was Done. It wasn't until 1891 that somebody had the seemingly obvious idea of designing scaffolding that places the bricks at chest-level, thus removing all of the unnecessary stooping that ruined bricklayers' backs and quartered their efficiency.

The person who had that idea was Frank Gilbreth, a bricklayer turned construction magnate turned engineer who made it his life goal to use scientific studies of job motions to design work environments which reduced stress and strain on labourers. It was not, however, until he married Lillian Moller in 1904 that he recognised that improving the worker's physical workspace was only half of the story. Frank Gilbreth understood, as nobody else did, the physical hardships of a task, but it was his wife who brought the psychology of industrial work into prominence. Together, they gave America's booming industrial scene the semblance of a soul by giving the physical and mental well-being of workers equal weight with profit and plant efficiency.

There was nothing in the first twenty-six years of Lillian Moller's life which suggested she would some day come to be known as the First Lady of Engineering. She was born to a wealthy family in the burgeoning city of Oakland, California in 1878. Her mother was prone to illness, and her older sister was a beauty, the darling of the family. And so Lillian grew up thinking that, since she wasn't pretty enough to be adored the way her older sister was, the only way to gain her parents' affection was through study. She curled up into a defensive ball of shy self-deprecation and hit the books, excelling in English, languages, and history, and was all set to go to the University of California when her father said that, as she was expected to be a help around the house and lady of refinement, he hardly thought college was appropriate for her.

And that might well have been the end of it, another promising mind cut off from education after high school for no other reason than the propriety of the thing. Fortunately, she talked her father round to her point of view in a rare display of self-assertion and studied English and psychology in college, writing her Master's thesis on Ben Jonson's play *Bartholomew Fair*. She was set to continue on to her PhD when she ran smack into a young construction booster named Frank Gilbreth.

Depending on who you ask, Frank Gilbreth was either a domineering bully who expected the world to wait on his pleasures and bend to his will, or a charming go-getter with stars in his eyes and the world's woes on his shoulders. By the same token, his marriage to Lillian was either a beautiful partnership of mutual support and surprise, or a one-sided act of domination that worked out well in the end only because Lillian was too allergic to conflict to defy Frank's will.

Frank was a dynamo, who grew up with a mother and aunt who made it their mission to see that his homelife was as undisturbed and tranquil as possible, attending to his every need, and instilling in him the expectation that such subservient care was the stuff of normalcy. He dashed off a dozen ideas a day for improving the way America and industry did business, and got so drunk on his own inspiration that he tended to expect that anything he set his mind to do was possible and undersell the difficulties he faced. When he and Lillian married, he informed her that he wanted a dozen children, and so they had a dozen children. He told her that the children were to be raised according to his principles of efficiency, and so they were. And he told her that he wanted her to be his partner in engineering, and so she started learning that trade from the ground up instead of continuing her work in English literature. As he expected the impossible from himself, so he expected it from her as well, and thus the profoundly shy girl who grew up believing that her purpose was to live a life of retired and unassuming academia was challenged to learn engineering, edit Frank's articles and books, raise a dozen children according to his stringent but progressive system, and accompany him on consultations with titans of industry. It was a schedule that gave her no rest for decades as she sought to prove herself equal to the plans he made for her. The result of that forced march was that, by the time Frank died, Lillian was herself recognised as an expert in the new field of scientific management, and as the originator of that field's psychological approach.

Frank's plan was that she would get a PhD in psychology, concentrating on its application to industry, while he showed the advantages of his motion theory work at a nearby factory, a task which would also generate additional data for Lillian to include in her study.

I am now going to tell you the name of the factory where Frank set up his ideal work plant and Lillian carried out her first studies, but you have to promise to be mature about it. Anybody caught tittering is going to have stay after class and clean the erasers.

It was the New England Butt Company.

Here Frank used motion pictures to break down the essential motions of each job and to find ways of cutting those motions down to their bare necessities, sometimes speeding up work by ten times as a result while actually lessening worker fatigue substantially. He invented plant flow charts to track products through their various stages and to determine how best to lay out the factory to avoid backtracking and personnel strain. Meanwhile, it was Lillian's job to study how industrial processes affected people mentally.

It was revolutionary work. Just two decades previously, the only studies done on industrial workers had taken as their starting point that problems with production were problems with worker motivation, that low numbers came from lazy workers. It was the insight of Frank Gilbreth and a few others in the early history of management that poor output was, more likely than not, the result of poor management and organisation, of not thinking rigorously about how individual jobs were to be done and how they should flow from one to the next. But those studies, as much of an improvement as they were, still felt that, once you'd found the One Best Way to do a job, the work was done. What Lillian Gilbreth found was that lining out a series of One Best Ways to do work, even if they made the work easier to do, could still result in workers who were unhappier than if you left them alone under the old inefficient system.

A popular example of her approach is large laundry centres, where women worked hunched over their ironing boards in close quarters, and had to, at the end of each load, haul their completed piles over to another room for further processing. Using Gilbreth motion study and workplace layout methods, a way was found so that the women could do their work from a more comfortable position, with better light, and with a better system for moving finished work out of the way that didn't require them to haul it out themselves.

And the workers *hated* it.

Yes, they were in a more comfortable position, but to do that the stations had to be separated off from each other, which meant the workers were robbed of the consolation of being able to talk to each other during the work. Yes, they didn't have to inefficiently move piles of finished clothes all over the place, but that also robbed them of the time when they mentally reset themselves between loads. Those trips let them approach a new pile of laundry with a fresh mental slate, something that Gilbreth's efficient delivery system didn't. Lillian, therefore, had to study how different types of jobs impacted workers mentally, and had to coordinate with Frank on ways of designing factories that kept his motion design improvements but that didn't hurt emotional well-being in the process.

From there, Lillian worked on studying how different types of jobs affect different types of people, and on developing methods of putting people into work that they were temperamentally suited for, instead of throwing people into jobs they couldn't perform well at and then punishing them when they didn't. After Frank's death, she continued for decades in developing, against the wishes of reluctant industries that thought of the Gilbreth methods as 'pretty but impractical,' the methods of motion analysis and psychological study, and of extending them into new areas. She designed layouts for homes that made kitchens at last sensible, and suggested farm practices that would cut substantially down on transport waste and agricultural worker fatigue.

During the Second World War, when the soldiers were overseas and American industry was faced with the challenge of increasing production with a workforce conjured from America's women, disabled, and elderly populations, who could not be readily replaced if injured, Lillian Gilbreth, though in her late sixties, sped from one end of the country to another, visiting factories and finding new ways to motivate workers to change their methods, to reduce fatigue and therefore injuries, and to compel owners and unions to see the mutual benefit that could be obtained by adopting scientific management principles. The field that had been considered outlandish, eccentric, and too pricey to be of value during Frank Gilbreth's life was now a vital tool in the drive to make America ready for war, and Lillian Gilbreth was proud to play her role, not only in improving the lives of workers, but in finding ways to allow veterans who had suffered losses of limbs to perform jobs they had been excluded from before.

The woman who had started her academic career as a student of Ben Jonson ended it as a professor of management (a position created more or less expressly for her) at Purdue University, a silver-haired old woman who nonetheless could regularly be seen at the gymnasium leading teams of workers in their morning exercises, and putting not a few to shame with her stamina. She advised presidents and was universally recognised as the world's authority on industrial psychology, all while financing college educations for her eleven surviving children from the fees she drew as an industrial consultant. The times were not always easy as a woman in engineering. She had twice been turned away from meetings to which she had been expressly invited because they were being held at clubs that refused entrance to women. Her early articles were rejected because publishers would not even consider printing engineering pieces written by a woman. But, pushed by the manic drive to succeed that Frank Gilbreth beat into everybody he came into contact with, she weathered those storms and became the most renowned woman engineer of her age, and a person to whom everybody who has ever been given an adequate chair, a set of tools in a sensible location, and proper lighting while on a job suited to their abilities owes not a little thanks.

FURTHER READING: *Frank and Lillian Gilbreth: Partners in Life* (1949) is the first important Gilbreth biography, written by the Ur-historian of women in science, Edna Yost. Before Yost, the study of women in science as a historical field was simply not a thing. Her books, *American Women in Science* (1943), *American Women of Nursing* (1947), and *Modern Women of Science* (1966) are the foundational works for anybody with an interest in studying women in science, and feature not only a fascinating array of scientists, but a writing style that is fresh and engaging still. Of course, they are a bit harder to come by, and *Partners for a Life* was written a full twenty years before Gilbreth's death, so if you're looking for something more modern and complete, you'll probably want *Making Time* by Jane Lancaster (2004). The *Cheaper by the Dozen* films (one with Myrna Loy, the other with Steve Martin) are also based on books by the Gilbreth children about their experience growing up under Frank's system.

Chapter Nine

Filled With People: The Teeming Mental Spaces of Melanie Klein

We are never alone. From our first connections with other human beings, we start filling ourselves with them, melding not only their actions and expressed expectations, but our imagined interpretations of their motivations, with our own natural mental machinery until finding the boundary between what is 'legitimately' us and what is 'imported' material formed of a mixture of outside reality and imagination is an all but impossible task. We are us but we are also, to some extent, a little bit everybody we've ever met, and once you realise that the question then becomes, how does that chaotic swirl of self and pieces of others possibly function in the real world?

In the realm of psychoanalysis, nobody pushed that question further, and into more uncomfortable places, than Melanie Klein (1882–1960). She is among the most controversial figures in the psychoanalytic pantheon, an individual who has been lauded as a thinker of an originality and depth that surpassed that even of Freud himself, and decried as a perverted hack who wrote her own damaged psychology onto a generation of hapless and innocent children. In the six decades that have passed since her death, the tempers and partisanship that her ideas inspired have just begun to settle, allowing us to finally approach the question, just what was Melanie Klein?

Klein's most reductive chroniclers place a great amount of emphasis on the unique familial constellation of her youth and early adulthood, and not entirely without reason. Born in Vienna, she was the daughter of Moriz and Libussa Reizes. Moriz had broken from his more orthodox Jewish family to study medicine, but was a generally ineffectual figure who spent more time airily polishing his intellect than effectively providing for his perennially impoverished family. The real powerhouse of the household was Libussa, whose business sense compensated for Moriz's ethereality, and whose resolute self-assuredness as to what is best in any situation dominated the lives of her four children, of whom Melanie was the youngest, and her personal favorite. After the passing of Moriz, the siblings devolved into a perpetual game of mutual misinformation and emotional manipulation to pry what they could from the extremely limited financial resources controlled by Libussa. Most manipulative of all was her brother, Emanuel, whose poor health led him to the belief that he would die young, and thence to the decision that he might as well spend his life living the Byronic ideal of drugs, travel, women, and maudlin self-pity at the family's emotional and financial expense.

For a certain stripe of psychoanalytic historian, everything you need to know about Klein the thinker is contained in that paragraph – *of course* she would grow up to be a theorist of the primary importance of the mother, and the omnipresence of fantastic versions of other people in one's mental life, growing up in the toxic web of control, guilt, and deceit that she did. This has then provided the space for those who are uncomfortable with her more controversial claims to disregard them as the products of an unbalanced mind with an unhealthy and unnatural perspective on how familial relations work.

Ideas, however, rarely have such neat origin stories, and we are going to have to do a bit better than, 'Her childhood was complicated' if we are to get to the bottom of Klein's unique ideas. Returning to our tale, then, Melanie Reizes was an unusually intelligent and strikingly beautiful individual who was being courted by multiple men simultaneously, but who settled on the seemingly unpromising figure of Arthur Klein. He was smart, with good career prospects as an industrial chemist, and perhaps the promise of financial stability after years of wandering in the economic wilderness thanks to the irresolute diffuseness of her father played a significant role in her fateful choice.

Arthur and Melanie were not temperamentally suited for each other, and the constant interference by Libussa, who continued to direct the course of Melanie's life after her marriage, did not help the couple find a workable equilibrium. Arthur was emotionally distant, unfaithful, and was enraged by any deviations from acceptable behaviour he detected in his children, going so far as to lie in hiding to catch his sons in the act of masturbation so that he could punish them for it. Melanie for her part coped with her deepening depression by being around as little as possible, taking regular extended excursions to the homes of friends and relatives to avoid the chilly domestic atmosphere.

After a series of dislocations motivated by Arthur's employment, the family was able to settle for some time at Budapest in the 1910s, where Melanie sought therapy for her depression from Sándor Ferenczi, who was one of Sigmund Freud's most important acolytes in the psychoanalytic cause. In 1914, Libussa died, and in her absence Melanie felt a swirling and contradictory mixture of relief, guilt, and the terrible burden of freedom. She worked through her grief in part by throwing herself into the theory of psychoanalysis, reading Freud's *On Dreams* and finding there a ready and deeply compelling explanation for the multifarious and complicated forces that seemed to be ever pulling at her mind. With Ferenczi's encouragement, she began the process of analysing her own children along Freudian lines.

The problem, however, was that her son Erich was younger than Freudians generally believed was useful or even responsible to subject to psychoanalytic analysis. Lacking the more sophisticated psychological machinery that Freud had said developed after the age of 6, Klein's critics said, young children don't have the capacity to understand what is being told to them about their impulses, or the ability to form the complicated links with their therapist (known as 'transference' in the Freudian literature) that were crucial to a successful analysis. Klein proceeded anyway, and found what she thought were decidedly fruitful results. So, let's talk about those....

Or rather, let's get *ready* to talk about those, because Klein's ideas about the mental world of children are of a sort that people tend to react to with instinctual disgust, or mockery, and while some of her conclusions are ones that you'll probably end up rejecting, what I recommend is to go into them with as much of an open mind as possible for as long as possible, for decorating the path through her life's work are these glimmering insights into the often tragic interconnectivity of humans that are well worth the having, and that you don't get if you reject the entire Kleinian system out of hand upon contact with some of its more shocking elements. That said, let's go.

For Klein, as her thinking would develop over the course of the two decades following her first exposure to psychoanalytic concepts, the critical moment of an individual's life comes when he or she is weaned from their mother's breast. Suddenly, the young brain has to scramble to make sense of a world where something purely good has been definitively taken away from it, and in the process of trying to piece together what existence is now about, the child takes into itself nebulous ideas about good and bad as represented in the person and body of the mother which go on to become long-term mental residents. The mother as a good giver of comfort and life becomes supplanted by mother the denier, who either keeps all the good things in her body to herself, or, worse, seemingly shares them with the father (this was a time when children routinely slept in their parents' rooms throughout early childhood and thus witnessed parental sex on the regular).

Internalising this conception of Mother as a denying and hostile force, a child develops sadistic impulses, impulses directed against their mother, and subsequent anxieties as they attempt to reconcile their conceptions of mother as a good and giving force with those of her as a selfish and castrating one. These anxieties, Klein hypothesised, come out in children's play, and it is the job of the therapist to notice them, and inform the children what tensions lie underneath their actions, as a way of helping them confront what underlies their anxieties and overcome them. This is where the discomfort really starts creeping in for most people experiencing Klein for the first time – the idea of taking a young child who has behavioural or emotional problems and telling them, for example, that the reason they cut the coal out of a toy train was that, subconsciously, they want to cut the feces out of their mother, or another who is sucking on the lamps of a toy carriage that it is because they want to suck on their father's penis is, to put it lightly, the last straw for most.

I get that, but let's put a pin in the carriage penises for now and look at another direction in which Klein took her thought in later years. Because at a young age we incorporate into ourselves both positive models of those around us, and threatening ones constructed from our anxieties and fears, we have at the same time conflicting notions of ourselves as good and whole individuals living in a loving world, and as individuals carrying in our deepest selves something rotten, something that threatens both the good in the world, and the good in ourselves. We live with the parts of people we have injected into ourselves, and with the fantasies of people that we have projected from ourselves out into the real world, and our poor egos have to navigate all of that as they attempt to find Reality in the mix of perceived hostilities and early wounds that colour our approaches to the world. For Klein, in a series of influential papers that peppered the 1920s and 1930s, the difficulties of recognising the real and the fantasy in the plethora of individuals we have let into ourselves as both good and bad semi-permanent residents, and in the characterisations of people we have constructed and layered over our perceptions of actual people in the world, drive a number of our psychological problems, from manic-depressive states to suicidal drives.

After the end of First World War, the anti-Semitic regime that ultimately rose to power in Hungary compelled the flight of the Kleins to Berlin, where Klein's increasing heterodoxy paved the way for a frosty reception among psychoanalytic circles there. Freud had held that the father was the driving force in creating early anxieties. Klein argued for the primacy of the mother. Freud had said that children probably couldn't be profitably analysed until about their sixth year. Klein included cases of children as young as 2.5 in her works. In fact, at this period, Klein and Freud had much more in common than their divergences, but in these early days, any small deviation from orthodoxy was grounds for concern, and in spite of having psychoanalytic luminary Karl Alexander as her mentor in Berlin during this time, these were not good years for Klein. Her marriage disintegrated, her ideas were regularly challenged, and her financial situation declined in the absence of Arthur's support.

Fortunately, what was heresy for Berlin was progress for London. Psychoanalysts there, led by Ernest Jones, found Klein's ideas promising, and invited her to lecture on them in 1925, the success of which led Klein to move to England in 1926, there to remain for the rest of her career. The British Psycho-Analytical Society chafed under the relative prestige of the Berlin and Vienna chapters, and considered the acquisition of Klein as a major Get. For at least the first half decade of her time in London, Klein's ideas were actively supported by the majority of the Society, and Klein grew under these positive influences to develop some of the deep insights into childhood and the processes of introjection and projection that we noted above.

As with Freud before her, however, Klein was to experience what happens when a fertile but controversial new path forward in the understanding of human psychology gathers a following. She found herself having to defend her ideas from those who increasingly vocally disagreed with her ideas, including the persistently brutal attacks of her own daughter, Melitta Schmideberg, herself a psychoanalyst, and those of Anna Freud's partisans, who had a very different approach to child psychology. In addition, she felt compelled to protect her thought from those who agreed with her concepts and wanted to push them beyond the borders she originally laid out. She became as controlling as Freud of the strict correctness of her ideas as stated in her foundational writings, and not a few supporters found themselves flung from her good graces on the grounds of minor differences of opinion.

In 1938, Sigmund and Anna Freud arrived in England, fleeing Nazi persecution, as part of a wave of continental psychoanalysts whose safe passage to England and the United States was orchestrated by Ernest Jones. Klein was worried about the arrival of her great rival, even to the point of being critical of Jones's refugee efforts, but in the end a tentative truce known as the 'Ladies Agreement' was laid out in 1946 which prevented the British Psycho-Analytical Society from irrevocably splitting into sniping factions by making space for both the devotees of Anna Freud and those of

Melanie Klein. Klein for her part continued to develop her thought into her seventh decade of life, focusing on how early internalisations and frustrations continue to have ramifications in adult life that manifest themselves in how humans experience envy, grief, and the fragmentation of an individual's cohesive sense of self.

Klein died in 1960, never having reconciled with her daughter Melitta, who did not attend her funeral, and having witnessed her grip on the British Psycho-Analytical Society gradually erode as medically trained and biologically oriented individuals rose through the ranks, making her insightful and intuitive approach to analysis seem increasingly like whispers of psychoanalysis's primal past. She had, however, decidedly left her mark on the development of psychoanalytic theory, and today, though individuals might disagree about the content of early psychological development, and the particular ratios of outside influences versus internalised individuals that determine our behaviours and approach to reality, there is no longer a question of ignoring these ideas altogether. Klein showed us that we are filled with other people, and that the sooner we make ourselves acquainted with them, the more we can understand who we are and why we do what we do.

FURTHER READING: Melanie Klein is perhaps the most written about woman in the history of psychology, so in a welcome change we find ourselves not having to *make do* with the sources at hand, but having to choose *which one* to approach first. The one that online book vendors tend to throw at you is Hinshelwood and Fortuna's *Melanie Klein: The Basics* (2018) but the structure of the book (which might just be the decreed structure for all *Basics* titles) I find adversarial to grasping the flow of her life and work. If you want a good summary of decent length, the section in Janet Sayers's *Mothers of Psychoanalysis* (1991) is a good place to start, or if you want to just dive right in I like Phyllis Grosskurth's *Melanie Klein: Her World and Her Work* (1986) quite a lot as a thorough account from a time when Klein was deceased but still very much a living presence, which takes seriously the genesis and development of her ideas while also detailing the idiosyncrasies of Klein as a person and intellectual. If you want to read Klein herself, you can still get copies of the excellent four volume *The Writings of Melanie Klein* series put out by The Free Press in 1984.

Chapter Ten

Helene Deutsch, As-If Personalities, Adolescent Friendship and the Art of the Quiet Revolution

Helene Deutsch (1884–1982) was a fantastically successful clinical psychiatrist and psychoanalyst who felt guilty about nothing quite so much as her fantastic success. As she grew older, and her circle of friends began contracting as her own position in the psychoanalytic tradition came increasingly under fire, Deutsch increasingly came to doubt the value of the family sacrifices she had made in the name of her early career even as she knew that, given her early life's story, she could hardly have chosen other than she did.

The woman who, of all that robust roster of early psychoanalysts, did the most to attempt to reconcile Freudian theory with the actual lived experience and development of women, grew up in circumstances that could not have been more Freudian in scope. Helene Rosenbach was born in 1884 in a far-flung formerly Polish corner of the Austro-Hungarian Empire to a respected lawyer father who looked upon her lovingly as the realisation of high intellectual hopes beyond the capacity of her feckless older brother, and a mother who, by Helene's account, hated everything about her, and would spring into violent rages almost at the sight of her. The dynamic of wanting to please her idealised father through professional achievement, while at the same time wanting to separate herself from the respectable expectations of her mother, drove Deutsch into the arms of medical study in the first instance, and into those of an older married man in the second.

Herman Lieberman (1870–1941) was a socialist-leaning Polish attorney who both devoted all of his waking hours and resources to the cause of workers, and privately despaired of workers' ability to ever truly raise themselves above their reduced station in life. Some fourteen years Deutsch's senior, and married, he was the object of her early adolescent fantasies, and formed a romantic attachment with him at the age of 16 to the undying shame of her mother, a connection that gave Deutsch unique insights into the grey regions between fantasy and action that she would so famously later elucidate in the first volume of her *The Psychology of Women*.

While frustrating her mother, and breathing in the heady atmosphere of early Austrian socialism and Polish nationalism, young Deutsch had to decide for herself where to apply her intellectual gifts. Her first choice would have been to follow her father into the legal profession, but as a legal education was entirely out of the question for a woman of her place and time, she settled upon medicine, attending the University of Vienna beginning in 1907, and early switching her field of study from pediatrics to psychiatry. For the next four years, she managed balancing her studies with Lieberman's demands to perform various errands for him in the city, and to meet him surreptitiously in a scattering of safe locations when he had the opportunity to slip away from his professional and family responsibilities. In 1910, she scored an 'excellent success' on her first round of medical examinations, thereby convincing her mother that her studies were something more than an act of rebellion, and represented an actual possible career path.

In 1911, she ended her relationship with Lieberman, realising that he would never leave his wife, and that she deserved something more than the intense but occasional scraps of attention he was capable of giving her. She soon found Lieberman's antithesis in the form of Felix Deutsch, an emotionally supportive, single, and stable doctor her own age who held out the prospect of a long-term and loving relationship that she craved after a decade of tumultuous life as The Other Woman, but that she would denigrate, as the years passed, as a poor substitute to the flawed but passionate relationship she had given up. They were married in 1912, and as much to escape a Vienna that had so many complicated memories as to deepen her knowledge base of clinical psychiatry, Helene left Felix behind in that city while she went to Munich in 1914 to work with Emil Kraepelin, a foundational figure in the establishment of a scientific psychiatry that had as its underlying premise the biological origin of most psychiatric disorders.

Kraepelin had been, like so many key figures in early psychiatry and psychology, a student of Wilhelm Wundt, and was skeptical of the work of Sigmund Freud, which Helene had been introduced to a few years prior to her arrival at his clinic, and which favorably impressed her. Kraepelin set her on a research project which had as its goal the disproving of one of Freud's assertions, and on familiarising herself with the work he was doing studying cases of *dementia praecox* (what we now call schizophrenia). Initially bursting with enthusiasm at working in a new city, with one of her field's greatest minds, in an institution that had scientific standards much more rigorous than those prevailing in Vienna, Helene's notes home to Felix soon expressed disillusionment. Next to the monolithic sureness of Freud, the scientific research process reigning in Munich seemed 'fumbling and searching.' If she was to become a great psychiatrist (for she was not yet contemplating becoming a psychoanalyst), she would need to find a working clinic that brought before her regularly the whole spectrum of psychiatric disorders, in their full diversity and complexity, rather than burying herself in a specialised research institution.

To realise this new ambition, Helene returned to Vienna, where the outbreak of the First World War would afford her a unique opportunity to oversee the entire women's division in the psychiatric clinic of future Nobel laureate Julius Wagner-Jauregg (1857–1940). Wagner-Jauregg was also not convinced of the validity of Freud's work, but knew of Deutsch's growing Freudianism and affably tolerated it, particularly as the personnel drain wrought by the war made un-draftable trained psychiatrists as herself so valuable. Privately, Deutsch's study of, and belief in, psychoanalysis was growing, as publicly she received plaudits and admiration for her deft and sure handling of a clinical division in wartime, and particularly for her nuanced and sympathetic approach to those suffering psychological damage as a result of their war experiences.

In 1918, Deutsch made the fateful decision to officially join the psychoanalytic movement, becoming a member of the Vienna Psychoanalytic Society and undergoing an analysis with Freud himself (albeit a truncated one) while taking on her first patient, another psychoanalyst by the name of Victor Tausk. It was an unfortunate first case, as she ended up squarely in the middle of the cannonades of jealousy and accusation volleying back and forth between Freud, who was uncomfortable around Tausk and felt that the younger man was trying to steal his ideas, and Tausk, who was desperate for Freud's approval but also felt the great man had stolen *his* ideas. Tausk used his sessions to try and talk to Freud through Deutsch, an untenable situation which Freud brought to an end after only three months. Later that year, Tausk committed suicide.

Deutsch's star, meanwhile, was on the rise. She filled her schedule with psychoanalytic work, and in 1923 left Vienna again to undergo analysis in Berlin with Karl Abraham, remaining there a year while Felix raised their 6-year-old son, Martin (born 1917). Felix had a natural way with children, and his bond with Martin was always a strong one, and a source of both guilt and jealousy for Helene, whose instincts and priorities simply didn't lie along those lines, as much as she might wish they did. While in Berlin, she carried on a brief affair with Sándor Rado, who brought some of the Lieberman-like zest and dash that she had been missing in the less sexually adept Felix, and wrote *Psychoanalysis of the Sexual Functions of Women*, the first full length book by a woman psychoanalyst, which was published the following year.

This was not her first published work in the psychoanalytic tradition – three years previously, in 1921, she had written an interesting paper, 'On the Pathological Lie,' which investigated people who compulsively lie, even when there is no advantage to themselves in doing so. Rather than the opportunistic lie of the charlatan, the deceptions of the pathological liar she described as creative acts that were either sources of pleasure as instances of spontaneous imaginative play, or sources of defense as the creation of a public alter-ego character that is free of attachments to the subject's real past traumas. *Sexual Functions*, however, was the work that put her well and truly on the map, presenting an image of women, psychoanalytically considered, that stayed within the technical confines of Freudian theory, while exploring realities of biology and circumstance thereto neglected by traditional Freudians.

Sexual Functions kept some of Freud's basic categories about masculine and feminine traits, while quietly passing over others, like penis envy, in order to get at Deutsch's real interest, which was the question of how women's biological development shapes their psychological destiny. Given that so many uniquely feminine biological events happen in the course of a woman's life, she argued, it is natural to assume that they have unique impacts on how women interpret the world around them, engage with it, and defend themselves from it. Like Melanie Klein, she argued that the role of the mother should be given a larger place in psychoanalysis than it had traditionally been in Freud's more patriarchal scheme. She also looked at the birth event as one that has unique psychological meaning for women that are not fully accessible to men, arguing that a mother's child is an ego ideal for the mother in a way that it qualitatively can't be for the father. Further, she was some decades ahead of her time in focusing on menopause not as a tragic boundary and end to normal and healthy biological functioning, but rather as a time of potential rebirth for women, a chance to rewrite the script of their lives and choose exciting new directions unavailable to them before.

Upon her return to Vienna, Deutsch assumed the directorship of the Vienna Training Institute, a position she would retain until her relocation to the United States in 1935. This was an attempt to import to Vienna the successful training regimen for prospective psychoanalysts that had been erected in Berlin. Freud wanted it led by a person with experience in both clinical practice and psychoanalytic theory, and Deutsch, with her time at Kraepelin's and Wagner-Jauregg's clinics, was a natural choice for the position. She played a key role in shaping the Institute along minimally invasive lines. While others argued that prospective analysts should have set rules to follow for the situations that they might encounter, Deutsch felt that analysis was a much more creative and artistic process that each individual had to approach on their own terms and with their own strengths, figuring out their own style along the way without too much pressure to fit one particular mold. In 1930 she published *Psychoanalysis of the Neuroses* as an extension of her Institute work. It was a collection of lectures designed to show, through examples, how different situations that arise in analysis could be observed and interpreted, without laying out strict and monolithic protocols for practice.

As the 1930s dawned, Deutsch's satisfaction in Vienna waned. The old guard clustering themselves around Freud's faltering body were engaged in bitter mutual battles for recognition and status, while Freud himself was positioning his daughter Anna as his heir apparent, with responsibilities that gradually crept into Deutsch's realm of responsibility. Through contacts with the psychoanalytic community in Boston and New York, and her own travels there, Deutsch was inspired by the United States as a land free from the musty tradition and frantic back-biting of Europe generally and Vienna particularly, where somebody of her talents could construct a new and engaging professional life. Against Freud's wishes, she moved to Boston in 1935, the year after she presented the Vienna Society with her first paper on the 'as-if' personality.

Next to *The Psychology of Women*, Deutsch's development of the 'as-if' character ranks as one of her most enduring achievements. Presented in its final form in 'Some Forms of Emotional Disturbance and Their Relationship to Schizophrenia ('As-If')' (1942), the As-If individual is characterised by their capacity to act as normally, intelligently, and carefully as a situation requires, but without any actual connection to what they are doing. Such people, due to the lack of an early identification with and internalisation of their parents, are prone to changing the structure of their personality on a dime to fit whatever the dominant paradigm around them happens to be, and while performing competently whatever they are given, never perform with genius or inspiration. They are masters of acting As If they are emotionally engaged with, and satisfied by, their actions in life, but no activity, no relationship, is able to do more than glance off their surface as they remake themselves in the image of the strongest personality in the room.

With the success of her As-If paper at her back, and her long experience in analyst training to recommend her, Helene was an instant hit in Boston psychoanalytical circles, even as Felix never fully found his feet in the new nation that didn't know what to do with somebody who was half physician and half psychoanalyst. Her greatest achievement of her American years was undoubtedly the two volume *The Psychology of Women* (1943, 1945), which delved into aspects of women's lived psychological existences still untapped after nearly a half-century of psychoanalytic practice. Eschewing the speculation about the psychological forces shaping children as infants which the public so associated with Freudian theory, Deutsch concentrated on the unique challenges she had seen girls have to overcome as they navigated adolescence on their way to motherhood. One of the most interesting aspects of these volumes was the light they shone on a hitherto entirely neglected aspect of adolescent girls' development, the role of strong friendship.

As young girls assert themselves to gain independence from their parental figures (and particularly their mothers), friendship takes on a new weight, whereby an adolescent girl is keenly aware of the precise level of connection she has with her best friends, and feels profound pain when she senses any diminutions in those relationships. Through her fantasy life, which can include masochistic elements she must learn to overcome, and through her social life, which provides her with alternate role models she can attach herself to as she attempts to navigate the seas of coming adulthood without the liferaft of her parents' rejected example, she attempts to establish herself as worthy of adult status and freedom. It is observations like these, of important but hitherto ignored aspects of women's early lives pulled from her decades of experience with both her own patients and those of the analysts she was training, that gave *The Psychology of Women* its staying power in the face of attacks (primarily by Karen Horney) that Deutsch did not go far enough in repudiating Freud's entire approach to gender.

After the Second World War, Deutsch turned increasingly away from her focus on women's psychology, and produced influential work on narcissism, which she characterised as the result of individuals having experienced a shattering blow to their self-image which compelled them to then place all of their sense of value in how others view them. In 1963, Felix died after years of steady decline that saw him drummed out of the Boston Society as mentally unfit to continue his practice. Deutsch lived on until 1982, publishing her memoirs in 1973 and being elected a Fellow of the American Academy of Arts and Sciences in 1975. For years, her reputation remained largely where Karen Horney had left it – as an important early Freudian whose legacy was tarnished by her unwillingness to split from Freud, but she was in truth much more than that. Disgusted by the in-fighting of the early psychoanalysts, she became an expert in absorbing the best of their thought, and folding into it the fruits of her own keen insights that challenged old conceptions without blowing those challenges into dramatic civil wars, quietly discarding what she didn't need, while advancing into the unexplored darkness of her predecessors with a sure-footed but understated instinct that moved psychoanalysis forward without breaking it apart.

FURTHER READING: For biographies, Paul Roazen's *Helene Deutsch: A Psychoanalyst's Life* (1985) is your best full-length treatment, though I've always thought it was written under some sort of sudden deadline pressure – the beginning is expansive in its treatment of every single detail of Deutsch and Lieberman's life, but the last forty years of her life, including some of her most important works, are rushed through at a gallop with *The Psychology of Women* stuffed somewhat unceremoniously into the Epilogue. It's just odd. Deutsch also makes up a quarter of a book I'll be bringing up again later, Janet Sayers's *Mothers of Psychoanalysis* (1991) which for my money does a better job of talking about *The Psychology of Women* though it necessarily gives much less detail about her early life. So, best probably to play it safe and get both. Meanwhile, for Deutsch's own writings, *The Psychology of Women* is available in more expensive modern printings, but you can also pretty easily find cheap copies of the mass market paperback 1973 printing, so I'd say start there.

Chapter Eleven

Speaking Culture to Psychoanalysis: Karen Horney's Gender Revolution

How much of womanhood is a matter of biology, and how much one of culture? Prior to 1929, Freudian psychoanalysis had closed rank determinedly around a biological position: the lack of male genitalia locked girls into a series of unavoidable psychological states that fundamentally plotted their course as women. Penis envy and the Oedipal complex were foundational dicta of early-twentieth century psychoanalysis which only the most heretical of apostates would even begin to doubt.

One woman, however, trained as a doctor and immersed in psychoanalytic theory during its formative years, looked at the field's reigning assumptions about women and dared to ask big questions: is this a description of women as they fundamentally are, or merely as a particular bracket of post-Victorian European males have chosen to interpret them? Is it enough for a group of men to produce brilliant analyses of the psychology of boys, and then just assume that everything is the opposite for young girls? And most fundamentally, are European women the way they are because that is how women must be, or have they been made that way over the course of centuries and millennia of social conditioning – conditioning that, once uncovered and explained, might be halted and even reversed?

Karen Danielsen (1885–1952) was, since youth, a poser of fundamental questions. Her father was a ship's captain who spent most of his life at sea but who, during his short spans at home, oppressed the household with his religiosity and his stifling moral strictures to such a degree that her mother, to escape the gloom of her domestic lot, took the unprecedented step (for a woman in the Kaiser's Germany) of leaving her husband and applying for divorce. In the midst of this homelife, Karen was pushed to confront several massive topics: What good was religion, if it bred tyrants like her father? What good was love, if it only led to misery?

Karen found her satisfaction in studies, and especially in the freedom that being a student allowed her. Attending a school in a neighboring town allowed her to slip the watchful eyes of her community, to experiment with the forbidden thrills of walking with boys without a chaperone, of observing the bustling life of the prostitution sector, and of bravely walking into a man's apartment alone to listen to him sing excerpts from Wagner. Her diaries from the period are filled with accounts of her romantic and fantasy life, micro-analysing her motivations and those of the people around her.

By the time she decided to, in defiance of all German tradition, set her sights upon becoming a doctor, two parallel threads of her life had been set: the drive towards challenging intellectual work that would make her one of the world's most famous psychoanalytic researchers by the time of her death, and a need for love coupled with a drive for independence that dominated her emotional, and sometimes professional, life.

After passing her *Abitur* exams, she enrolled in the University of Freiburg in 1906, one of the few institutions in Germany that had opened its doors to women. There she made progress in her studies while obsessing over a complicated romantic triangle involving two friends: a movie-handsome medical student called Losch and a down to earth but emotionally complex PhD student by the name of Oskar Horney. She could go on adventures with Losch, but Horney was the one she would always write to afterwards to work through what those adventures *meant*. She noted in herself the tendency to push away love objects as soon as they had demonstrated their devotion to her, but knowledge of one's faults is not always enough to overcome them, and Karen in due course rejected Losch and embraced Horney, marrying him in 1909 only to embark on a series of affairs that opened up new triangles and complications.

Those trials were in the future, however, when Karen and Oskar moved to Berlin in 1909 and thereby placed themselves squarely in the centre of a radical new intellectual atmosphere. Communism, transvestitism, drugs, free love – everything was on the menu during these, the closing years of the Kaiser's Reich in this city which permitted itself everything. And here was gathered a group of intellectuals devoted to a new school of psychiatry born in Vienna from the mind of one of the twentieth century's most towering figures: Sigmund Freud.

If Vienna was the birthplace of psychoanalysis, Berlin was where it experienced its heady adolescence. Karen Horney, who had never been satisfied by the traditional answers to humanity's neuroses, found herself drawn into the orbit of Berlin Freudianism, enticed by the potentials for curative self-discovery that the new discipline seemed to offer. Though she would come to differ with Freud on many points, on many basic assumptions she would never cease to extol his insight: the power of the unconscious, the significance of dreams and accidental utterances, the importance of childhood, the utility of analysis for revealing deep psychological truths, and the need to recognise women's sexual desires. For a person of such keen powers of self-observation as Horney, psychoanalysis seemed uniquely suited as the primary vehicle of her medical practice.

To become a psychoanalyst, one had (and has) to go through analysis one's self, and Horney's analysis began in 1910 under Karl Abraham. The experience was alternately illuminating and frustrating as Horney came hard against early Freudianism's tendency to paper over individual trauma and neuroses with blanket statements referencing childhood genital embarrassment or inadequacy. She began to question the universality of certain Freudian structures: did shame and frustration at lacking a penis *always* play a formative role in a girl's psychological development, and was resentment of the mother *always* at the core of a woman's adult behaviour, or were these merely things that male researchers guessed ought to be true lacking firm evidence to the contrary, and subsequently made canonical?

Horney's great revolution lay two decades in the future when she first began her analysis, however, and in the interim there was enough triumph and suffering to fill her time. Her psychoanalysis practice bloomed while Oskar was making gobs of money from his position in one of Germany's most important wartime industries. They had servants and wealth, and soon three children entered the picture, but Karen and Oskar's work kept them from home, and Karen's tendency to throw the children into a series of mutually contradictory experimental education programs made them resentful and miserable. Professionally satisfied, her guilt over her missteps as a parent and her seeking after alternate outlets for affection through affairs made of these years a decidedly mixed bag.

Until 1923, however, there was at least enough money about to keep the family afloat. While other families struggled to find food, Oskar's work kept him insulated from the gnawing depression of post-war Germany until his employer's various speculations led the firm to ruin and Oskar found himself suddenly bankrupt and, in a doubling knife twist of fate, ill from meningitis as well. These final blows were too much for the family to withstand, and Karen and Oskar separated three years later. Karen kept the children and entered bravely into a world of emotional and economic hardship, but also one of intellectual recreation as she finally put to print the doubts that she had been feeling about the conservative orthodoxy of Berlin psychoanalytic circles.

In 1926, the year of her separation, she wrote *The Flight from Womanhood*, in which she argued that the very terms in which the analysis of womanhood had been couched were tied to the assumptions of masculine culture: 'How far has the evolution of women, as depicted to us today by analysis, been measured by masculine standards and how far therefore does this picture fail to present quite accurately the real nature of women?' Men place an overbearing emphasis upon the nature of their phallus, and therefore when they write theories about women, this phallus is located at the centre of their psychology. Men take themselves as the model for success, and therefore label anything that doesn't correspond to their core imperatives as inadequate or substandard. And so it has come to pass that women, who possess the ability to create *life itself*, have been told that they are psychologically crippled for life because they don't possess male genitals. Might it not, Horney suggests, be possible that men, lacking a womb, are far more psychologically crippled by that realisation of biological extraneousness than are women confronted with their lack of a penis.

Beyond these considerations, there is the matter of culture and society. If women are seeking to abandon their femininity and become 'masculine' in modern society, might that not be because of the socio-economic realities of modern Europe, and the necessity of labour in industrial society? Has not perhaps millennia of social and economic subordination pushed women into a sense of inferiority, a skewed sense of self-worth, that has much to do with power structures and less with childhood sexual drives?

To be clear, Horney is not saying that childhood sexual exploration and fantasy do not exist and play important roles in psychological development. In fact, she wants in this and other papers to push the vaginal stage of development significantly earlier than traditional Freudianism would have allowed. But the idea that biology and culture had to be considered together when evaluating the neuroses and psychological struggles of the modern woman was, as reasonable as it seems today, rank heresy to the psychoanalysts of the 1920s.

The papers that followed shone, if anything, a more intense light upon the shortcomings of Freudianism's concept of women. She pointed out that psychoanalysis had not yet attempted a thorough-going cross sectional sample of womanhood, but had rested content with its sampling of well-off intellectual European women who displayed extreme neurotic symptoms. Why, Horney asks, if so much of our theory is based on the traumas and acts of self-redirection of little girls, do we not do studies of them? Why don't we study middle and lower class women who lead 'normal' lives so that we can evaluate more clearly the universality of our statements about women? Maybe then we'll discover how much of woman's psychology is due to the basic facts of being biologically a woman, and how much to the specifics of a child's upbringing.

For these radical departures from established theory, Horney was fast earning a name as a heterodox practitioner. Had she remained in Berlin, the consequences would have eventually been sure ejection from the heart of psychoanalysis. With the rise of Hitler, however, the predominantly Jewish intellectual makeup of early psychoanalysts would increasingly work against the growth of the movement in continental Europe. Horney herself was not Jewish, but between the antagonism of her colleagues and the darkening political scene, a move to new environs was welcome, and in 1932 she immigrated to America at the invitation of Franz Alexander to work at the new Chicago Institute for Psychoanalysis.

For the next two decades, Horney's life was a mixture of massive success amongst the lay public and diminishing status amongst the professional psychoanalytic community. Her books, such as *New Ways in Psychoanalysis* (1939), *Self-Analysis* (1942), and *Neurosis and Human Growth* (1950), were wildly popular, selling over 500,000 copies and driving a steady stream of patients to Horney's practice. In them, she advocated for a new approach to psychoanalysis, one which focused on a warmer engagement with the patient, and on resisting the temptation to relate Everything the analysand says to a pre-formed childhood trauma theory. Simultaneously, however, her professional position was steadily eroding as concerns about her heterodoxy and teaching style drove her from America's most important psychological institutions.

The years 1932–1952 present a frustrating and complicated tale without clear villains or heroes. In 1941 she was ejected from the New York Psychoanalytic Institute for a tangled web of reasons. From the Institute's point of view, she was an instructor who seemed totally unwilling to let students absorb the basic principles of psychoanalysis before attempting to convert them to her particular interpretation of it. From Horney's point of view, the Institute was just trying to squash any approaches to psychoanalysis that deviated from their strict Freudianism.

Incensed, she formed her own institute and no sooner had she founded it than she ejected one of its most important teachers (and Horney's erstwhile lover), the philosopher and psychologist Erich Fromm, for what seemed to many of the student body and faculty as insufficient reasons. It appeared to them that Horney had left the New York Psychoanalytic Institute in the name of intellectual freedom only to erect an institution that enforced Horneyean Orthodoxy with the same narrow rigidity. Her institute cracked in two, and then cracked again. She began a romantic relationship with one of her younger students, at least the second in her career, and an embarrassing breach of professionalism. She took longer and longer vacations and neglected her practice in favor of her writing, which was itself moving towards notions of authentic self-actualisation that sounded to the psychoanalytic establishment more like Zen-infused self-help tracts than serious scientific research.

But Horney by the late 1940s was beyond such criticism. After its fumbling start, her Institute was a runaway success heading into the 1950s, and she could point to a string of patients and readers whose lives had been changed utterly by her common sense, jargonless approach to the detailing of their mental lives and the struggles between actual self and ideal self that kept them chained to perverse cycles of self-loathing. In the last year of her life, suffering from a lung cancer she didn't yet know she had, she visited Japan to learn more about the Zen beliefs that had so captivated her, and there enjoyed the greatest intellectual and personal adventure of her life, visiting Buddhist shrines and coming face to face with the truth of her old belief that a psychology that neglects culture can never truly understand people.

Karen Horney returned from Japan in the Summer of 1952, was admitted to a hospital after an attack of what seemed to be pleurisy in October, and died on 4 December at the age of 67.

FURTHER READING: Susan Quinn, who wrote my favorite Marie Curie biography, is also my favorite biographer for Karen Horney, her *A Mind of Her Own: The Life of Karen Horney* being an exhaustive account, backed by Horney's diary entries and correspondence, of the complex inner world of one of early psychoanlysis's most compelling if frustrating figures. Horney's books are still available through Norton publishing and your choice of which to read will be guided by whether you are more interested in her as a challenger to traditional Freudian conceptions (*New Ways*), as a populariser bringing psychoanalysis within the grasp of everybody (*Self-Analysis*) or as a creator of new conceptions of the internal struggles that keep people chained to cycles of dissatisfaction and self-torment (*Neurosis and Human Growth*).

Chapter Twelve

Bringing Science to Psychoanalysis: The Many Survivals of Sabina Spielrein

The life of Russian psychologist Sabina Spielrein (1885–1942) began in emotional and physical abuse, and ended with the murder of herself and her two daughters, and in between those two dim poles she experienced professional and personal character slander at the hands of one of the century's most famous intellectuals, the systematic destruction of her family at the hands of the Soviet state, and an almost constant poverty at the hands of her own intellectual standards.

And yet, in spite of the almost constant misery that attended her ill-starred existence on this earth, she managed in her fifty-seven years to produce clusters of works that pointed to a new, scientifically-centred approach to psychoanalysis nested in biological fact and Darwinian theory that was decades ahead of its time, and a psychoanalytic approach to children that directly influenced many of the twentieth century's most prominent child development specialists. She eschewed the tribalism of early psychoanalysis in search of higher syntheses in a way that guided the future of that movement but did little to help her in her own fractious time.

Synthesizer and innovator, Spielrein's insight into the human condition was perhaps so well developed because her experience of its darkest depths was so foundational to her person. Born to a Jewish family in Rostov-on-Don, a section of the Russian Empire where Jews were permitted to live and work, Sabina experienced physical beating and emotional humiliation from her deeply troubled father, a man of great intelligence and natural curiosity but also profound depression and explosive rage. Alternately pressured to perform at the highest possible level of academic achievement and beaten for no other reason than the shifting state of her father's suicidal instability, Sabina developed a robust array of obsessions and neuroses that determined the course of her young life.

Slowly broken by her youthful experiences, she developed a keen hunger for punishment and humiliation, and a sexual interest in the idea of people eating while defecating, all accompanied by splitting headaches, pains throughout her body, and uncontrollable urges that grew into full neurotic episodes after the death of her sister. In 1904 her parents sent her to the Burghölzli mental hospital in the hope that the new, progressive psychological techniques practiced there by Eugen Bleuler might do her some good.

At that time, there was on the staff of Burghölzli a young psychologist by the name of Carl Jung who was one of many professionals charged with recording Spielrein's behaviour and progress. At the time, Jung was just at the start of a career that would lead him to international fame in spite of a character that was, let's be frank here, *execrable*. A serial philanderer who was not above using the power of his connections to bury fallout from his bad behaviour who went on to hitch his star to the rising Nazi Party in order to shove his Jewish colleagues out of psychology so as to make more room for himself, Jung was a person of minimal personal morality and a bottomless need for praise and worship, and Spielrein had the great misfortune of being placed squarely in his path during one of her weakest moments.

She fell wildly in love with the handsome Christian psychologist who played the role of tormented genius to the hilt. As he had before, and would do multiple times in the future, he ignored any considerations of professional ethics and allowed their relation to grow into a romantic one, all while mining her case for material to put in his own early papers, all be it in a form that distorted the facts to fit his own ideas. Spurred on by the discovery of this great love, and not yet aware that she was just one in a series of victims, Spielrein's condition improved to the point that she became less of a patient at the hospital and more of an assistant, developing her knowledge of the nascent theories of psychoanalysis as promulgated by Freud and filtered by Jung.

After a year's time at Burghölzli, Spielrein applied to medical school at the University of Zürich, which had one of Europe's longest traditions of admitting women to its departments. Jung was a lecturer there, and Spielrein his enthusiastic student, and of course the difference in power and position made not the least difference to him as the relationship continued under the nose of Jung's wife and his growing family of children. These were the times of the great psychoanalysis turf wars, when Sigmund Freud was fighting to root out alternate practices from his own, and his heir apparent, Jung, was slowly developing his own idiosyncratic, occult-infused version of psychoanalytic theory based on universal archetypes.

Spielrein's family was aware of her relationship with Jung, and while her father generally accepted the idea, her mother Eva was concerned, particularly after she received an anonymous letter (possibly written by Jung's wife) in 1909 warning her that if she allowed her daughter to continue seeing Jung it would only end in her ruin. Eva wrote to Jung expressing her concerns, at which point Jung, panicking that his behaviour might impact his career, wrote to Sigmund Freud to attempt to pre-smear Spielrein's name, saying that she was 'kicking up a vile scandal' in revenge for his refusal to impregnate her. He cast her as a deranged former patient who had abused his virtuous friendship terribly.

Jung and Spielrein would go on to reconcile, but throughout the rest of their correspondence, Jung's tone was one of condescension as to her own intellectual efforts, and flat-out irritation any time she mentioned the value of theories that weren't his own. In 1911, she finished her studies at the University of Zürich on the strength of a paper which represented an early attempt by psychoanalytic methods to understand schizophrenia, and the first by somebody possessing an actual medical doctorate. The next year she moved to Vienna, where Freud held court over the course and development of psychoanalytic theory.

Here, she placed herself on the intellectual map with the reading of her paper 'Destruction as the Cause of Coming into Being' before the Vienna Psychoanalytic

Society. In it, she proposed that evolutionary theory and biological knowledge be employed in the analysis of humanity's psychological drives. She posited the foundational importance of living beings' reproductive drive as a biological imperative that percolated into the psychological realm, which was a challenge to the primacy of the pleasure drive as theorised by Freud. She wanted to investigate how the biological facts of reproduction and death as honed by millennia of evolution might colour individuals' experience of sexual intercourse, and the neuroses that spring from the suppression of the sexual drive.

As an attempt to import biological knowledge into the practice of psychoanalysis, the paper was decades ahead of its time, and though the idea of a death drive (though in a far different form) would make its way into Freud's later theory, the reaction of the assembled notables was not a favorable one. Psychoanalysis was to be a pure practice based on observation and subjective interpretation of individual cases, the reigning minds declared, and any attempt to drag biology into it was essentially heresy.

Undeterred, Spielrein would continue to work out her theories, about the competing biological claims of the need to reproduce and the need of the individual to survive, and how our brains negotiate between the energy needed to maintain our individual integrity and that needed to raise a new generation. The ensuing years represented a time of professional creativity but financial and personal hardship. She married a Jewish doctor in 1912, though she spent most of their marriage separated from him, tending to their daughter Renata while attempting and failing to build a psychoanalytic private practice for herself in Berlin and Switzerland, perhaps largely due to her unwillingness to declare Freud, or Jung, or Adler *absolutely* right and the others *absolutely* wrong, which hurt her standing in the psych wars of the 1910s and 1920s. Reduced to living off the handouts of friends, but still maintaining her professional if unremunerative commitments that included a new interest in the development of language in children, Spielrein's life was intellectually rich if materially meager during this time.

Working with another rising intellectual star of the century, the young Jean Piaget, she began developing theories about how to apply psychoanalysis to children to help them recover from traumas such as those that marked her own childhood. She developed games and other forms of play that would draw the children into healthily revealing the roots of their extreme behaviour and safely expressing their angers and frustrations, all while developing ideas about how babies acquire language which would be influential to her more celebrated successors, Anna Freud and Melanie Klein.

In 1923, at the insistence of her family, and in anticipation of a better life promised by the apparently pro-science Soviet state, she returned to Russia and lived a fruitful and satisfying life so long as psychoanalysis was an officially accepted branch of psychological theory. With the death of Lenin and the fleeing of Trotsky (who was a primary supporter of psychoanalytic methods in the Soviet Union), however, and the subsequent denigration of pedology (which Spielrein advocated as a union of biological knowledge with educational theory, and which you can read more about in the feature of Siberian doctor Anna Bek in the first volume of this series) as ideologically suspect, Spielrein found her career prospects radically curtailed.

One by one, members of her family who had once had positions of intellectual promise were targeted by Stalin's regime. Her brother Isaac was executed by firing squad in 1937 for his position at the head of industrial psychology in the Soviet Union, followed by her other brothers Jan and Emil in 1938. Her husband had died in 1937 of a heart attack, and her father, after having been arrested and tortured in 1935, died in 1938 following the loss of a third son to state-sanctioned murder in the span of a year. Spielrein, alone with her two daughters, one in her twenties and the other a teenager, carried on as best she could until the invasion of the Soviet Union by the German army in 1940. As the army approached Rostov, she was urged to flee with her children, but for some reason she did not, and in 1942 *Einsatzgruppen* descended upon her town and began rounding up its Jewish residents. She, aged 57, and her two daughters, aged 28 and 16, were loaded into vans which were taken to great trenches outside of the city, where they were shot and buried along with some 20,000 other Jewish residents of the city.

Carl Jung, who had used the machinery of the Nazi state to remove his Jewish rivals from their positions and protections, lived until 1961. In his autobiography he left an account of Spielrein that referred to her only as a nameless 'talented psychopath' he had once treated.

FURTHER READING: Sabine Richebacher did a huge amount of the legwork in bringing Spielrein's obscure last years to the light of day, and you can find the results of that in her *Eine fast grausame Liebe zur Wissenschaft* (2008) but if you are looking for an English source, then you'll probably want *Sex vs. Survival: The Life and Ideas of Sabina Spielrein* (2014) by John Launer. If you want further reasons why Carl Jung was just a terrible, terrible person, and a not much better thinker, I'd recommend Walter Kaufmann's *Freud, Adler, and Jung* (1980).

Chapter Thirteen

Tsuruko Haraguchi: The Strenuous Road to Becoming Japan's First Woman Doctor of Psychology

(1) Do Not Interact With Men Easily
(2) Do Not Dance
(3) Go to Church Every Sunday
(4) Spend Money From Home Wisely
(5) Do Not Become as Aggressive as an American Girl
(6) Mind Table Manners Well

Such was the advice bestowed upon the 21-year-old Tsuru Arai (1886–1915) by her English travel companion when the two parted ways in Vancouver, and Tsuruko headed out on her own into the vastness of the United States, where no aspiring Japanese woman graduate student had ventured before. As counsel goes, it was on the harrowing side, suggesting to the already nervous young woman that the land she would be studying in was a lawless and immoral wilderness that would swallow her whole unless she remained ever vigilant against its seductive wiles.

Over the next five years, as Tsuruko worked towards her PhD at Columbia University, she came to find that all of the fears lurking behind Ms. Philips well-intentioned advice were, in fact, illusory, and that the example of American women, far from being a cautionary tale, was an inspiring one, with many aspects worth importing to her own country, including their capacity for independence, practical self-direction, open-hearted honesty, and spiritual engagement. Unfortunately, she would have but precious time to instill these newly internalised lessons in the aspiring women scholars of her home country, as upon returning to Japan in 1912, she had but three years to live until her tragically early death at the hands of tuberculosis at the age of 29.

Though gone in body, however, Tsuruko Haraguchi remained vibrantly alive in spirit, and scanning the names of Japan's first generation of university-trained women scientists, it is hard to come across one who does not mention Tsuruko's story directly as a primary inspiration in deciding their life's course. That story began on 18 June 1886, just under two decades into Japan's Western-leaning Meiji era, when Hirosaburo and Tane Arai welcomed into the world their second daughter, Tsuru. Hirosaburo was what was known as a 'muko,' which was a man who married into his wife's family and took up their name, which was necessary to carry on family lines when that family lacked male heirs. Her mother died in childbirth when Tsuru was 6 years old, and understandably Hirosaburo gave his children more leeway in their youth and adolescence than was traditional, doting on them and encouraging their intellectual and personal growth. In 1902 Tsuru began preparations to enter Japan Women's University, and in 1903 matriculated there, with a focus on English literature.

At the time, Japan Women's University boasted among its professors the nation's most renowned psychologist, Matataro Matsumoto (1865–1943), who had earned his PhD at Yale in 1899, and spent time studying with that wellspring of modern psychology, Wilhelm Wundt. Matsumoto was uniquely well-disposed towards women's education, and women's equal participation in academic life. Inspired by the presence of an eminent individual who not only held the opinion that women should expand their minds, but expressed that opinion readily in public, Arai switched her field of study to psychology, graduating in 1906.

That degree represented the extent of what Arai could expect from the Japanese educational system, and if she wanted to pursue her education further, she would have to travel to a foreign country. She decided upon Columbia University, located in the archetypal 'American' city of New York, and featuring such psychological luminaries as James Cattell (who was, like Matsumoto, a former student of Wilhelm Wundt and a foundational figure in American psychology), Robert Sessions Woodworth (a former student of William James and a pioneer in psychometrics), and a young up-and-comer by the name of Edward Thorndike, who would play an important role, a decade later, in encouraging Tomi Kōra to follow in Haraguchi's footsteps and work towards a psychology degree.

In 1907, accompanied by a Ms. Philips, Arai set out from Japan on the twelve day journey to the Americas. Parting ways in Vancouver, Arai boarded a train that took her through the great breadth and ecological diversity of the land she would call home for the next half decade, before depositing her at last in New York. In order to get a sense of how American women behaved and thought before plunging into her studies, she signed up to participate in the Young Women's Summer Camp at Altamont, where she experienced sleeping in tents (which she enjoyed), hay-rides (which she did not), and the startling diversity of character exhibited by American women. Instead of the loud, uncouth, materialistic beings she had been warned against, she found them companionable, welcoming, philosophically reflective, self-sufficient and deeply curious, and began to wonder what other stereotypes about American culture she had absorbed from Japanese and European culture might crumble upon closer inspection.

The camp having concluded, Arai reported to Columbia, where she met with Thorndike himself to go over her potential schedule. She believed that she wasn't smart enough or fluent enough in the language to try for anything more than a Master's degree, but Thorndike encouraged her to try for a full PhD, and when the departmental administration called him to question her qualifications, he insisted that she be allowed into the program. Thence began her dual education, one in the ways of experimental psychology, and the other in the ways of American culture. As to the latter, she soon found that dancing was an entirely respectable occupation, well-organised and supervised, and indeed essential to social functioning in a society that didn't partake in arranged marriages, that one-in-the-morning secret sandwich making parties in her fellow students' dorm rooms were pretty fun actually, that sleeping on a roof in a snowstorm will not kill you, and that Americans will talk about anything, at any length, at any time.

As to her academic path at Columbia, Arai began in 1909 to conduct the research that would lead to her PhD three years later. Reading through her dissertation today, the punishing and monotonous routine she subjected herself to in the name of experimentation is well-nigh inconceivable. Her main area of research was mental fatigue, and for the first three major experiments of her study, she had nobody but herself to act as a test subject. To test the effects of mental fatigue, of course, requires a person to become mentally fatigued, to which end Arai would, for *eleven hours* at a stretch, mentally multiply four digit numbers by each other (reasoning that this task had no accompanying physical component that might muddy the causal waters), measuring changes in her temperature, heart rate, solution time, and accuracy as she did so, and *then*, as if eleven hours of mental multiplication were not enough, she would then subject herself to further hours of mental challenge performing translation and memorisation tasks to see if the fatigue accrued in one task transfers over to other tasks, or if the change of task type 'resets' the brain in some significant way (spoilers: sometimes it does, sometimes it doesn't).

Having straightened out the kinks in her procedure, it was time to open up the experiment to other test subjects, and in 1910–1911 Arai gathered information about mental fatigue from twenty-seven poor hapless souls who themselves had to experience the mind-numbing hours upon hours of mental mathematics and memory testing required for her to probe the outer edges of human mental strain. In 1912, she defended her thesis successfully before a committee that included Thorndike and Cattell, and thereby became the first Japanese woman to earn a PhD in any subject. She graduated on 5 June 1912, and went straight from the PhD ceremony to her wedding, marrying Takejiro Haraguchi, a graduate of the Hartford Theological Seminary.

The couple returned to Japan in 1912, where Arai (now Haraguchi) published her thesis in Japanese in 1914, gave birth to a son in 1913, and a daughter in 1914, and prepared a Japanese translation of Francis Galton's *Hereditary Genius* (an 1869 volume that was among the first to attempt to scientifically investigate the heredity of genius), and a book recounting her experiences in America, *Tanoshiki Omoide (My Happy Memories)*, both of which were published in 1915. All the while, however, her health was in a state of precipitous decline that puts one in mind of the tragic fates of Anandibai Joshee or Srinivasa Ramanujan following their return from foreign study. On 26 September 1915, she succumbed to tuberculosis, exiting her life of often self-imposed mental strain, and entering Japanese history as a prodigious example of what can be accomplished by a woman possessed of a desire to know the world better, and a community supporting her in that journey.

FURTHER READING: In 2006, Yoko Kamei translated *My Happy Memories*, and it is just a wonderful book. Most of the middle section is devoted to informing Japanese readers about the habits of American men and women, and what their society and institutions are like, which are interesting but less personal, while the beginning and the end represent valuable resources for how she navigated American academic and social life. Her book *Mental Fatigue* is available online for free through Google Scholar. In Japanese, Izumi Ogino's 1983 biography of Haraguchi is the wellsource from which most biographical information seems to flow, but best of luck finding a copy!

Chapter Fourteen

Of Gifted Children and the Banality of Menstruation: The Educational Psychology of Leta Hollingworth

What do you do with a gifted child? A child who learns new concepts three or four times faster than their contemporaries, often withdraws from social interaction, and who brings unsettling intensity to both their passion and apathy.

How do you even identify one?

In the early twentieth century, while Anna Freud worked with traumatised children, and Maria Montessori with the very young, it was Leta Hollingworth (1886–1939) who devoted half of her all too brief career to sounding the profound riddles and contradictions of the gifted educational system. She had been, herself, a gifted child tossed constantly about on the waves of youthful tragedy. She was born on the hardscrabble Nebraskan frontier in a town where a gun fight was the accepted means of settling debates. Her mother died when she was 3, and her father, who was only occasionally present in her life up to then, turned tail and ran when faced with the task of raising three daughters on his own.

Leta and her sisters were left in the care of grandparents, and just when all seemed to be going well, who should return but their banjo-strumming, whiskey pounding father with a new wife on his arm. He insisted on taking the girls in to live with him and, once they were settled, left them in the care of a stepmother who routinely beat and abused them while he was gone for months at a time on adventures that would be dashing if they weren't built on such hopeless suffering. So it was that, at age 10, Leta made herself a solemn vow to skip the rest of childhood and proceed directly to adulthood. It was a decision determined by the hard grind of daily tragedy and a growing sense of nascent powers stirring.

Like Margaret Mead, her polar opposite in almost all other respects, her initial ambitions were centred on literature. Her poetry was filled with the yearning of a clever and emotional girl stranded in an intolerable life, and rings with an honest intensity that couldn't have been more out of touch with the poetic climate of the early twentieth century. After a series of failed attempts at securing publication, she gave up on the notion of a career as a writer, though not on writing privately for herself and her friends.

She turned instead to psychology, a field that was just finding its feet in the United States. Getting her undergraduate degree at the University of Nebraska, she moved with her husband, Harry Hollingworth, to New York City, only to find that her status as a married woman prevented her from obtaining a teaching position. State law at the time prevented the hiring of new married teachers, and only permitted established teachers to get married and continue working until they had children, at which point they were compelled to retire.

Stranded and withering, relief only came with acceptance to Columbia University as a Master's student in psychology. Her chosen field of early research was a provocative one, aimed directly at the most cherished gender theories of her advisers at Columbia. She sought to prove statistically the invalidity of two oft-cited theories about the mental inferiority of women. First, that women showed less variability in features than men, a sign of their lesser capacity for brilliance and lesser evolutionary importance. Second, the theory of functional periodicity, which held that women, during menstruation, were so diminished in their capacities that any intellectual or professional work that required persistent competence was beyond their ability.

She aimed first at the variability theory, gathering and analysing a list of 20,000 physical measurements taken at the baby ward of a local hospital. The result demonstrated unequivocally the same level of variability among male and female infants, and easily buried the variability theory. Next, for her dissertation research, she arranged for a series of men and women to perform a group of set tasks at a set time every day, and measured their performative variation. For both physical and intellectual functioning, she reported no significant alteration of performance during menstruation, and another centuries-long myth was sent scurrying for the corners.

The results were important, but they are not what we remember Hollingworth for. The breakthrough work of her life was performed from 1916 to her death in 1939, and centred on the problems of identifying and providing academic aid to special needs students, both the highly intelligent and the intellectually hindered. When she began her studies, psychometrics was in its infancy, but was roaring into prominence on the strength of Lewis Terman's Stanford-Binet IQ test.

In our age, when standardised testing is threatening to strangle an entire generation's self-motivated love of learning, it's hard to realise how exciting and important the development of these diagnostic exams was. For the first time, educators had something more than the instinct of their variously-trained teaching staff to identify students in need of particular support. Leta Hollingworth was an unabashed proponent of these exams, though she argued strenuously that they must never be administered in a group setting, but only one-on-one, with the educator following up on the results via interviews with parents and supplementary diagnostics to evaluate alternate intelligence types (our current acceptance of multiple intelligence types is an advance at least partially of Hollingworth's doing).

The crowning achievement of Hollingworth's career was the establishment of the Speyer School, an experiment in educating children with both very high and very low IQ results. The press centred on the gifted aspect of the program, the first thorough-going experiment of its kind. The gifted children were encouraged to meet in committees to decide amongst themselves the topics that they'd like to investigate and report on, with the teacher acting as facilitator and guide rather than lecturer. In place of simply accelerating the students through the expected curriculum, Hollingworth designed a schedule that permitted a quick gathering of the basics, and then extra time for broadening exercises and expeditions. Constant field trips, to factories and museums, were the order of the day, supplemented by the students' self-guided work on researching related topics of interest.

Of particular interest to Leta were the issues affecting the hyper-advanced students, those with an IQ of over 180. Incredible statistical rarities, Hollingworth only found twelve of them in her decades of research, but her posthumous work detailing the particular challenges they face in learning and socialising is still a standard text in the field.

That she contributed foundationally to the discipline of gifted education is beyond question, and that her role in combating the gender prejudice against female education at the beginning of the century ought to be more celebrated, likewise.... Which brings us to an unpleasantness.

For, having been educated in psychology in America in the early twentieth century, and in particular being a devotee of psychometrics and genetic explanations of intelligence and character, Leta Hollingworth was an unabashed eugenicist. As against the egalitarian and democratic psychological theories of her colleague, William H. Kilpatrick, she emphasised the deterministic role that superior breeding stock plays in bringing about exceptional children, and argued for the enforced sterilisation of the mentally deficient.

Yep, the story's going in *that* direction now.

She held it to be inconceivable that a superior child could come from sub-standard parents, and had no patience with social programs that held the contrary. It was a waste of time to educate everybody the same way, she asserted – cruel for the slower of intellect who were thrown at the same topics again and again only to fail again and again, and cruel for the exceptional children who were weighed down by the sluggish pace of their comrades. While we can dispose without hesitation of her views about the creation of exceptional children, there is some truth yet in the idea that inflicting the same educational regimen on all children regardless of ability is a form of cruelty, one we've been slowly correcting through advances in differentiated instruction.

Generous to a fault with members of her family and friends, her lifelong dogmatic adherence to the tenets of eugenics caused her to lash out at colleagues and students who dared to question her assumptions. To her, the facts were the facts, and anybody who disagreed or tried to add nuance to her views was simply hurting science out of foolish soft-heartedness.

Nobody is a hero in all things. In dozens of ways, the vista of world education has been enriched and improved by Hollingworth's stubborn adherence to the content of her collected data and devotion to specialised education for those requiring specialised learning environments. Multiple intelligence types, differentiated education, student-driven learning, individually focused testing for special needs identification, and the non-concomitance of intellectual precocity with social or artistic genius, were all ideas either originated by her or promoted heavily thanks to the prominence given them by her school experiments.

As against that, she held horrendous if common for the time beliefs about social engineering that are only forgivable from the context of her having died before the Second World War showed the all-too-real result of such airy theorising. She was a member of the Heterodoxy Club and other early feminist groups aimed at gaining greater social, professional, and educational access for women, but was firmly against Franklin Roosevelt's programs to provide security for the unemployed and elderly. Surviving so much childhood tragedy had hardened something inside her – if she survived so much pain, then everybody else ought to as well and should stop asking for help to cover their failure. If anybody can earn the right to such a grim and inhuman view of humanity, I suppose, it was she, and if anybody has profited from the simmering misery that pushed her work and world-view, surely, it is we.

FURTHER READING: Leta's husband, Harry, wrote a biography of Leta but in spite of some beautiful passages, his view of her and her work is tempered by his relentless conservatism which saw in Leta the virtues and beliefs it wanted to see. Far better is Ann G. Klein's *A Forgotten Voice: A Biography of Leta Stetter Hollingworth*. Klein is a professor who has worked in gifted children's education for decades, and her insights into Leta's continued significance are thoroughly worth the search.

Chapter Fifteen

Come Together? Inez Prosser and the Psychological Impact of Mixed Schooling Systems

Many Hopes Lie Buried Here.

These words, etched on the tomb of Inez Prosser (1895–1934) express an entire constellation of grief and frustration in the space of five short but devastating words. Just 38 years old, and one year out from becoming the first Black woman to earn a PhD for a psychology dissertation, Prosser had the world before her, and a bevy of challenging questions in her head that she sought to solve, when all of that potential was brought to a sudden halt on a lonely Louisiana highway.

Though the exact year of her birth is a matter of debate, with answers ranging from 1894 to 1897, no matter what date you choose it places Prosser's youth firmly in the shadow of the Supreme Court decision of *Plessy v. Ferguson* (1896), which ruled for the legality of 'separate but equal' facilities for the white and Black populations of the United States, including education. Had she been born in the North, her educational opportunities would not have been so impacted by *Plessy*, which only said that separate but equal facilities were legal, not that they were compulsory, but in Texas of the late 1800s segregation was very much the order of the day, and Prosser's family moved several times to maximise the limited educational opportunities available to their children in the state.

Born in San Marcos, Texas, which at the time had a population of some 2,300 individuals, when it came time for her to begin attending school, the family moved to Yoakum, a bustling new town which had been created from scratch in 1887 as a hub along the San Antonio and Aransas Pass Railway, and which by 1900 already had a population of 3,500. Prosser attended the 'colored' schools available to Black youths at the time, graduating as valedictorian of Yoakum Colored School in 1910. Her family only had resources to send one of their eleven children to college, and nearly settled upon her older brother, but Inez's dedication to education as her profession won them over, and she began attending Prairie View State Normal and Industrial College, graduating with a teaching credential in 1912.

For the next decade, Prosser split her time between teaching at elementary and high schools in the vicinity of Austin, Texas, and studying first for her Bachelor's degree from Samuel Huston College (a historically Black college opened in 1900, and known today, after its 1952 merger with Tillotson College, as Huston-Tillotson University) and then via correspondence and over summers for her Master's degree in education from the University of Colorado, graduate degrees not being available to Black individuals anywhere in the state of Texas. Her studies included both educational and psychological topics, and as they took place during the 1920s, what later historians of psychology would dub 'The Decade of Testing', it was all but inevitable that her interests would turn towards the intersection of psychological standardised testing and education.

Her Master's thesis, completed in 1927, was on the reliability of a series of English grammar tests she had designed and evaluated herself. Following the completion of her Master's degree, she took up a faculty position at Tillotson College, which had been founded in 1877 as a co-educational institution, but which in 1926 had become a purely women's college. She was noted by students and colleagues both for her dedication to the teaching profession, and the lengths she went to in bringing new and unique opportunities to her students, rising through the ranks to become the college's registrar and dean in addition to her duties as a psychology professor.

In 1931, she began the most promising, and most controversial, phase of her career when she received a $1,000 grant to study at the University of Cincinnati, where she focused on the psychological impact of voluntary segregation on Black children. In her 1933 doctoral thesis, Prosser made a distinction between mandatory segregation, a fundamentally anti-democratic structure created by white elites to minimise mixing with undesirable races, democratic anti-segregation, which held that all schools ought to be mixed-race in the name of the country's basic ideals, and what she termed voluntary segregation, which was the creation of specialty schools that catered to one particular, often discriminated against, segment of the population. Employing a battery of personality and character tests, including the Burdick Apperception Test, Lehman's Play Quiz, Attitudes SA Test, Personality Attitudes Test for Younger Boys, Woodworth-Cady Questionnaire, and the Personality Adjustment Test, she sought to determine, from the point of view of character growth, self esteem, and personality, whether students at a nearby mixed school were better off than those at a nearby Black-only school.

What she found was that students at the all-Black school scored better than their mixed counterparts on measures of sociability, social stability, relationships with faculty, and breadth of occupational and recreational interests, though the relatively small size of her sample prevented her from making general statistical claims about her results. She explained the more positive personality growth for students at Black institutions in terms of access to sympathetic faculty, and the lack of a need to navigate relations with a majority white student body, which drove black students in mixed schools towards introversion and social disengagement. It was an interesting result that was bound to make just about everybody upset, with anti-segregationists worried that Prosser's results about voluntary segregation would be used as ammunition by proponents of mandatory segregation, and segregationists upset with her declaration that making racial separation the law of the land was fundamentally anathema to democratic practice.

Prosser received a PhD in 1933 for her work, which brings up the question, 'Was Inez Prosser the first Black woman to receive a PhD in psychology?' It's usually the first thing about her that comes up in online articles about her life and significance, but it's not *quite* true. Though she was the first Black woman to receive a PhD for a psychological dissertation, she received a PhD in education for her work, not one in psychology. The first PhD earned by a Black woman in psychology, then, goes to Ruth Howard (whom we'll meet later), who earned hers the next year, in 1934, from the University of Minnesota.

Whether she was the 'first' PhD or not, her 1933 dissertation was a promising beginning to a fearless career that aimed at following the truth wherever it might lead, regardless of what political feathers were ruffled in the process. One year later, however, on a trip back from a family gathering in Texas to Mississippi (where she had been employed since 1930 as registrar and professor at Tougaloo College), her car collided head-on with another vehicle, sending her flying through the front windshield. Had the accident happened just three years later, when cars sold in the United States were required to use safety glass, the injuries she sustained might not have been so severe, but as it was she was taken to the hospital on 28 August, 1934, and died there on 5 September. She had used her earnings to send six of her siblings through college, her organisational skills and personal charisma to lead a generation of Black students to higher educational possibilities, and her fearlessness as a researcher to bring the best (if subsequently controversial) analytic tools of her age to bear on some of the darkest questions of her time, and suddenly, at age 38, she was gone. If there is a silver lining to be found in any of this, it is to be located in the person of a young woman, just entering Howard University the year of Prosser's death, by the name of Mamie Phipps Clark, who would go on to expand Prosser's investigation of the psychological impact of segregation in ways unavailable to Prosser, and to finish her work in eliminating once and for all mandatory segregation from the face of the United States.

FURTHER READING: R.V. Guthrie's classic *Even the Rat was White: A Historical View of Psychology* (1976) is a good starting point for the history of Black psychologists, while Wini Warren's 1999 *Black Women Scientists in the United States* contains a couple of paragraphs about Prosser. A better source is Benjamin, Henry, and McMahon's 2005 article 'Inez Beverly Prosser and the Education of African Americans' from the *Journal of the History of the Behavioral Sciences*, and Prosser's own dissertation, 'Non-Academic Development of Negro Children in Mixed and Segregated Schools.'

Chapter Sixteen

Children are People: The Life and Science of Anna Freud

Humans have a profound genius for generating terrible ideas. Slavery. Theocratic government. But there is one particular idea we hung onto for an unfathomably long amount of time before finally questioning, and that is the notion that Children Are Property, and therefore may be treated more or less however we please. Our appreciation for the importance of their early environment, and the responsibility we bear for positively structuring their earliest years, is of incredibly recent provenance, and rests to a significant degree on the work of a woman who has spent much of the last three decades languishing in semi-obscurity as a casualty of war: Anna Freud.

The Freud Wars, a twenty year exercise in discrediting Sigmund Freud's work and denigrating his humanity that was waged in the late twentieth century, brought with it the side effect that the remarkable work of his daughter in clinically investigating the developmental stages of children was held likewise in suspicion. She was cast as the reactionary, psychologically damaged protector of Sigmund's lamentable legacy, and her work sloughed off as unscientific and dated. Fortunately, cooler heads eventually prevailed and historians came around to the view that Sigmund Freud is indeed a primary figure of intellectual history whose cultural situation led him astray on a number of points, and that Anna's contributions to child therapy were foundational to modern programs like Head Start, while her Hampstead Clinic was a crucial model for the development of American child psychopathology research.

Anna was born in 1895, the last of Sigmund and Martha Freud's six children. Being the youngest daughter brought with it a weight of cultural expectation. Often, in Viennese Jewish circles, while the older daughters were given leave to marry, and the sons sought their fortune, it was expected of the younger daughter that she would stay behind and help nurse her parents through their old age. That role was confirmed when she showed from an early age an originality and depth of mind that made her, unique among her siblings, an intellectual companion and partner for her brilliant but intensely private father.

She was a daydreamer, given to constructing wild fantasies that astonished her parents for their bold intricacy, with a way of naively but incisively describing her world that more than once made it into her father's correspondence with his colleagues. Growing up, she eventually settled on the idea of pursuing a career in teaching while her father instructed her in the techniques of psychoanalysis. Now, part of being trained as a psychoanalyst is a requirement to perform a self-analysis under the guidance of a trained practitioner. It is an emotionally intense experience which requires absolute honesty with one's analyst in the reporting of dreams and urges, and total comfort with reporting any associations those might unearth.

That Anna had as her analyst her own father strikes us today as perhaps uncomfortable, even creepy, but then we are after all a generation largely allergic to genuine intimacy in our pursuit of anonymous approbation from online strangers. That said, it's hard not to feel a little proxy embarrassment, reading along as Anna analyses her dreams of being beaten and her masturbatory urges in vacation letters home to her father.

Or perhaps it's envy of that level of familial comfort masquerading as embarrassment.

In any case, unlike most of the historical accounts of women in science we've seen in the previous volumes of this series, where the would-be scientist has to fight every inch of the way to be recognised by a chauvinistic power structure, Anna was received enthusiastically into the psychoanalytic fold. Anna's interest turned quickly to children, especially as the destitution wrought by the First World War had left so many children in desperate need of guidance and support. Before Anna, there was interest in providing materially for the well-being of such children, but very few people were actively engaged in studying exactly HOW extreme environments impact the psyches of children, and how effective therapies might be developed to help them return to a somewhat normal life.

Anna Freud came into her own in the 1930s in spite of the gathering clouds of anti-Semitic conservatism in Austria and her father's steadily deteriorating health at the hands of cancer. Just before being forced to emigrate by the arrival of the Nazis, she established the Jackson Nursery, where her decade of experience working with troubled youths allowed her to construct environments that would ease anxiety and help the children rehabilitate after the traumatic events of their childhood. She kept index cards on all the social interactions, behaviours, dietary choices, and hygiene preferences of each child, collating them into a master system that would one day become the towering Hampstead Index, a treasure trove of day-to-day data on regular and arrested child development.

Then came the Nazis. The Gestapo arrived at the Freud house and took Anna away for a day of questioning. Sigmund paced the floor, consumed by worry, and decided that they must leave Austria as soon as possible after her return. Ernest Jones, who would later write the first standard biography of Freud, arranged for the emigration to London of the Freud family, as he did for so many other Austrian and German psychoanalysts who faced anti-Semitic persecution under the Nazi regime and its willing collaborators, like Carl Jung.

Sigmund died within two years of arriving in England, leaving Anna alone in very hostile territory. For London was where Melanie Klein held court. Believing herself to be Sigmund Freud's true heir as well as the reigning expert on Freudian child psychology, she was upset that Ernest Jones would offer her rival asylum. There were many sticking points between Anna Freud and Klein that shaped the debate about childhood psychosis in the fifties and sixties, the residue of which is our inheritance of common parenting wisdom today. As we saw in the portrait of Klein above, her focus was on the cavalcade of internalised and warring demons resulting from the loss of the mother's breast that create in us ideals of good and bad that affect how we view ourselves and the world around us.

Anna Freud criticised Klein's theory for being totally untestable, and for insufficiently taking into account the child's larger environment. She argued for a largely harmonious early development story which could be subverted by stress from the environment. Children who are the focus of aggression at home, she discovered, will identify with the aggressor as part of their defense mechanism. They will seek opportunities to enforce The Law upon others, or direct their acquired aggressive instincts against themselves, all resulting in a suspension of the normal integrative path of development.

During the Second World War, she had a chance to expand her knowledge of these defense mechanisms, running the Hampstead War Nursery, a haven for children during the Blitz who couldn't be removed from the city. She had to develop therapies to help ease children through the pain of losing their parents, and divided her charges up into small groups of families headed by a therapist to allow them to once again know the comfort of being loved, and avoid the regression to earlier developmental stages that often comes when one's stable love objects are no longer present. It became clearer and clearer to her that each child had their own unique path through some regular developmental stages, that a good child psychologist must not attempt to foist a universal origin myth upon the child, but rather must follow as closely as possible their life story to find factors in the environment that deflected the child into self-damaging behaviour.

Her work at the Hampstead War Nursery bled into the development of a fully staffed, permanent clinic which featured not only therapy and support for children exhibiting neuroses, but also a nursery for non-crisis children, observation of which served as a baseline for normal developmental psychology that had not existed previously. The titanic records kept by the nurses, teachers, analysts, and staff formed the basis of a publishing bonanza in the 1960s that spurred the blossoming of a new child psychology renaissance in America, leading not only to governmental programs like Head Start, but public mental health initiatives to educate new parents about their role in shaping the psychological health of their children.

Instead of the property that children were assumed to be in the nineteenth century, Anna Freud instructed us, we need to think of them as fragile psychological beings who absorb all of the anger we direct at them, and inflict it ten-fold upon themselves and the world. Abandonment, violence, belittling, all of these cause the child to employ a variety of defense mechanisms that interfere with normal development, trapping the child in repetitive activity or overpowering neuroses, and must be treated not with renewed harshness, but with a redirection of energy to substitutive activities, to play and art. Some of her ideas of normalcy sound reactionary to modern ears (in particular her stance on homosexuality as something that could and ought to be cured), but her overarching contributions, of giving children their due as people, and informing the world of their extreme sensitivity to the least of our actions, have made us better parents, and our world a more mutually supportive place.

She continued her work, carrying on her struggle against renewed outbursts of Kleinianism and desperately attempting to do justice to her father's memory in the teeth of his over-popularisation until her death in 1982.

The Hampstead Clinic where she gave birth to modern child clinical psychology was renamed the Anna Freud Centre in 1984.

FURTHER READING: Elisabeth Young-Bruehl's *Anna Freud: A Biography*, now in its second edition, is pretty much your go-to book for learning more about Anna and the dizzying array of phenomenal talent that rushed to psychoanalysis in the twentieth century. It is a lovely book motivated by a sincere and profound desire to do justice to Anna Freud's great productivity and constant emotional struggles. Alex Holder also has an interesting book out on the conflict between Melanie Klein and Anna Freud, and how that conflict continues to play itself out today, for those intrigued in fleshing out Klein's contributions a bit more than Young-Bruehl does.

Chapter Seventeen

Neuroembryology in Wartime: Rita Levi-Montalcini and the Discovery of Nerve Growth Factor

It is 1942, and Allied bombs are raking the city of Turin, wreaking a thudding vengeance for Il Duce's cynical alliance with Nazi Germany. Amidst the panic and carnage, a woman carefully gathers her most precious items, a microscope and a set of pain-stakingly prepared slides, before heading into her basement to wait out the attack. A Jew, ejected from her university position in 1938 because of racial laws, she has been continuing her research in a makeshift lab set up in her bedroom, entirely unaware that this work will, one decade hence, rewrite everything we thought we knew about early neural development.

Rita Levi-Montalcini's youth prepared her well for this life of social isolation. Her parents were Jewish, but her father was a devoted secularist who scoffed at holidays and practices that, as he saw it, had their basis in violence and vengeance. When other children asked her what religion she belonged to, Rita didn't have an answer, and went to her father for advice. 'You children,' he said, 'are freethinkers. When you reach twenty-one, you'll decide whether you wish to continue as before or whether you prefer to belong to the Jewish or the Catholic faith. But don't worry about it.'

Would that all children had parents that put such trust in their spiritual self-determination.

Born in 1909 in Turin, Rita looked at the absolute control her father had over her mother's life and decided early that the raising of a family was not going to be her lot. While the other girls oohed and ahed over whatever babies happened to be tossed their way, Rita set down to the task of what to do with her mind. Her sister and brother both had marked artistic abilities from an early age, none of which she possessed. It wasn't until witnessing the slow, gnawing death of a dear family friend that she felt a glimmering of purpose – to devote her life to medicine.

Fortunately, a handful of brave women like Maria Montessori had already paved the way in the Italian university system for her and, after an intense course of self-study during which she picked up four years' worth of mathematics, Greek, and Latin in six months, she entered the Turin School of Medicine in 1931. The great histologist Giuseppe Levi taught there, and under his often brusque guidance she became a first rate practitioner at using silver to render nerve systems more visible through a microscope.

Gradually, she found her way to the research that would consume most of her 103-year long life, the study of neuroembryology, which is to say of early nervous system development, first through an examination of developing chick embryos, and later through the sophisticated in vitro methods that Levi had been a pioneer of.

When she first began, neuroembryology was a jumble. Experiments by the Austrian Paul Weiss with limb grafting seemed to suggest a high degree of adaptability in nerve growth and development, while his student Roger Sperry arrived at precisely the opposite conclusion. In one of Sperry's tests, the motor nerve system servicing the left leg of a rat was switched to the right leg, and vice versa. When said rat received a shock on the left paw, the right leg twitched, and in fact the poor animal's nervous system was never able to rewire itself to compensate, suggesting a deterministic, genetically programmed view of neural development.

Levi-Montalcini entered that world at first unenthusiastic about being able to make a substantive contribution to such a murkily defined and underappreciated field. She began her tests, on chicken embryos because of their relatively low nerve count and quick incubation time, carrying out amputation experiments along the lines of work carried out previously by Viktor Hamburger to determine how growing nerves react upon arriving at an excised limb bud. She noted that embryological nervous systems underwent a hurried growth stage followed by a massive die-off once they reached the amputated limb, suggesting a fierce competition for some missing, vitally important resource.

Those crucial studies were carried out in steadily worsening conditions throughout the late 1930s and early 1940s. Italian fascism did not embrace anti-Semitism as a central principle to the degree that Nazism did, but when Mussolini wanted to earn the support of Hitler, he realised that an offensive against the Jews was a cheap means to that end. The racial laws ejected Rita and Giuseppe from their university positions, but did not stop their work. Rita set up a new laboratory in her bedroom, and continued working there even through the bombings that tore through Turin in the early forties, until finally, for the sake of her family, she moved to the countryside, working as best she could with makeshift dissection instruments forged from sewing needles until Germany's invasion of Italy following Mussolini's resignation.

Travelling under fake documents, she and her family fled to Southern Italy, carrying on a false life for two years while waiting for war's end. Research during that time was out of the question, so Rita turned to nursing and document forgery to pass the time and help the Jewish community evade the German army.

That war did finally come to its dreary conclusion, leaving behind an Italy split by shame and hobbled from hubris. Bad memories and lack of funding for science led Levi-Montalcini to accept an offer from Washington University in St. Louis, from Viktor Hamburger himself, to continue her work in the United States. Her association with the university would last three more decades, and include the work for which she would win the Nobel Prize in 1986.

While there, she received a note from a fellow embryologist about some abandoned research he had done previously on grafted mouse tumors. Trying the experiment for herself with her chick embryos, she was astonished to see nerve fibers, ordinarily so well behaved and predictable, branching out like mad in the presence of the grafted mouse tumors. The nerves had a greater volume and chaotically reached out hungrily in the direction of the tumor. To Rita, this suggested the presence of something in the tumor that induced nerve growth, a factor that determined direction and growth rate which could over-ride the programming of normal development.

Continuing her work in vitro, and by a chance suggestion with snake venom and mouse oral secretions in place of the tumors, she found the effect validated again and again – a halo of neural tissues stretching out in the direction of the source. She, with Stan Cohen, was able to isolate the protein responsible, and found that its effects held even when heavily diluted. It was a major discovery, a bomb set off at the heart of everything people thought they knew about nerve development, and yet the importance of which wouldn't be generally realised for another three decades. In place of the dignified march of nerve growth under the standard model, Levi-Montalcini uncovered a mad race for growth factor, in which all newly developing nerves participated, and most died, cut off from their desired protein by their over-ambitious neighbors.

While the therapeutic possibilities of this insight struggled into the light, there was still other work to be done. In the sixties, Levi-Montalcini spent half of her time at St. Louis, and half at a new Research Centre for Neurobiology in Rome, of which she was made director. There, the mysteries of nerve growth factor continued to be examined, along with questions fundamental to the inflammation response. Though officially retired in 1979, she continued her work and, at the age of 82, she and her team continued to make discoveries about the function of mast cells and their regulation. This work, combined with her earlier breakthroughs with nerve growth factor, suggested new analgesics and new therapies for neural degenerative disease and the regrowth of Schwann cells, many of which are just now making their way to clinical trials.

In 1986, she finally received the Nobel Prize in recognition of the importance of her nerve growth factor work, and, more remarkably, in 2001 she was made Senator for Life in the Italian Senate, a post she held until her death in 2012 at age 103.

FURTHER READING: Levi-Montalcini is the author of one of the great scientific autobiographies of all time, *In Praise of Imperfection* (1988), a fascinating look at Italian politics at the height of fascism, and at the development of neuroembryology in the fifties and sixties, mixed with poignant personal reflections of people who lived tragic lives far from the eye of history. It's one of those rare science memoirs that speaks as much to its own particular field of scientific interest as it does to the dazzling, frustrating, chaotic continuity of human experience.

Chapter Eighteen

Mary Ainsworth, Infant Anxiety and the Case of the 'Strange Situation'

We tend to think of babies as, psychologically, relatively uncomplicated creatures. They are happy when clean, warm, and fed, and angry when wet, cold, or hungry, and that's more or less the extent of it. Concurrent with that sense of the basic simplicity of babies is the conclusion that, as long as *someone* is attentively seeing to that baby's drying, warming, and feeding, the child will be fine, and not turn out evil.

By the 1930s, this notion of infant psychological simplicity was being challenged from two sides, by the Freudians on the Continent, who theorised the existence of profound and disturbing relationships between the infant child and its parents, and by developmental psychologists in the United States and Canada, who became interested in how different amounts and types of parental (and particularly maternal) attention correlated with different infant behaviour patterns. This American school of infant psychological studies, which properly began with the insights of William E. Blatz, culminated in the work of Mary Ainsworth (1913–1999) that established attachment theory as a major hypothesis in developmental psychology.

Born Mary Salter in Glendale, Ohio on 1 December 1913, Ainsworth was the oldest of three siblings, and the one who displayed the earliest predilection for intellectual pursuits, learning to read at the age of 3. When she was 5 years old, her family moved to Toronto because of her father's job, and ultimately became Canadian citizens when he was given the presidency of the Canadian branch of the company he had been working for. For the next twenty years, until the outbreak of the Second World War, Toronto would be the centre of Ainsworth's world, a comfortable nest where she would remain even when the common academic wisdom urged her to flee.

Her path through elementary and high school was an easy one, the way paved by the family tradition of going to the library every week and checking out as many books as their five library cards would allow, with just one bump in the road when Ainsworth affected a disinterest in academic pursuits to raise her social standing (something I think many of us nerds have experimented with at least once in our lives – for me that was sixth grade, when I hung out with the skateboard kids and tried to drum up an interest in Alice In Chains to be *cool*). When she was 15, and a senior in high school, a book came her way which determined the course of her life – William McDougall's *Character and the Conduct of Life* (1827). To Ainsworth, it was a revelation – through psychology, humanity had the chance to determine how emotions and behaviours originated, allowing humans to understand themselves and why they act as they do, instead of throwing up their hands in resignation as the mere vessels upon which irresistible emotions and outside forces act.

In 1929, she matriculated at the University of Toronto, where she directed herself as soon as university requirements would allow to the honors psychology track, which featured a battery of instructors filled with a quasi-messianic zeal for psychology and its potential to change the world which Ainsworth caught in due course. While an undergraduate, she took courses from William E. Blatz (1895–1964), who was developing the security theory which was the predecessor to Ainsworth's attachment theory. Blatz believed that, as an individual grows and develops, their ability to courageously and competently navigate the world around them flows from the possession of a series of secure relationship bases, first provided by the parents, then by a friendship group, and ultimately by a spouse, and that individuals lacking this dependable harbor of emotional safe return have a tendency to develop a less exploratory approach towards their life options and goals.

Ainsworth remained at the University of Toronto for her Master's and PhD work, rejecting the general advice that graduate school should be about seeking new opportunities and new points of view. She did her dissertation work on Blatz's security theory, developing self-reporting tools that would allow researchers to gather consistent and scorable information about children's evolving sense of stability at home and with their friends. In 1939, she received her PhD, just as the world was tumbling headlong into the Second World War. With most of her professors involving themselves in war-related work, Ainsworth had the opportunity to remain yet longer at the University of Toronto (along with her colleague Magda Arnold, about whom more later), keeping the department afloat for three years, until the call to action became too insistent and, in 1942, she joined the Canadian Women's Army Corps.

With her training in psychology, and particularly her ability to turn the results of personality tests into measurable information about the behaviour and aptitudes of individuals, it was almost a foregone conclusion that the CWAC would steer her towards personnel selection as her primary occupation. This was both a rewarding experience, in that it gave her an appreciation for how much could be learned in a clinical, rather than academic, setting, and a frustrating one, as for the most part the army was really only interested in infantry soldiers, and so generally ignored most of her department's recommendations for where to place people so as to best employ their abilities.

After the war, she continued with Rehabilitation Services for the Department of Veterans' Affairs, but after a year returned to the University of Toronto, where she taught classes on personality assessment while continuing work with Blatz on how best to quantify individuals' sense of self security. In 1950, she married Leonard Ainsworth, who was still working on his PhD, and whose lesser status meant that decisions about where the new couple would work were largely based around what places would hire him. Together, they moved to London, where Mary was able to work with John Bowlby and James Robertson on separation theory, i.e. on how different young children respond to separation from their parents, and on what evolutionary or ethological explanations might lie at the heart of those responses. She was at the verge of beginning a longitudinal study of how mothers and infants interact with each other when Leonard finished his PhD and decided that he wanted to work in Africa, which meant Mary had to say goodbye to her planned research and find some paying and relevant work to occupy herself with in Uganda where Leonard found a job as a research psychologist.

As it turned out, she not only found financing, but a golden opportunity, to observe mother-infant dyads in Uganda that would provide a crucial baseline for her later work in attachment theory, and the results of which she would publish in the mid-1960s. The primary contrast she observed between Ugandan mothers, and the middle class Western mothers she would later study, was in the pure amount of time that mothers were in physical contact with their children in Uganda, and the tendency to let children set the cadence for the meeting of their needs, instead of attempting to rewire the children to fit into the cadences of the household.

The Ainsworths remained in Uganda from 1953 to 1955, when, once again, the availability of a job for Leonard dictated the couple's next move, this time to Baltimore, where Mary found work as a lecturer at Johns Hopkins's evening classes, and as a psychologist at Sheppard and Enoch Pratt Hospital. The dual responsibilities meant that, during the late 1950s, Mary's research career was at a virtual standstill, and did not pick back up until after her divorce in 1960 and her decision to stop working at Sheppard and Pratt in 1961, events which cleared the decks for her to begin the work which her name will forever be associated with, the invention of the Strange Situation, and its use in the development of attachment theory.

Beginning in 1962, Ainsworth set out on a program of observation to determine how different types of mother-infant interactions lead to different types of behaviours in infants vis-à-vis their relations to other people and the space around them. To probe this issue, she created the Strange Situation, an experimental setup that featured eight stages, wherein a caretaker and her child were introduced to a new room, filled with toys, and where psychologists observed the interaction between the mother and the infant, the infant and a stranger introduced into the room, the mother, the infant, and the stranger, and the infant left by itself. From her observations, she determined that babies ultimately belonged to one of three categories in terms of how they perceived their mother and the world. The majority of babies, at around 70 per cent, were Type B, whose mothers regularly engaged in physical contact with them and were generally highly responsive to their infant's needs. These babies, she hypothesised, had developed an inner model of their universe based in security, in the expectation that their needs would be met, and that their caretaker could be generally relied upon. As such, they tend to be more confident in exploring their surroundings, using their mother as a kind of base to return to, but not requiring constant physical contact to reassure them of her presence, are more cooperative, less prone to anger, and are willing to engage with strangers so long as their caretaker is around.

Type A children, then, representing some 15 per cent of the total, possessed caretakers who were actively uninterested or flat out rejective of contact with their infants, being generally unresponsive to their vocalisations, and unwilling to engage with them physically unless absolutely necessary. These 'avoidant' children, Ainsworth believed, have in response to habitual rejection built up a defensive indifference that manifests in a lack of interest when their mothers leave the room, an ambivalence when being left alone with strangers, and a greater tendency to angry outbursts. Type C children, meanwhile, make up the remaining 15 per cent of babies observed, and represent the middle of the road, being the children of parents who are responsive to their needs, but inconsistently so. These children have learned neither to completely rely, nor completely give up on, their caregivers, and so have a greater feeling of anxiety that manifests in clinginess and an unwillingness to explore, a finding in agreement with the basic principle in reward theory that, if you want to keep a person pushing a button, you need to give them rewards only intermittently, to keep their expectations from solidifying into sure knowledge about what the result of each push will be.

For the next three decades, Ainsworth fine-tuned her attachment theory, often in the face of feminist criticism that her results about children's apparent need for a single dependable figure whose time was given over to their development ran against the entire spirit of women's liberation, which held that babysitters were just as effective as parents in giving infants what they fundamentally need. Ainsworth herself always claimed that her findings were not essentially anti-feminist, that more research needed to be done on the possibility that multiple attachments could take the place of a single maternal one, or that fathers could fill that role just as adequately as mothers would. Her point was to highlight the hitherto undocumented complexity of expectations formed by individuals even as infants, and how the fulfillment of those expectations has an early impact on their engagement with the world, while leaving it to the next generation to determine how best to meet those expectations within the confines of each individual family's economic situation, while for the moment suggesting at the very least the creation of government programs to aid couples in understanding the developmental needs of children, and accessing resources to best serve them in providing those needs.

In 1975, Ainsworth transferred to the University of Virginia, which had a more robust roster of developmental psychologists to collaborate with than Johns Hopkins did, and where she remained until her full retirement in 1992, the same year she was elected a Fellow of the American Academy of Arts and Sciences. She died of a stroke in 1999.

FURTHER READING: A wonderful source for Ainsworth's life and the inspirations behind her research is the autobiographical sketch she wrote in 1983 for *Models of Achievement: Reflections of Eminent Women in Psychology*, edited by Agnes O'Connell and Nancy Russo, which is just a remarkable resource generally, featuring seventeen autobiographical accounts from prominent women psychologists of the early to mid-twentieth century. Beyond that, her 1978 book *Patterns of Attachment: A Psychological Study of the Strange Situation* is an exhaustive account of the development and results of attachment theory, and if you can't wait for that to show up, her papers 'Infant-Mother Attachment' (1979) and 'Attachments Beyond Infancy' (1989) are nice and digestible summaries of her main points, with the latter in particular fulfilling Blatz's original goal of tracing the impact of different levels of security throughout an individual's life.

Chapter Nineteen

Separate: Mamie Phipps Clark and the Psychology of American Segregation

Separate But Equal.

Of all America's variously Orwellian brandings, few have wrought as much human suffering as those three words. It is a phrase wrapped in a dire, folksy simplicity, an Aw Shucks veneer of common sense coating a core of pure racist malevolence that cut a deep gash through American society for a half-century and beyond.

In 1896, in a decision on everybody's shortlist of that body's most disastrous rulings, the Supreme Court declared in *Plessy v. Ferguson* that racial segregation was legal, so long as equal facilities were made available to all races concerned. It wasn't the first time the Supreme Court had thrown its weight behind the South's attempts to undo the work of Reconstruction – the Civil Rights cases of 1883 had gone far in removing crucial federal safeguards for Southern Blacks a decade before – but in terms of gifting the South a plan of action for the perpetuation of antebellum race expectations and roles, it was the court's dark masterpiece.

Its surface reasonableness made it difficult to combat – who could argue against equality, after all? Who could argue that, sometimes, separation wasn't good for all concerned? It was necessary to show, not through anecdotes, not through abstract reasoning, but rather through research and statistics, exactly what the human toll of Separate But Equal was, if *Plessy* was to be overturned. And as it so happened, just as a cadre of lawyers led by Thurgood Marshall and William Hastie were preparing their case to take down *Plessy* in the late 1930s, a woman from Hot Springs, Alabama was carrying out the psychological research to show definitively how segregation affected the minds of the young children raised within it.

Mamie Phipps Clark (1917–1983) has largely fallen through the cracks of history. Her research, which was used in *Brown v. Board of Education of Topeka* to strike down segregated schooling, was attributed to her husband, while her thirty years of nurturing a mental health services program for the children of Harlem in the teeth of administrative indifference remained solidly at the level of local news. For one whose recognition was never on the order of her accomplishments, and whose every advance was scratched out of the hard stone of professional racial prejudice, however, her childhood was by all accounts a happy one.

Her father was a physician, and her mother helped him at his work. Phipps took to school like a hamster to a wheel, intoxicated with the thrill of learning things, and upon graduating high school was offered several different scholarships. She was interested in majoring in mathematics and minoring in physics, because Mamie Phipps never half-stepped anything that she did, and enrolled in Howard University. Once there, however, between the uninspiring nature of the mathematics faculty, and the fact that her boyfriend, Kenneth Clark, was urging her to transfer to his department, psychology, she decided to change fields. Phipps was to be a psychologist.

In 1938, she received her degree (magna cum laude, naturally), married Kenneth Clark, and embarked on the research that was to change the shape of American society. Travelling to New York she met with Ruth and Gene Horowitz, whose work centred on self-identity in pre-schoolers. Their data on the self-perceptions of Black children was lacking, and Clark, who was simultaneously a secretary for one of the lawyers involved in building a case against *Plessy*, saw a perfect opportunity to do field research in psychology while at the same time building up data to support the repeal of segregation.

She traveled to an all-Black nursery school and performed a series of tests to determine when children are aware of their racial identity, and how they feel about that identity. The most famous of these were the 'doll' tests which Clark would spend the next years developing to reveal ever finer layers of children's self-perceptions. In these tests, children had an array of dolls of different skin colours to choose from. Clark proceeded to ask them questions about which dolls they thought represented them, which ones they thought were 'nice' or 'bad', and so by steps to reveal their feelings about race and themselves.

The results, presented in 'The Development of Consciousness of Self in Negro Pre-School Children', were devastating. Clark found that children had a sense of race identity by the age of 3, and that overwhelmingly they used 'negative' terms to describe anything connected with being Black. Separate But Equal was not producing minorities with a robust sense of self-worth, but rather a generation filled with unconscious self-denigration beginning from the first moments of racial self-identity.

Her husband Kenneth became interested in her research and together they obtained a Rosenwald grant to extend her early work. For the next five years, as Mamie added the responsibilities of raising two children to the demands of her jobs at the American Public Health Association and the United States Armed Forces Institute, the Clarks produced a stream of studies showing the depth of segregation's impact on the self-identification of Black children. When it came time for Thurgood Marshall to prepare his case against segregation for the Supreme Court, Kenneth included their results in a brief sent to the NAACP which was ultimately referenced in the final 1954 *Brown v. Board of Education of Topeka* decision.

The court stated unequivocally that it was convinced that segregation impacted children in ways 'unlikely ever to be undone' and in one grand sweep of judicial might crushed segregation as a legally defensible institution. Mamie had begun the research, and had carried it through in spite of great domestic and professional responsibilities, and yet in the few later publications that mentioned this contribution at all, credit was unilaterally given to Kenneth for submitting the results to the NAACP rather than to Mamie for having originated them.

Mamie, however, had more pressing concerns than posterity's distribution of credit. Her PhD advisor (wrongly) felt that Black individuals did not have the same mental capacities as white ones, and made it clear to her that, as a Black woman, the best she could hope for was to become a teacher at a Black school, but that carrying out an actual career in research was thoroughly out of the question. She applied for jobs only to watch white people with less experience than her get hired, and what work she could get involved tasks far beneath her skill and training.

From 1944 to 1946, Clark drifted, frustrated and under-utilised, until at last she had an experience that pointed the direction for the next three decades of her life. Upon becoming a psychologist at the Riverdale Home for Children she was brought hard against the realisation that urban minority children were woefully lacking in basic access to mental health services. She tried to get other social service institutions interested in this problem, and when they responded with deafening indifference she gathered money from her family to start her own organisation in the heart of Harlem, the Northside Testing and Consultation Centre.

It had a rocky start – the community looked with suspicion upon a service that had theretofore not existed, and was wary of the stigma that might attach to their children if they went to Dr. Clark's Centre. That reticence, however, was overcome when parents discovered how Clark could be an important ally in fighting one of the great tools of *de facto* segregation: special class placement.

The idea was devastatingly simple: if you had a school district that was technically against school segregation, but you didn't want Black people in your classes, an easy way to solve your problem was simply to designate Black kids as 'retarded' and shunt them over to special needs classes where they wouldn't get in the white children's way. Parents trusted the school's placement and trickled into Clark's centre, worried about what they could do for their child who seemed fine but was apparently mentally deficient.

Clark re-tested the children brought to her, discovered in case after case that these children were in fact performing perfectly at grade level, and worked with parents to reinstate their children back into normal classes. Her efforts earned her and her Centre the trust of the community, and she served as Northside's executive director from 1946 to 1979. That Centre, which Mamie began with a $1,000 loan from her father amidst an overwhelming lack of enthusiasm from the reigning social care providers of the day, survives still, providing tutoring services, nutritional guidance, parent training, and psychological consultations for the community.

In later life Mamie Phipps Clark was critical of psychology that kept itself bottled in university or clinical research, refusing to apply its knowledge and methods for the good of communities. For four decades, she was a living example of how research and social aid could walk hand in hand with no diminution to either, how advocacy led by psychological insight could empower whole cross-sections of a community in ways that generated long-standing social good. She served on the board of directors of nearly a dozen organisations and companies, from ABC to the Museum of Modern Art, forming bonds between swaths of society that had little knowledge of each other's problems and methods, and through it all had the satisfaction of knowing that her life's work had formed one of the nails driven into the coffin of American segregation.

Mamie Phipps Clark died of cancer in 1983.

FURTHER READING: Wini Warren's *Black Women Scientists in the United States* (1999) is an indispensable book generally, and in the case of Clark puts together material from the three main sources we have about her work, Robert Guthrie's *Even the Rat Was White* (1976), a feature on her work in the *Ebony Success Library*, and Agnes O'Connell's classic *Models of Achievement: Reflections of Eminent Women in Psychology* (1983).

Chapter Twenty

Virginia Satir and the Art of Family Communication

To many, Virginia Satir was an instinctive therapeutic genius, whose invention of family therapy has instructed untold millions of husbands, wives, and children how to better communicate with each other and improve their relationships by doing real and substantial work on building up their own individual senses of self-esteem and appreciation of each other's subjective value. To others, however, she was a character bordering on farce, whose naive sense of psychological theory and the sources of psychological distress, reliance on vaguely defined pop categories and awkward analogy, and crude attempts to systematise the practices that worked best for her represent an ungainly, almost accidentally effective, contribution to therapeutic practice.

That's quite the opposition of opinion, and finding the true Virginia Satir (1916–1988) within the binary swirl of hagiography and condemnation that characterises a goodly number of discussions about her is a difficult task. But we'll try anyway. She was born the eldest of five siblings on a farm in Neillsville, Wisconsin on 26 June 1916. Neillsville was the county seat of Clark County, but was still very much a modest rural town with a population of around 2,000 which has grown all the way to 2,300 today. By her accounts, Virginia taught herself to read by the age of 3, and by 5 had settled on the resolution of being a sort of children's detective, investigating the mysterious lives of adults.

That decision might have been inspired by a religious conflict around the same time between her parents. Her mother, a Christian Scientist, was dead set against Virginia receiving medical attention when she complained of abdominal pains, and her father went along with it until her appendix ruptured, and he insisted she be taken to the hospital, where her life was saved. After a few months in the hospital, Virginia returned home and to the rural education that most country kids in 1920s America could expect – seven years in a single room school with the eighteen other children from the town whose parents bothered to educate them. Thanks to her voracious reading, Virginia was ready for high school by the age of 13, and the family moved to Milwaukee so that the children would have a better high school education than was available to them in Neillsville.

Satir graduated South Division High School just before turning 16, in the year 1932. This was, to say the least, not a propitious time to be heading towards a potentially expensive college education, with America firmly in the teeth of the Great Depression. She arrived at Milwaukee State Teachers College with $3 to her name, and assured the understandably concerned registrar that, if he let her sign up for classes, she would find a way to come up with the money needed for tuition and books. And she did, by virtue of taking up a job at Gimbel's (which most people today remember as the rival department store to the one Santa Claus worked for in *Miracle on 34th Street*) and another for Franklin Roosevelt's New Deal era Works Progress Administration (which employed some 8.5 million people from 1935 to 1943). In 1936, she received her BA in education, and went on to a teaching post at Williams Bay, Wisconsin, a small town of some 700 individuals at the time. Here she began the habit of accompanying a different student home each day, to spend some time with their parents and attempt to engage the whole family unit in student support. It was a great idea from an era when teachers weren't so bogged down with standardised testing pressures and administrative hoop-jumping that they had time to engage with the wider school community and interact with parents on a semi-regular basis, and it gave Satir crucial insights into how family dynamics work, and how different communication styles of family members contribute to the creation of tensions within the family unit.

Satir realised that, if she wanted to do more to help students and their families, she could use a wider familiarity with the current state of social work practice, and to that end she enrolled for summer courses at Chicago's Northwestern University, beginning in 1937, before transferring to the University of Chicago for graduate work, where her grades began to falter, either as a result of gender discrimination, or of her own preference for self-constructed pragmatic solutions over more academic theorising. Nonetheless, she received her Master's degree in 1948, a year before her first marriage, to a Second World War soldier, ended in divorce.

Her private practice (and second marriage) began in 1951, and it is here that her career as 'The Mother of Family Therapy' well and truly begins, as she followed her instincts and developed her insights into the processes that pull families apart. She insisted on seeing families as complete units during her sessions, rather than meeting with each individual separately, so that she could observe how they communicate, and show them how their communication styles impacted each other. She delineated five of these styles– the placater, the blamer, the distractor, the intellectual, and the congruent, the first four of which deflected real and equal communication between individuals, each dodging in their own unique ways honest engagement with, and acceptance of, others' emotional states. Placaters accept too much responsibility for how things are, and don't honestly communicate their own frustrations, blamers take too little responsibility and want problems to be somebody else's fault, distractors are uncomfortable with emotional communication and revert to irrelevant questions or jokes to change the topic, and intellectuals (or computers) do everything they can to reduce conversations to their logical content and deny the importance of what is happening emotionally in a given situation, leaving only 'congruent communicators' to acknowledge and defend their own worth while respecting the subjectivity and reality of the emotions around them.

I don't think anybody seriously argues against the value of Satir's methods for allowing family members to realise the shortcomings of their communication styles, appreciate their own worth, and work on recognising the existence of others as subjects in their own right. The tensions tend to come when we talk about Satir's work in the 1960s after moving to California. She began attending sessions at the Esalen Institute, a retreat in Ben Sur that catered to a wide range of New Age interests, and which still exists today as a place where wealthy people can go to discuss the flow of their energies, what animal they want to be reincarnated as, and their experiences with extraterrestrials (at the time of writing, I see I am able to sign up for a weekend workshop on 'Primordial Qigong: Vibrant Health, and Harmony with the Tao' for a mere $1,000 to $10,000, depending on whether I want to sleep in a bag or in the 'South Point House'). In later talks with Barbara Jo Brothers, Satir expressed the great change in her thought wrought by her time at Esalen, and it shows in her work. Her previous interest in lack of self-esteem and opportunities for self-actualisation as key players in family conflicts morphed into a model of energy flow whereby inspirational energy from the heavens and grounding energy from the physical world merge, propelling us along our life's path in a seed-like growth that taps into the overall energy of the cosmos. Likewise, the generally valuable instinct from her experiences of troubled families that a focus on 'potential over pathology,' i.e. on how individuals can build themselves into better persons instead of a hyper-focus on what was wrong with them, was a useful therapeutic practice, changed into a guiding philosophy that all humans are essentially good, and seeking to achieve the growth of their own personal energy seeds, and that all problems, therefore, including that of global-scale war, can be treated by centreing individuals in the overlapping energies around them, and achieving a 'third birth,' when they become the fullest versions of themselves and their own decision makers, on their way to a 'fourth birth,' when they join all consciousness. (By the by, in case you were wondering about the math of it, Satir's stated estimate was that, if 6 per cent of the global population were to adopt this approach towards the self and its relation to the energy of the cosmos, that would be enough to end war.)

These sorts of thoughts, as expressed by Satir and elaborated upon in almost messianic tones by her followers, made professionals from many different disciplines uneasy. Psychologists were not too keen about the wholesale dismissal of pathology, feeling that it misled people with very real and biologically grounded neural conditions requiring treatment down roads of false promise. Historians and social theorists decried the reduction of world conflict to a matter of self-esteem. Feminist theorists were uncomfortable with the waving away of deep cultural and economic factors in explaining the position that women found themselves with regard to their families. And physicists just plain didn't like the analogical appropriation of some popular explanations of quantum effects that tended to crop up in the Satir school of self.

However, these were never the people Satir was attempting to win over. She did not care about theory, only about what she observed in the trenches of practice, and the effectiveness of her techniques for those she attempted to help. In this, she was tireless, sometimes spending upwards of 350 days a year on the road, answering calls for help, holding teaching seminars and public demonstrations, and doing her level best to introduce her techniques to countries traditionally closed to input from the West. She met with tens of thousands of individuals over the course of her career, and energised a new generation of therapists with her unconventional ideas about how humans could be made whole, and families united, through a mutually realised path of self-discovery. There are reams of testimonials from husbands and wives, sons and daughters, speaking to the change to their families, the shift in their common perspective, brought about by just a single meeting with Satir.

All of which brings us back more or less to where we started – what are we to make of Virginia Satir, considered as a whole? Can we separate the important contributions she made – the treatment of the family as a unit rather than as individuals, the use of communication stances as a therapeutic tool, the highlighting of self-esteem as the issue underlying many familial surface problems, and the turn from pathological micro-analysis to constructive potential seeking – from the somewhat vaguely defined smorgasbord of New Age concepts she and her followers increasingly draped those core techniques in over the last two decades of her life?

The answer is, *of course we can*. As humans, we have plenty of experience in honoring our innovative pioneers, even those who drifted into realms of near self-parody later. Virginia Satir helped thousands of people, created a new approach to therapy, and did it all on her terms, for better and worse, and the sheer audacity with which she lived her life, and resoluteness with which she approached her purpose, combined with her real and true gift for identifying what works and what does not in opening the lines of actual and honest communication between individuals, will linger on in the annals of family therapy long after Esalen runs out of well-heeled rubes to bamboozle, and long into the age when we all feel well enough about ourselves and others that we do, in fact, have slightly less inclination to blow up each other (and ourselves) at the least provocation.

FURTHER READING: I'm not entirely sure what to tell you. *Well-Being Writ Large: The Essential Work of Virginia Satir* (2019) by Barbara Jo Brothers is one of the most comprehensive accounts of Satir's thought you can easily get, and if when reading the above, you thought to yourself, 'Gosh, Dale, you're being too harsh on the post-California turn in Satir's thought,' then this is probably a good book for you, as it very distinctly believes that turn represents peak Satir. *Virginia Satir: The Patterns of Her Magic* (1999), is useful for the presence of a verbatim transcript of one of Satir's therapy sessions, to give some insight into how she dealt with people face to face. Of her own works, *Conjoint Family Therapy* (1983) and *Peoplemaking* (1972) are easy to flag down copies of, and form a good direct introduction to her methods.

Chapter Twenty-One

More than the Sum of their Parts: Eleanor Maccoby's Studies of Children in Groups and the Adults they Become

When Eleanor Emmons left home to matriculate at Reed College in 1934, she had life pretty well figured out. Her family's Theosophy gave her an idiosyncratic but firm moral centre. Her high school pacifist and socialist activism gave her a sense that good could be wrought in the world so long as people of idealism existed and followed their consciences. She was the top student in her class, a bookworm over-achiever who had never known a moment's academic struggle. If anybody was headed for success, it was Eleanor.

Three years later, at the end of her sophomore year (she had to take a year off after her freshman year to earn enough money to return to college, her father's carpentry business having been decimated by the Great Depression), she was failing out of Reed and contemplating suicide, her intellectual and emotional life caught in a downward spiral touched off by her study of psychology, and particularly by the wrecking-ball-like effect that her discovery of behaviourism had on everything she had ever believed or known.

Eleanor Emmons Maccoby (1917–2018) was destined to become one of the greatest psychologists of the twentieth century, but her road to greatness was anything but direct. She grew up on a small, half-mile long island in Puget Sound, just across the water from Tacoma, Washington. She lived there with her father, who owned a small but profitable carpentry business, her mother, a gifted singer and guitarist, her grandmother, and her sisters. Her parents had abandoned traditional Christianity, which they felt did not adequately explain the world's various miseries that were visited upon some of its most innocent individuals, for Theosophy, a popular religious movement of the late nineteenth and early twentieth centuries. Theosophy taught the equality of all humans, the value of animal life, reincarnation, and the ability to communicate with the souls of those who have recently passed, and are awaiting their next round of rebirth.

Holding these idiosyncratic beliefs cost her family some esteem in the community, but also gave them a new group of thoughtful, idealistic, and sensitive fellow believers to discuss the nature of existence with, and the Emmonses were deeply involved with the global Theosophist community, hosting a variety of internationally renowned figures at their home when they came to speak in Washington, and actively pouring resources into the development of a Theosophist summer community on Orcas Island in 1927, which continues to this day.

Being a Theosophist made young Eleanor automatically a bit different from her classmates. They all ate meat with thoughtless abandon, whereas her belief in the beauty and sanctity of all animal life made her a vegetarian, a lifestyle almost unheard of in that era. They felt no doubts about the use of military force to achieve political ends, while she could not reconcile herself to shoveling human beings into the gnashing maw of war to achieve ends that could be attained just as easily through negotiation and mutual understanding. Her refuge in these years was books, and each week she would go to the Tacoma library, return the pile of books she had devoured, and pick up a new stack in a ritual that I suspect is familiar to a number of you out there.

As an outsider, Eleanor was often compelled to think about structures and assumptions that other people took for granted, and in high school that led her into political activity. She formed a club to talk about the day's political issues, usually from a pacifist, left-leaning perspective that led her into engagement with labour and socialist leaders of her time, including her role in an anti-war demonstration that she and her fellow students snuck into the middle of an Army Day parade.

This, then, was the Eleanor Emmons entering Reed College – a smart, compassionate, politically engaged individual of whom much was expected, and who expected much of herself. Her first year went well – she readily found an attractive and well-respected boyfriend from the senior class, and did top notch work in her studies, and after a year of saving up, she headed into her sophomore classes full of high hopes. That was when she collided head-on with the classes led by Monty Griffith, a garrulous bull of an individual, who taught psychology from a behaviourist perspective. On one hand, it was intoxicating looking at humans from a totally different point-of-view from any she had considered before. On the other, behaviourism represented a complete repudiation of what she had previously believed. Instead of humans being individuals who crafted their destiny from free-will possessing souls that earned ever advancing places in the chain of reincarnation, she increasingly saw them as creatures who more or less mechanistically responded to stimuli in biologically coded ways that resulted in behaviour patterns that were predictable and environmentally determined.

It was a blow, and it caused her to wonder what the point of being a human was, if we were essentially being railroaded forward to a largely predetermined end. Added to this existential angst was guilt about becoming sexually active before marriage, and a general lack of interest in the academic work that used to form a central pillar of her sense of self. She stopped doing her work, contemplated suicide, and to add embarrassment on top of misery, Reed's only employee qualified to act in a counselor capacity was – Monty Griffith, who showed up unannounced to her dorm room to ask her why her grades were slipping. Eleanor was understandably too mortified to tell her psychology professor about what the psychological principles he had taught had done to her, and he left without having offered much by way of solving her problems.

Gradually, Eleanor's spirits returned, and over the summer she made up the work to pass her classes, and applied to the University of Washington to complete her undergraduate career. Here, she met and fell in love with Nathan Maccoby, a graduate student who was a teaching assistant in the psychology department. The pair married in 1938, and received their degrees (he a Master's and she a BS) in 1939. This was a propitious time to have a newly minted college degree, as the expansion of government programs ushered in by Franklin Roosevelt's New Deal meant a parallel expansion in the need for qualified professionals to organise those programs, and soon after their graduation, Eleanor and Nathan moved to Washington DC as government employees. Eleanor's specialty as it evolved was in survey design – how to take a program that the government wanted to know the effectiveness of, create a series of informative, non-leading questions to get that information, find a meaningful but non-biased sample in the desired community, hire and train people to conduct the survey, and then hire and train people to code the responses and crunch the numbers in a way that yielded quantifiable, actionable results. It was interesting work, Eleanor was good at it, it gave her interesting connections and skills that would be of use later, and it gave the family an extra income in what could have been lean times, but it wasn't precisely using her psychological training to the fullest, a situation she would have to wait until after the war to rectify.

In 1947, the Maccobys relocated to Michigan, to continue their graduate studies at the University of Michigan at Ann Arbor in the company of the waves of returning soldiers studying under the GI Bill. Here, Eleanor became more interested in experimental psychology, and particularly in problems related to perception and learning. In 1950, Nathan was offered a position at Boston University, which allowed Eleanor the chance to study at Harvard with B.F. Skinner (some online sources say that she studied with him at the University of Michigan, a confusion created by the fact that, though she received her PhD from the University of Michigan in 1951, she was in fact at Harvard at the time, where she finished her PhD remotely). Skinner was one of the world's most recognised names in psychology on the strength of his first book, *The Behavior of Organisms*, which in 1938 established his views on operant conditioning, including his use of the famous 'Skinner Box' to train rats and pigeons into exhibiting ever more complex behaviours in the face of stimuli.

Skinner was media savvy and tended towards the egotistical, but Eleanor learned a great deal from him about the use of technology in psychological experiments, while at the same time keeping one foot in the rival department of social relations, where she worked with Robert and Pat Sears to design questionnaires that would attempt to establish whether Freud's theories about child rearing methods being recapitulated in a person's later behaviour were valid. While the Searses worked on the study of the children, Maccoby (who was not a Freudian in any sense of the term) was tasked with developing the parent interview portion of the study. Crunching the numbers later, it turned out that there was no significant link between factors like toilet training strategy and later child behaviour, but the raw data generated by her interviews was important in later studies on the diversity of American approaches to parenting.

By the 1950s, the Maccobys were a professionally successful couple with a decade and a half of happy marriage behind them, but they were consistently unable to conceive a child, and so they took steps to adopt, first inviting the child of a friend's relative to come live with them, and eventually adopting a baby boy and baby girl. These were hectic days, as Eleanor had to balance the unique needs of adopted children, with her teaching responsibilities (including running an important course on field surveying methods), with her own research activities (among which was a neat study determining what characters in movies people preferentially give their attention to).

Eleanor's work on the parenting project appeared in 1957's *Patterns of Child Rearing*, while the new edition of Ted Newcomb's classic 1947 *Readings in Social Psychology* which she served as lead editor on came out in 1958, which represented two large chunks of work successfully completed, allowing her and Nathan to consider an offer from Stanford University to spend some time there working with Robert Sears on a project to determine how parents get their information in order to better design public information campaigns directed towards them. The Maccobys left for Stanford in 1958, and what was to be a temporary project turned into a permanent assignment as both Nathan and Eleanor were offered tenure positions at the university (though Eleanor's salary was, unbeknownst to her, the lowest of any offered to a full professor there).

It was here, at Stanford in the 1960s and 1970s that Eleanor did the work that made her name not only in academic circles, but in the public imagination. She was asked by the Social Science Research Council to put together a book which presented the best knowledge to date about the developmental differences between men and women. Acting as editor and as author on one of the chapters, she produced *The Development of Sex Differences* in 1966, which represented the beginning of her career in gender psychology that culminated in 1974's *The Psychology of Sex Differences*, co-authored with Carol Jacklin. The pair combed through all studies about sex differences, and then tracked down the authors for any data that they didn't publish because they didn't find any differences between the genders worth reporting. Eleanor churned through this mass of public data and long neglected but crucial unpublished results, and found that, statistically, the differences between the genders developmentally had been overstated. It was a landmark book in the history of Second Wave Feminism, and an important example of the utility of sifting through published studies not only for their stated results, but with an eye towards 'uninteresting' data that might have been discarded in the editing process.

These were also the years when Eleanor launched longitudinal studies of child development during their first six years of life, and produced a classic 1978 paper, 'Social Behavior at 33 Months in Same-Sex and Mixed-Sex Dyads' that brought attention to the understudied topic of how child behaviour changes when a subject is placed in a room with another child, and how gender seeps into that interaction from an early age, even when the children have been given identical uniforms that visually obscure gender information. The early 1980s saw Maccoby's interest turned towards designing studies that revealed as much as possible the intricate structure of the mutual influence of parents and children, including 'Socialization in the Context of the Family: Parent-Child Interaction' (1983), 'Sex-of-Child Differences in Father-Child Interaction at 12 Months of Age' (1983) and 'Children's Dispositions and Mother-Child Interaction at 12 and 18 Months: A Short-Term Longitudinal Study' (1984), while in the late 1980s, she began a collaboration with Robert Mnookin about the psychological impact of divorce and split custody on children that resulted in a series of papers that made the legal profession more aware of the toll different custody configurations can take on the children involved.

While all of this was going on, Eleanor was turning over some of the newer results on gender differences that other researchers uncovered, and that she found in her own studies, and was wondering if, perhaps, *The Psychology of Sex Differences*, as thorough as it had been in 1974, required a re-think in the light of new information. After writing *Dividing the Child: Social and Legal Dilemmas of Custody* (1992) with Robert Mnookin, detailing the results of their studies, she turned to this topic in her last book, 1998's *The Two Sexes: Growing Up Apart, Coming Together*. Here, she found that children's tendencies to group together with children of their own gender enforce certain conceptions of behaviour and goal formation that carry over into the social interactions and life strategies of adulthood in ways that are distinct. To some, *The Two Sexes* represented a betrayal of the egalitarian spirit of *Sex Differences*, while to others it represented a necessary complexification of the subject of socially determined gender development in the light of more sophisticated investigation of intricate early-life group dynamics.

By 1998, however, Eleanor Maccoby's position in the fabric of world psychology was all but unassailable. She won the Lifetime Achievement Award from the American Psychology Foundation in 1996, was made a member of the National Academy of Sciences in 1993, and has served as president of the APA's Division 7 (the developmental psychology section) and of the Society for Research in Child Development. She published her last psychological paper in 2007 at the age of 90, and completed her memoirs in 2017 at the age of 99. On 11 December 2018, the woman whose life saw the end of the First World War and the launch of the first iPhone, and whose career contributed to our knowledge of the role that group relations play in the development of children, and the variety and mutuality of their relations with their parents, passed away at the age of 101, and if the Theosophy of her youth turns out to be true, and her future existence is shaped by the good she did while she was with us, she must be living a most happy life now indeed.

FURTHER READING: In 1989, Maccoby wrote an autobiographical account in Stanford University Press's *A History of Psychology in Autobiography 8*, but it is much easier to find her more complete memoir, titled simply *A Memoir: 1917-2017*, which counts as one of the most engaging, honest, and charming examples of autobiography from a major psychological figure that I've ever read, from her accounts of her Tacoma youth, through her difficulties at Reed, and into the detailing of the mutual support system that Nathan and she developed to sustain each other, it is simply wonderful. If you can't wait, and must know more about her this very moment, *Reed Magazine*'s In Memoriam appreciation of her life and work, which is available online (and free, unlike the *American Psychologist* obituary of her, which they will only let you read after you hand over $18), is very good and should tide you over until her memoirs can arrive.

Chapter Twenty-Two

Building a Kingdom in the Brain's Unfashionable District: Brenda Milner's Century of Neuropsychology

There are scales and metrics you use to evaluate the lives of most neuropsychologists, and then there are those you have to invent in order to speak with any degree of justice about people like Brenda Milner.

She was born in 1918, four months before the end of the First World War.

She earned her BA degree in 1939, on the eve of Europe plunging headlong into the Second World War.

Her masterful study of the role of the hippocampus for memory consolidation was published in 1957, the year that the Soviet Union launched Sputnik into space.

Her major study of speech lateralisation was published in 1977, as movie-goers flocked to see *Star Wars* in the theater.

Her studies of brain imaging during tone perception for Chinese and English speakers were released in 2001, as the world changed in the wake of the World Trade Centre attacks.

And when this book is published, if the world decides, just this once, to be just, she will be but a few months away from her 106th birthday.

How much the world has changed over the course of Milner's life is just barely within the realm of comprehension, but what is more amazing still, is how much she contributed to that change, pushing our knowledge of the brain and memory forward by leaps and flashes, decade after decade, resolutely refusing to stem the tide of her discovery even as she entered the eighth, ninth, and tenth decades of her life.

She was born Brenda Langford, the daughter of two musicians, on 15 July 1918. Because her father was a music critic and pianist, it left him time during the day to spend with her, and up until his death when she was 7 years old, he took her education into his own hands, teaching her mathematics, German, and the arts. It was clear from a relatively early point, however, that she was not to be a musician like her parents, though she did seem to have a gift for literature and languages, which her teachers at school encouraged her to develop. She was more interested, however, in physics and mathematics, and, against all advice, decided to attend Cambridge to study the latter, matriculating in 1936.

As often happens with mathematics students, however, once she began her course of study she realised that there is a difference between loving mathematics, and being the sort of person who wants to be a mathematician as their career. She did not, however, want to switch back to humanities, believing that one could always study humanities independently on the side throughout one's life, but that science, once given up, is more or less gone for good, as to really *do* science means being part of a research team in a way that being a poet generally doesn't. She thought about switching to philosophy so she could study logic, which is close kin to mathematics, but her advisors at Cambridge assured her that a philosophy degree was effectively a receipt for unemployment, but that psychology might scratch a similar intellectual itch.

Importantly for her future work, Cambridge, unlike many British higher institutions at the time, was doing work with the connection between psychology and neuroscience comparable to that being done in North America, and Milner had the chance to work with Oliver Zangwill (1913–1987), who was interested in brain lateralisation (i.e. in brain functions that are more heavily associated with one hemisphere of the brain than the other) and the effect of brain lesions on brain functions. She received her BA in experimental psychology in 1939, and secured a scholarship to remain at Cambridge for graduate studies when the Second World War broke out, and her skills as a psychologist were recruited for the war effort. Like many psychologists in this volume during the Second World War, she was initially detailed to develop aptitude tests for armed services personnel selection, particularly with creating tests that would distinguish prospective bomber and fighter pilots. Eventually, however, she found her way to radar research, where she focused on the effects of different display types on radar operators.

It was while doing radar research that she met a young electrical engineer, Peter Milner (1919–2018). The pair hit it off, and when, in 1944, he was set to be transferred to Canada to help in the development of Canada's nuclear program, he proposed to Brenda, who was about to return to her Cambridge graduate studies, but decided to accept his offer instead and move with him to Canada, where her destiny awaited her.

Of course, destiny doesn't happen all at once, and her immediate concern on arriving in Canada was finding a job that used her skill set to at least some degree, and for seven years she taught animal behaviour and experimental psychology at the University of Montreal. It was good work but, she realised, if she wanted to get back in the game of research in North America, she would need a PhD, and applied to McGill University, where she hoped to work under Donald Hebb (1904–1985), who had just published his landmark *The Organisation of Behavior* in 1949. That book gave to the world Hebb's Postulate, stating that the connections between neurons are made more efficient the more that they fire together, and was a pioneering text in neuropsychology, i.e. the attempt to explain psychological phenomenon with reference to underlying neural structures and reactions.

Hebb's work stood out in a psychological landscape still very much in the throes of a behaviourism which viewed with great suspicion any attempt to relate outwardly recordable behaviour to inner mental or neural states, but Milner was fascinated by the work he was doing, and promised him that, if he took her on as a PhD student, she would see the work through, and not abandon it at the whim of her husband or the vicissitudes of his career. Soon after starting at McGill, however, Hebb received a request from Wilder Penfield at the Montreal Neurological Institute (or 'the Neuro') for a student to help him analyse his surgical patients, and Hebb put Milner's name forward.

Penfield, who students of psychology will recognise from the 'Penfield Diagrams' that depict, via a large-handed, large-mouthed human figure draped over the brain, the relative motor cortex spaced devoted to different parts of the body, and that are featured in virtually every introductory psychology textbook, was a masterful neurosurgeon who employed direct probing of the brain and EEG technology to hone precisely in on which parts of the brain he could safely remove without affecting the patients' quality of life. He needed Milner to study patients upon whom he performed unilateral medial temporal removals. His research had shown that these were regions of the brain most active during epileptic seizures, and that careful removal of part of the medial temporal lobe most involved in seizures (we have two, one on each side) could vastly reduce epileptic episodes for his patients. He was, however, cautious, and brought Milner on to exhaustively test patients before and after the procedure to determine what other effects the surgery might have produced.

At first, it was rather dull work. Penfield was careful only ever to remove one of the medial temporal lobes, so that the other one could compensate for any loss of function. As such, the patients Milner interviewed all seemed to be doing just fine, with improvements to their symptoms and no other loss of function. Then, one day, she gave a memory test to a patient known in the literature as P.B. The test involved patients' ability to recall details of short stories read to them. P.B. usually started well, but his accounts lost cohesion as he went along, and when asked a short time later about the same stories, he couldn't even recall having been read anything at all.

Something was definitely amiss, and when a second patient, F.C., turned up with similar memory problems, Milner and Penfield began to hypothesise about what the underlying issue might be. According to the most advanced theories of the time, memory was a global phenomenon in the brain, something spread uniformly through its structure, not localised to a single location. How could the removal of such a tiny section, then, produce such intense results, of individuals unable to form new memories? All the pair could do at that moment was theorise. They *believed* that both patients had happened to have damage to the other, non-operated-upon, temporal lobe, either through stroke or some congenital defect, and that, as a result, when their remaining good lobe was removed, they lost completely whatever ability lies in the temporal lobes, which, apparently, was the ability to solidify experiences into long-term memory.

Lacking our modern era's imaging capacities, however, they could not prove their assertions until Penfield happened to run into W.B. Scoville at a conference. Scoville worked primarily with psychotic patients, whose symptoms and behavioural problems he often treated with bilateral temporal lobectomies, i.e. with removal of *both* temporal lobes. Scoville shared with Penfield that his patients were suffering severe memory problems, and readily accepted Penfield's request to have Milner perform her memory tests with those patients who had undergone the procedure. Most of those patients were psychotic, and therefore difficult to examine with anything like certainty as to the origin of their problems, but one, H.M., was an individual who had suffered from extreme epileptic episodes his whole life in spite of heavy medication, and had elected for Penfield's procedure as a desperate last measure to give himself some manner of normalcy.

Twenty-nine years old, affable, psychologically 'normal', and with a measured IQ of 118, H.M. was an ideal research candidate. For decades, he reigned as a superstar in psychological circles, with researchers vying for time with him, as an example of a 'pure' injury to both medial temporal lobes not likely to ever come again. What Milner gathered from H.M., and from Scoville's psychopathic patients, was that damage to the hippocampus hindered a person's ability to form long-term memories, and the worse the damage, the greater the hindrance. She and Scoville collaborated on writing up these results in their 1957 paper, 'Loss of Recent Memory After Bilateral Hippocampal Lesions', which has just recently crossed the 10,000 citations threshold, ranking as one of the most cited and important papers in the history of neuropsychology.

Shortly after having revolutionised neuroscience with the discovery that memory consolidation is regionally located in the brain, and not globally distributed, Milner pulled off a second revolution after giving H.M. a new test, which involved tracing a star in a mirror. The expectation was that he would get better and better at the task, then lose all of his memory of how to do it, like he lost memory of everything else, and on the next day perform as well as he did when he first received it. What Milner found, however, was that, given the task the next day, though he did not remember ever having seen it before, he was, somewhat to his amazement, performing as well as he was at the end of the previous day, and after three days was performing it perfectly. He had remembered *how* to do the task, even though he didn't remember ever learning to do it. The implication was staggering: the brain possesses different ways for processing and storing different types of memory, what we now think of as implicit and explicit memory systems, with damage to the centres responsible for the one still allowing the other to take place normally.

Milner's studies of memory were so far-reaching in their impact, so unexpected in their conclusions, that some historians give them credit for putting cognitive neuroscience on the map as a respectable field of study, at a stroke saving the discipline from the years of disdain heaped on it from the behaviourist camp. For her part, Milner would continue checking in on H.M. through the 1960s, but by the 1970s had largely handed that role over to MIT, where Suzanne Corkin (whom we shall meet in more detail soon) would continue the work she had begun so spectacularly. Milner, meanwhile, was on to the next thing, which was her study of the frontal lobe. In her words, 'When I first arrived at the Neuro, the frontal lobes were being debunked and were wildly unfashionable.' The studies that had been done on them were of poor quality, and tended towards the conclusion that this region of the brain didn't do anything particularly important.

Milner suspected otherwise, particularly in the light of research being done at the University of Wisconsin on frontal lesions in monkeys. She decided to import one of the tests being developed there to the Neuro, and try them out on humans with frontal lobe lesions. This was the Wisconsin Card Sorting Test, first developed in 1948. This involves giving subjects cards to sort, which have one of four different ways that they can be sorted (based on colour, symbol shape, etc.). The examiner only tells the subject if they have sorted a card correctly or not, leaving them to figure out what the rule is. The trick comes with the fact that, at every tenth card, the examiner changes which sorting rule the subjects need to employ to get 'correct' answers. Milner found that patients with frontal lobe lesioning had a much harder time than control subjects in switching to new rules, evincing a deficit in their ability to reverse previous learning, resulting in characteristic problem solving rigidity. She published her results in 'Effects of Different Brain Lesions on Card Sorting: The Role of the Frontal Lobes' (1963), which has been cited some 3,300 times since, and which was an early step in the rehabilitation of the frontal lobe which we are now very much in the midst of.

Again, Milner could have kept riding the new interest she had created in the frontal lobes, just as she could have kept riding the interest she had stoked for the temporal lobes a decade earlier, but though she kept updating her previous work (writing about frontal lobe lesions and their impact on cognitive and behavioural functions in 1982 and 1984, respectively, and updating the world on H.M.'s amnesiac syndrome in 1968) she also kept pressing forward into new fields of interest, including the lateralisation of language and handedness with interesting results on how early lesions to the left hemisphere can cause redistribution of faculties between the hemispheres (1977), and imaging studies of bilingual brains performing word generation tasks aimed at discovering whether second languages are stored in the brain differently from how primary languages are (1995) and at how general tone perception differs between speakers of tone-based languages (like Chinese) and non-tone-based ones (like English) (2001). She continued teaching, advising, and researching past her 100th birthday, which was celebrated by the psychological community in 2018 with a symposium featuring an appreciation of her ten decades of life and seven decades of pioneering neuroscientific work. She consistently ranks 'curiosity' as the number one trait that she looks for in graduate students, and credits it as the quality that has kept her pushing forward, never resting on her abundant laurels, but instead always thrusting her nose around the next corner, spending the long life she has been given in the acts of Noticing and Inquiring, of finding those pieces that don't quite fit the neurological structures as we know them, and excitedly devising ways to plumb the contradictions at large. One curious brain has been the gateway through which we have learned about all our brains, as they move through life attempting to observe, learn, and remember.

FURTHER READING: Because of her longevity and importance, we are fortunate to have a number of in-depth interviews with Milner where she reflects, with an accuracy and precision at 90 years of age that I have never possessed in my entire life, on the course of her career, the people she worked with, and the challenges she faced, studying against all counsel those parts of the brain deemed unimportant by other scientists, and striving, against vociferous opinions that it was a waste of time, to bring psychology and neuroscience into closer connection. For her and Penfield's part in studying H.M., you can pick up Luke Dittrich's *Patient H.M.* (2017), with the caveat that, when it's good, the book is very good, and when it isn't, it very isn't. The parts about Milner and Penfield are among the best, and are based on interviews with Milner, but I wouldn't recommend it as a good source about Suzanne Corkin, for reasons we'll get into in her portrait later.

Chapter Twenty-Three

Edith Graef McGeer, the Great Neurotransmitter Race and a Glimpse towards the End of Alzheimer's

The brain can be its own worst enemy. In a host of those diseases, the merest whisper of which is enough to send a streak of black dread through a formerly happy family, the instigator is not a virus, or a bacterium, but rather the brain's own protection mechanisms which, when kicked into overdrive, wreak havoc throughout our nervous system. Sixty years ago, there was little a family could do for a loved one diagnosed with Parkinson's or Alzheimer's disease besides make them comfortable and wait for the dark times to come, but then there arose a dedicated phalanx of neurochemists, emerging from the 1960s rush to identify and lay out the pathways for the brain's various neurotransmitters, whose steady and brilliant work in delineating the chemical systems involved in neural disease have placed us tantalisingly close to understanding and treating some of the most heinous examples of our brains' autodestructive tendencies.

At the centre of that vanguard of neurochemists for the last half-century and more lie two figures, a married couple who participated in the great neurotransmitter revolution, and then gave us our first paths towards understanding the chemistry of Alzheimer's and aging, Edith Graef McGeer (b. 1923) and Patrick Lucey McGeer (1927–2022). Edith was born in New York on 18 November 1923 to Charles and Charlotte Graef. Her father was an ear, eye, nose, and throat specialist who could afford to send her to private school for her elementary and high school years. She attended Swarthmore College on an open scholarship, though when she attempted to register for classes with the chemistry department, its head informed her that she was wasting her own and the department's time, since women did not make good chemists. With the draining of the department's manpower in response to the Second World War, however, Edith became a much more valued component of the student body, and received her degree in chemistry in 1944.

As difficult as her Swarthmore years were, however, they provided her with an excellent education centred around small discussion groups instead of traditional lectures, and when she moved to the University of Virginia for her graduate work she found that she was so in advance of the other students, and of the expectations of the faculty there, that she was able to wrap up a PhD in organic chemistry by 1946. As a PhD chemist, she had little difficulty finding work in the post-war industrial scene, and soon landed a job with DuPont, which at the time was such a powerhouse of chemical research that it alone was responsible for hiring *half* of the entire country's PhD chemistry graduate population. Edith worked in the intelligence division of the chemicals department, which was tasked with thinking up new ideas for chemicals and doing ground-level research for chemical lines in development.

It was while working at DuPont that Edith Graef met Patrick McGeer. He was a former regional basketball star and amateur pilot who had attended Princeton during its Golden Age, when its faculty was flush with the displaced cream of the European universities including John von Neumann and Albert Einstein. His interest in science was fired in that environment, and in 1951, fresh from his PhD work, he began working at DuPont in the polychemicals department. The two met when they moved into apartments across from each other in a block of flats DuPont had constructed for its workers. Neighborliness soon turned to romance, and by 1954 the couple were married and had moved to the University of British Columbia, where Patrick had been accepted as an MD student. These were the heady days when neurobiologists were honing their focus on the chemicals of the brain, and their differential concentrations within different neural structures. Prior to the 1940s, the reigning theory had been that neurons talk to each other through electrical signals, but as we have seen by the late 1940s Marthe Vogt had discovered acetylcholine's potential as a neurotransmitter, and with new studies in the early 1950s about the effect that different antipsychotic drugs had on neurochemical concentrations, the race was on to catalogue the constellation of chemicals that neurons use to send signals to one another.

The McGeers were fortunate at this stage of their careers to work in the lab of William C. Gibson (not to be confused with William Gibson, the author of the classic cyberpunk novel *Neuromancer*, or with William Clyde Gibson, the serial killer – they are all *very* different people), a neuroscientist with a deep interest in the underlying chemistry of psychological phenomena who had studied with Wilder Penfield, the brain probing surgical superstar whom we met in the tale of Brenda Milner. For some years, Edith worked with Gibson in a part-time capacity as she split her time between research and the raising of three children, born in 1957, 1958, and 1960. While the McGeers were honing their skills, Arvid Carlsson was studying the impact of dopamine levels on Parkinson-like symptoms, Marthe Vogt was reporting on the unusual levels of noradrenaline in the hypothalamus, and a small army of researchers were finding regions of the brain with heightened serotonin concentrations.

By 1960, the McGeers, armed with new radioactive isotope tagging techniques, were ready to join the neurotransmitter hunt, which aimed not only to identify what chemicals were and were not neurotransmitters, and where they were most concentrated, but to lay out their life-cycle, i.e. the chemicals that triggered their production, the chemicals that they were formed from, the enzymes that aided those formations, the regions of a neuron they interfaced with, and the mechanisms of their eventual degradation. Knowing more about the different chemicals involved in the life-cycle of a neurotransmitter was additionally useful because some transmitters were unstable in a way that made them difficult to measure, but some of the proteins involved in their production or break-down were entirely stable, and thus an easier source of information about what transmitters existed where, and in what quantities. Researchers at the McGeer lab used choline acetyltransferase, an enzyme that catalyzes the creation of acetylcholine, as their guide to the eventual complete mapping of the cholinergic system, which plays a role in memory and learning, and the degradation of which is a key event in Alzheimer's disease.

Similarly, when they wanted to map which neurons use the inhibitory GABA neurotransmitter, the members of the McGeer lab used GABA transaminase (GABA-T) as a guide. Unlike choline acetyltransferase, which is an enzyme which synthesises acetylcholine, GABA-T is an enzyme that *breaks down* GABA, and thus its presence or absence can also be taken as a sign of GABA's presence or absence in a region, and what's more, GABA-T can interact with certain dyes that would then produce a beautiful and definite map of where GABA is and isn't in the brain.

And as if mapping the GABA and acetylcholine neurons in the brain weren't enough, the McGeers also oversaw the confirmation of glutamate's role as a neurotransmitter, a particularly difficult task given how many different roles glutamate plays in the body, creating an obfuscating omnipresence which is analytically tricky to penetrate in order to catch glutamate in one particular role. Tricky, but for the McGeers and their crack staff, not impossible, and in 1977 they were able to announce to the world glutamate's neurotransmitter nature.

By the 1980s, the great neurotransmitter hunt was largely wrapped up, and the McGeers, after being challenged by the leader of an Alzheimer's support group to use their knowledge and expertise to do something immediately useful for people suffering from brain disease, took up the challenge, and began investigating the role that autotoxicity, or neuroinflammation, played in neural diseases like Parkinson's and Alzheimer's. They looked at the complement system, which usually forms part of our normal immune response, creating a sequence of compounds called cytokines which enable us to attack the cell membranes of pathogenic cells.

What the McGeers found was that the membrane attack complex created by the complement system was richly abundant on the damaged neurons of Alzheimer's patients. Something about the brain's own immune system was being weaponised against it, with inflammatory cytokines apparently playing a central role. Edith and Pat wondered if anti-inflammatory drugs might offer protection against the worst aspects of Alzheimer's, and began pulling together statistics from rheumatology clinics (where anti-inflammatory medication is often prescribed) about the prevalence of the disease among their patients, and followed up a lead about leprosy patients in Japan who seemed to have very low Alzheimer's rates, ultimately tracing that effect to the drug dapsone that they were prescribed, an anti-inflammatory medicine that seemed to cut Alzheimer's incidence from 6.5 per cent among those who went off the drug down to 2.9 per cent for those currently prescribed it.

In 2012, the McGeers founded Aurin Biotech, a private company centred around the promise that aurintricarboxylic acid (ATA) seemed to show as a medicine to inhibit the complement system, and with it the inflammatory cytokines that attack neurons directly, and weaken the blood-brain barrier, allowing peripheral immune cells into the brain, augmenting neural inflammation further.

Delineating what among the many contributions of the McGeer lab can be attributed to Edith, and which to Patrick, is a difficult task, made somewhat easier by the knowledge that, from 1962 to 1986, Patrick was heavily involved with a political career as a member of the legislative assembly, and therefore had to leave much of the running and directing of the lab to Edith when the assembly was in session, indicating a major role for her during precisely the era when the lab was doing its most fundamental work in mapping neurotransmitters. Beyond that, we have to rest content with the assertion that theirs was a partnership of equals, resulting in over 500 publications, including some truly foundational discoveries about the chemical workings of our brain and the diseases it is prone to. For their decades of service, the McGeers were Officers of the Order of Canada in 1995, Fellows of the Royal Society of Canada in 2002, and members of the Order of British Columbia in 2005, but their greater legacy will likely lie in that future day when families only have to read of Alzheimer's as a thing that happened to people once upon a time, in that age before the McGeers and their colleagues used the full measure of their chemical genius to illuminate the first steps on the way to safeguarding our minds through the treacherous valleys of late life.

FURTHER READING: Relative to what they accomplished, there is not nearly enough out there about the McGeers, but one of your best sources is the autobiographical sketch they wrote for the third volume of *A History of Neuroscience in Autobiography*, edited by Larry Squire, a critically important resource that encompasses some twelve volumes that are all, incredibly in this day and age, free to access, so I'd head over there and give it a read before they change their mind!

Chapter Twenty-Four

Virginia Johnson and the Development of Effective Sex Therapy

In 1955, if you were a man who suffered from premature ejaculation, or a woman who had never known an orgasm, your choices were few and far between for constructing something like a sustainable and satisfying sex life for yourself and your partner. Physicians of the era were willfully and shockingly unknowledgeable about the physiology and psychology of sex, while even those professional figures who did profess a deep interest and expertise with sex, the Freudian psychoanalysts, could only offer a course of analysis that would take years to complete, cost tens of thousands of dollars, and in the end only result in single-digit percentage success rates.

In spite of Alfred Kinsey's groundbreaking surveys of American sexual habits during the late 1940s and early 1950s, the United States was still stuck in the sexual Dark Ages, possessing more knowledge about *what* people got up to, but profoundly ignorant about how any of it worked. Finally, however, one obstetrician based out of St. Louis, Missouri decided, against all advice, to turn his gift for observation and research to the problem of detailing, physically, how sex worked. His name was William Masters and his early experiments of 1955 were daring but did not reach anything like their full potential until 1957 when he was joined by the woman who would be his research and life partner for the next thirty-five years, Virginia Johnson.

Together, Masters and Johnson (for so their names were always ordered in the press) revolutionised how first America and then the world looked at sex, and the potential for meaningful sex therapy. Through direct observation and measurement of some 14,000 orgasms, they gathered the knowledge of how bodies work during sex that overturned both millennia-old sexual myths and many of the modern theories of the Freudian school, launching thereby a sexual revolution that has informed spousal expectations and advertising practices ever since.

Johnson was born Mary Virginia Eshelman on 11 February 1925, in Springfield Missouri and, except for a brief period when her father tried to make a go of a career as a groundskeeper in Palo Alto, California, Missouri was to be home to her for the rest of her life. Her adolescence was spent in Golden City, Missouri, a sleepy town of some 867 souls then (and 656 today!) where nothing much ever happened and nobody ever went much of anywhere. Virginia's mother expected more for her life than to just settle down with a local farm boy and spend the rest of her days within a 10 mile radius of Golden City, and pushed her to attend Drury College in Springfield, and to sing for political functions in the state capital of Jefferson City to make herself better known among what passed for the Missouri elite.

The life of a singer soon won out over that of a student, however, and Virginia dropped out of college to focus on her vocal career, as well as on a series of short-term sexual relationships with soldiers she performed for at USO shows who were due to be sent overseas. Uniquely for her time and upbringing, Virginia didn't feel a particular pull to 'save herself' for marriage or to deny herself attractive sexual opportunities that might be forthcoming, and enjoyed a variety of casual sexual partners until her 1947 marriage to Ivan Rinehart, a lawyer some twenty-one years her senior. That marriage was soon over, however, foundering on the rocks of Rinehart's unwillingness to start a family, and in 1950 Virginia married George Johnson, an engineer who also happened to lead a band that toured semi-extensively. Unlike her first marriage, which Virginia entered into at least partially as a means to the end of getting out from the controlling reach of her mother, with George she at least shared the central interest of music, and in the beginning, all was well. She performed with the band, travelled, and was well-received by audiences, but with the birth of the couple's two children, some fundamental changes needed to take place in their lifestyle, changes which George was not willing to entertain at that point in his career, and so, in 1956, Virginia obtained a divorce, maintaining custody of the children.

Thirty-one years old, twice-divorced (or thrice, if you count a rumored two-day marriage to a Missouri politician, the certificate for which has yet to surface), with two children to support, Virginia decided that the only way to make a reasonable living for herself and her kids was to obtain the college degree she had been deflected from a decade earlier, a resolution that brought her fatefully to the steps of Washington University, where she signed up for classes, and for a job as an assistant at the obstetrics and gynecology department, where she was hired after a brief interview by the department's star surgeon, William Masters.

The early days of Masters's sex research were filled with all the innovation, poor judgment, and risk that come with entering a field of study that none before you have tread. Reasoning that prostitutes were a good source of information about sex as it is actually practiced, Masters cut a deal with the local police to ignore the activities of certain volunteer prostitutes who agreed to allow Masters to watch them through a peephole as they serviced their clientele. The experience opened Masters's eyes to the variety of sexual behaviour that existed in the world, but any tear in the veil of secrecy he wove about his investigations threatened to end his research and his career. Added to the danger of his activity during these early studies was the fact that he simply didn't understand a good deal of what he was observing – when one prostitute asked him if, for his studies, it would be useful for her to fake an orgasm like she often did with her clients, he simply couldn't wrap his brain around the idea, and she delicately suggested that, perhaps, if he really wanted to compile an accurate account of how sex worked, he should bring on a woman as a partner.

Here the tale gets somewhat twisted. In some versions, Masters (who was married with two children) reached out to Johnson about becoming his assistant, starting with having her take down case histories, and gradually revealing to her the full extent of his studies (which had moved from prostitutes as the main subjects and onto student and local volunteers), which over time morphed into the (somewhat mutual) decision to practice the sexual behaviours they had witnessed on each other. In others, Masters made it clear from the first that, were Johnson to take the position as his assistant, she would be expected to make herself sexually available to him as part of the research. Both versions are, of course, problematic, even considering the era, but I am hard pressed to decide which reflects worse upon Masters – either drawing Johnson into a project, allowing her to identify with it and receive much of her sense of self through it, and then loading on sexual demands when she was too committed and dependent on him to refuse, or the equally creepy forthright declaration of future sex as a prerequisite for receiving a much needed job.

However the partnership started, the presence of Johnson was the essential ingredient that pushed the project forward. Her insights as a sexually experienced woman, and her ability to recruit new volunteers, make them feel comfortable and even somewhat heroic in a potentially embarrassing situation, were invaluable in moving the project from Masters's determined but essentially blind fumbling through the world of women's sexual physiology, and into the diverse and robust study it became. Though eventually all but kicked out of Washington University when their colleagues saw the extent of the work they were doing (including, horror of horrors, the use of dildos with cameras attached at the end to film actual footage of what happens inside a woman's body when orgasm occurs), they continued their efforts at a nearby location, and in 1966 compiled their results in the runaway best-seller *Human Sexual Response*.

The book was a thunderbolt aimed directly at everything Americans thought they knew about sex. As opposed to the reigning image of a confident, manly male thrusting his way triumphantly in the missionary position, providing maximum satisfaction through Freud-approved vaginal intercourse, what Masters and Johnson found was that male sexuality was a somewhat fragile and meager thing, with erections easily lost, premature ejaculation common, inability to have an orgasm not unknown (particularly in those raised in religious households) and with men subject to refraction periods of up to an hour after orgasm before being able to engage anew in sexual activity. Women, by contrast, were capable of multiple orgasms, with clitoral orgasms being in no way inferior physiologically from vaginal ones, suffered from no refractory periods, and, by the numbers, seemed to have better orgasmic responses from self-stimulation with a machine than with a male partner. Further, Masters and Johnson informed an America still reeling from the dethroning of men as the stallion-like powerhouses in sexual relations, sex was still possible and pleasurable for individuals in their sixties and seventies, with those who had it more often being less prone to having troubles with it in later life.

Lambasted by conservatives for changing sex from a spiritual union between a man and a woman into a physiological process between a woman and her chosen electrical device, and by Freudians for ignoring the deep mental aspects of sex, and the superiority of vaginal orgasm, the book was nonetheless eagerly embraced by a generation that had always suspected that the New Victorianism of 1950s America had sacrificed the full measure of life's pleasures in a head-long push for normalcy and material prosperity, and increasingly by a medical establishment that had historically rejected the need to know the details of sexual physical processes so long as the end results were more or less clearly established.

For Masters and Johnson, the physiological data obtained over a decade was the starting point of the development of new sexual therapies, and if Masters was the lead partner during the data gathering phase, Johnson took to the helm during the development of their famous 'dual therapy' approach to sexual counseling. In an age when psychoanalysis couldn't claim much more than 10 per cent efficacy in helping men with erectile problems or premature ejaculation, or women whose vaginas contracted to the point that vaginal sex was not an option for them, and when regular doctors didn't have much advice beyond, 'Try getting drunk,' the techniques developed by Virginia Johnson boasted an 80 per cent success rate. Her approach was to get an individual's sexual partner involved with the counseling process from the start, with first the husband talking to Masters and the wife talking to her before they switched and went through the process all over again, each mining for the frustrations, expectations, worries, hopes, and practices as expressed by the wife and the husband in their own words in a judgment-free space. Following the information gathering phase, Masters and Johnson would then assign the couple 'homework' which involved mutual exploration of each other's bodies without sexual expectations, to familiarise each other with the terrain, how things worked, and in general to lower the anxiety-inducing mystifications that Americans brought up in the 1940s and 1950s had internalised as part of their upbringing.

They published the results of their sex therapeutic research in technical form in *Human Sexual Inadequacy* (1970) and in a more popularly accessible form in *The Pleasure Bond* (1976). Once the pariahs of American sexual research, by the mid-1970s Masters and Johnson were hailed as the two individuals who, more than anyone else, had given American couples a road to mutual sexual fulfillment, guided by scientifically rigorous understanding of each other's physical beings, and respectful, practical advice about confronting problems of sexual performance as a mutually supportive couple, dedicated to working exploratively through their problems together. Professionally respected, internationally renowned, in 1971 the pair took the final step to combining their life interests together when they married after Masters informed his wife of his decision to seek a divorce from her.

On the surface, all looked well, but for Johnson, these were years of steadily growing crisis. Masters had only asked for her hand in marriage after it seemed like she was going to marry another man, and thereby potentially distract her from her work with (and sexual engagement with) him. Though well suited as work partners, they were hardly suitable as life partners, with Masters's idea of a good time being to sit in his underwear and watch football, while she wanted nothing so much as to go to parties thrown by their famous friends and socialise with celebrities and big name politicians. Further, Masters was obsessed by the idea of completing a third great study of sexuality, centred around homosexuality and possible 'cures' for it, which resulted in the publication of *Homosexuality in Perspective* (1979), which Johnson did not support, and which featured claims of successful homosexuality conversion therapy that strayed far from the rigorous, data-driven studies that had put Masters and Johnson on the map in the 1960s, and which opened the doors to decades of far-right religious conversion therapy programs in America's heartland that have ruined untold lives.

Over the 1980s, then, Masters and Johnson, who were once America's go-to couple for advice on sexual life, grew gradually out of touch with the leading lights in sexual theory, as their marriage grew colder, their finances dwindled through poor business planning, and their new publications, including *Crisis: Heterosexual Behavior in the Age of AIDS* (1988) and *Masters and Johnson on Sex and Human Loving* (1988) failed to capture imaginations or engage the scientific community as their classic work had. In 1992, Masters announced his intention to divorce Johnson so that he could marry a sweetheart of his early youth whom he had recently discovered was now single, and by 1993 Johnson found herself a single woman, again. The remaining two decades of her life saw Johnson increasingly frustrated by her inability to find somebody to share her late life with and provide the same happiness that Masters had found with his lost love, upset with the opportunities that she had turned down because of her need to cater to Masters's needs and demands (including a college degree), and unable to find a pursuit that would resurrect her name from the damage it had received over the 1980s and 1990s. She died on 24 July 2013, two months before a new Showtime series, *Masters of Sex*, based on her and Masters's complicated life and work, made a whole new generation aware of how much she had done in introducing ordinary Americans to the workings of their own bodies, and laying out means by which sexual problems even decades in the making could be gradually overcome, creating a happier world for those concerned, with orgasms for all, and malice towards none.

FURTHER READING: Thomas Maier's 2009 *Masters of Sex* (upon which the Showtime series is based) is a great book which does the work of tracking down all of the people who observed Masters and Johnson over the years, attempting to nail down the plethora of conflicting accounts of how their work evolved, what their relationship was like, and who was responsible for what in their creative output and therapeutic techniques. I'd also recommend Paul Robinson's *The Modernization of Sex: Havelock Ellis, Alfred Kinsey, and William Masters and Virginia Johnson* (1989) not only for how it puts M and J in the context of early sex experts, but for Robinson's delicious roasting of their unfortunate prose style. Used copies of *Human Sexual Response*, *Human Sexual Inadequacy*, and *Masters and Johnson on Sex and Human Loving* are easy to come by with the earlier works being more historically important, but the later works being more pleasant to read, so take your pick.

Chapter Twenty-Five

Beyond Nature vs. Nurture: The Enriched Heredity of Marian Cleeves Diamond and the Evolutionary Psychology of Leda Cosmides

In 1964, two publications announced the beginning of two roads out of the centuries-long quagmire represented by the Nature Versus Nurture debate among philosophers, psychologists, and social theorists. The first, 'The Effects of an Enriched Environment on the Histology of the Rat Cerebral Cortex', was the opening move in a multi-decade program by Marian Cleeves Diamond (1926–2017) to demonstrate that it's not a question of innate vs learned, but rather of how the two interface, with environment producing measurable changes in the baseline physical characteristics of the brain. The other, 'The Genetical Evolution of Social Behavior' by W.D. Hamilton, started the ball of modern evolutionary psychology rolling by positing how different degrees of altruism could be explained as a mathematical maximisation of gene survival whereby individuals might give up their right to reproduce and even their very lives if it served the purpose of ensuring their kin's overall genetic survival.

The analysis of social behaviours through the lens of evolutionary theory would become the hallmark of Leda Cosmides (b. 1957), who with John Tooby would edit and contribute to the classic text *The Adapted Mind* (1992), which sought once and for all to bury any remaining vestiges of the limiting Nature vs. Nurture debate, replacing them with a larger interest in how our common neural architecture came to be that places the focus on our species-wide commonalities instead of on questions of the micro-tuning of our common features through individual experience and environment. Taking it as given that research like Diamond's showed what it claimed to show, and that environment can play a role in nudging basic architecture a bit this way or that way, Cosmides then pushed to the much wider problem of how that architecture came to be, and whether it is based on general structures that we adapt to particular situations, or a myriad of finely tuned neural circuits each of which has been calibrated over 10 million years to deal with one very particular situation.

Marian Cleeves was born in Glendale, California, in 1926. Her father was a British doctor who built something of an idyllic paradise on 20 acres of land, constructing playsets and tennis courts for the children to romp on when they weren't sampling liberally from the fruit trees or tending to the chickens. Her mother was also highly educated, having studied German literature, Greek, and Latin at the University of California, Berkeley (or Cal), an institution which would play a central role in her daughter's academic life. Cleeves knew she wanted to go to Cal, but decided to attend

Glendale Community College first to extend the amount of time she could spend with her family. She attended Berkeley starting with her junior year, graduating in 1948, and enrolling straightaway in the graduate program there, earning her Master's in 1949 for research she did on referred pain (a phenomenon whereby you feel pain in a location other than that where an injury is sustained).

While in graduate school, Cleeves lived at the International House (which is still there today), where she met an athletic young aspiring physicist by the porn-ready name of Dick Diamond, whom she married in 1950, and subsequently followed to Harvard in spite of not having an academic position lined up for herself either there or anywhere nearby. She received her PhD in 1953 the same month she had her first of four children, and eventually found her way to a part-time research project at Cornell with Marcus Singer that turned into a three year instructor's position when she was hired at the last minute to fill in for Singer, whom the university sacked for refusing to divulge the names of other professors with whom he had attended a Communist Party information meeting.

While teaching at Cornell, one day she happened to read in *Science* magazine about an experiment that had been performed at Berkeley by David Krech, Edward Bennett, and Mark Rosenzweig which had demonstrated that a rat strain that had been bred for hyper-competence at running mazes showed markedly higher acetylcholine levels in their brains than those who had been bred for their hyper-incompetence at running them. It was a tantalising result that suggested that learning ability had a chemical basis, and in her autobiographical sketch, Diamond reported immediately being struck by the possibility that not only the chemicals of the brain, but the very anatomy of the brain itself, might change in response to learning. She would get the chance to put that theory to the test when, in 1959, as a result of her husband's acceptance of a job at Cal, she was in a position to approach Krech, Bennett, and Rosenzweig directly with her research proposal.

Her experiments compared the brain anatomy of rats who had been brought up from their 25th to 105th days of life in a stimulation rich environment full of toys and other rats as against that of rats who had been raised in a solitary environment with access to just food and water, and in enriched rat after enriched rat she found a consistent 6 per cent increase in the thickness of the cortex, a bombshell of a result which showed learning and enrichment having a direct, measurable, and physical impact on the properties of the brain itself. She published those results in the 1964 paper we mentioned at the beginning of this piece, and to dispel criticism that what she had observed was just the hastening of the pace of brain maturation, she spent seven years laying out the groundwork for the baseline cortical development of male and female rats against which all future results could be compared.

One of the most significant results to come out of this baseline work was the establishment of gender distinctions in brain structure, with male rats featuring a thicker right cortex, which in old age settled down to a greater symmetry between the hemispheres, while female rats tended towards hemispherical symmetry their whole lives. Follow-up experiments in the 1960s and 1970s on the neural impact of enrichment found not just an overall thickening of the cortex in enriched rats, but greater dendritic branching, higher glia to neuron ratios, larger synaptic junctions, and perhaps most comfortingly for us, that even elderly rats, when exposed to enriched environments, were capable of thickening their cortices some 10 per cent more than their unenriched brethren, a most welcome result suggesting that some capacity for neural plasticity remains into old age.

In the 1980s, Diamond divorced her husband in order to marry Professor Arne Scheibel, an emotionally expressive widower with a great and open curiosity for the world, and embarked on a new direction in her research, tracing the possible connections between neural activity and the stimulation or inhibition of the immune system. At the same time, she co-authored the massively popular *Human Brain Coloring Book* (1985) and wrote up the results of her three decades of research on the neural impact of enrichment for a popular audience in *Enriching Heredity: The Impact of the Environment on the Anatomy of the Brain* (1988).

As Diamond was winding down her incredibly impactful career, just across the San Francisco Bay at Berkeley's great rival, Stanford, Leda Cosmides was wrapping up her postdoctoral work and was busily constructing the foundation of a new approach to evolutionary psychology which would not shy away from the deep and treacherous waters of the evolutionarily derived neural circuitry at the basis of social transactions. By 1985, when Cosmides published her dissertation, *Deduction or Darwinian Algorithms? An Explanation of the 'Elusive' Content Effect on the Wason Selection Task*, the intellectual descendants of W.D. Hamilton had already advanced the idea of evolutionary psychology from the realm of insect altruism into that of human behaviour, with Donald Symons's 1979 *The Evolution of Human Sexuality* representing a particular milestone in the revitalisation of the field. Evolutionary psychologists (EPs) believed strongly that the diverse capacities of the brain developed as adaptations which solved particular problems of survival deep in humanity's past, and that the human mind, then, is the result of the interlinked running of highly specialised sub-modules, rather than of the specialisation through experience of a general brain that comes pre-loaded with only a few rational or logical functions that it applies to different situations as they arise.

In one of Cosmides's memorable examples, she points out that humans need carbohydrates to survive, and dung beetles need poop, and that as a result one species developed an instinctive circuit that responds positively to sugar but resolutely not to poop, and the other a circuit that responds positively to poop and, well, likes sugar all right but wouldn't build a house for the kids out of it. Ancestors to humans might at one point have had a circuit that drove them to find poop delicious, but they likely died out very quickly, leaving behind a group of pre-humans with all more or less the same mental mechanism that responds to the taste of sugar with a, 'Yes, Good, Let's Eat More of That Please,' which has allowed us to consume the calories we have needed to survive as a species.

Studies of infants in particular lend credence to the ideas of the evolutionary psychologists. Instead of the blank slates of the Empiricist tradition, babies come prewired with a number of astonishing abilities and ideas, the result of basic neural modules that proved useful throughout humanity's long past, including instinctive attraction to faces, a basic sense of causality, a distinction between animate and inanimate objects, and ideas about objects as solid and unitary. Humans are not beyond animal instincts from an EP perspective, they have just made higher order intellectual categories instinctual through the development of specialised neural circuits.

Cosmides extended these ideas into the realm of social interactions through a famous adaptation of the Wason selection task, which presents a logical statement and then a series of cards related to that statement. For the statement, for example, I could say 'If you cast Fireball, then you are a Fire Mage.' I would then present you with four cards, which have a spell on one side and a fantasy class on the other, and only show you one side of each, like [FIREBALL], [FIRE MAGE], [ICEBOLT], [LIGHTNING MAGE], and ask you which ones you have to turn over in order to discover if the rule that I gave you was *violated*.

People are not very good at this, as a rule, with only something on the order of one out of every four people doing it correctly (by the way, in this example, the two you would need to flip are FIREBALL and LIGHTNING MAGE). Spotting situations where violations might occur requires logical thought that it can be difficult to work our brains around to, particularly if we haven't taken a logic or geometry class in a while, but Cosmides believed that, since humans must come with a whole battery of specialised social circuits in order to navigate the complicated web of expectations of social existence without getting cheated, people might perform distinctly better on the Wason task if the statements and the resultant cards had a social exchange dimension. If instead of talking about mages and fireballs, we say, 'If you watch the baby, then you get $20' and present cards which have on one side an action performed, and on the other side a payment, like [GOT $20], [WATCHED TV], [GOT $0], and [WATCHED BABY], and ask people to spot which cards need to be flipped for violations of the implied social exchange, instead of 25 per cent of people identifying the correct cards according to logic, the number leaps upwards to 65–80 per cent, but, and here's where things get interesting, that accuracy is related to evaluating violations of fairness rather than violations of logic. Logically, we should check GOT $0 to make sure the other side does *not* say WATCHED BABY, and WATCHED BABY, to make sure it does *not* say GOT $0. But if you have social circuitry which has been trained in the hard evolutionary school of constant social interaction, what you're interested in isn't logic, it's catching cheaters.

One of the great problems of social cognition is how to build a neural circuitry that creates a functioning society which prevents people from just bluffing their way to all of the community's resources constantly without doing any of the work. If you end up on the Sucker end of the stick all the time, you will not be long for the world, and so we have built up strong bits of mental machinery that lurch towards Cheater Detection, and in this case drive us towards the GOT $20 and WATCHED TV cards, because we want to know right away, if somebody got $20 for not doing a job (even though, logically, within the rule we gave, you're allowed to get $20 for doing other things) or if they got $20 for Watching TV (even though, again, according to the rule, there's nothing against that). We HAVE to find the cheaters, and are VERY good at making the card selections that will weed them out, while ignoring the ones that won't.

This is just one example of how Cosmides has probed our latent social circuitry to determine how evolution built brains that allowed us to build civilizations, the tip of a deep iceberg of necessary algorithms which generate the decisions we use instinctively to rate the trustworthiness and dependability of those we have to regularly work with. In 1992, together with her husband John Tooby and Jerome Barkow, Cosmides edited and contributed to *The Adapted Mind: Evolutionary Psychology and the Generation of Culture*, a fascinating walk through the evolutionary challenges and neural solutions that had to be faced and created to go from our deep primate past into our rich apex species present, in which mere questions as to what ratio our individual genetic inheritance combines with our environment to produce our individual behaviour become dwarfed by the larger issue of how we came to have the rich multiplicity of baseline instinctive functionalities that we do, and that bind us all.

FURTHER READING: Diamond's autobiography can be found in the sixth volume of *The History of Neuroscience in Autobiography* which, like the memoir from McGeer in that series mentioned earlier, is also available online. Her 1988 book *Enriching Heredity* is pretty easy to find copies of, and is engagingly written, clearly communicating her excitement for her topic and the importance of its results for every stage of life, three decades into her project. A good introduction to Cosmides's work on the neural machinery of social exchange can be found in the chapter she and Tooby wrote for *The Adapted Mind*, 'Cognitive Adaptations for Social Exchange' (1992), and while you're waiting for that to arrive, her staff page at UC Santa Barbara will lead you to her engaging introductory essay, 'Evolutionary Psychology: A Primer,' which entertainingly distinguishes what evolutionary psychology is, and what it is not, what it has accomplished, and what the future holds.

Chapter Twenty-Six

Brief Portraits: The Early Twentieth Century

Harriet Babcock (1877–1952):

Columbia educated Harriet Babcock was a key figure in the movement to prevent the fragmentation of psychology into its experimentalist and public practitioner wings. She felt that each needed to keep itself aware of the work of the other, and that only by doing so could some of the extreme challenges of her era be met. For her part, she devoted herself to the study of dementia praecox (schizophrenia), seeking for a physiological/biological explanation for the disease that would in turn allow for the development of an effective treatment. She was deeply critical of psychiatric/psychoanalytic attempts to ascribe the disease to the interrelation of various abstract mental categories, which she felt to be woefully unscientific.

Babcock also worked with lobotomy patients, measuring the gap between how they perceived themselves as changed by the operation, and how they actually were. She found that though post-lobotomy patients evaluated themselves as more confident and mentally energised, nonetheless when their mental aptitudes were tested they were found to be significantly reduced in terms of memory and focus performance.

Babcock is also known as the co-creator (with Emily Burr) of the Babcock Test of Mental Efficiency (not to be confused with the regular Babcock Test, which is a way of determining the fat content in milk). One of the things that frustrated Babcock was the lack of a good way to determine quantitatively how much a patient's mental ability changed after an illness or procedure, relative to their previous 'normal' state. She reasoned that vocabulary, being one of the earliest measurable forms of learned content acquired, would be minimally affected by most mental diseases and thus most useful for establishing a stable baseline for general mental aptitude, which could be used to calculate a mental 'age' as calculated from a normed table. The subject would then be given a series of tasks that focused on mental speed and new learning to determine their current mental efficiency (for example, recalling their year of birth or the current governor, or performing map-based tasks), which score would then be correlated to a current mental age. The baseline age, calculated from stable vocabulary, and the current age, calculated from mental efficiency tasks, would then be subtracted from each other to get a measure of mental deterioration. In an evaluation of the Babcock Test performed at Elgin State Hospital by Phyllis Wittman on 245 patients and 26 non-psychotic subjects, it was found that alcoholism resulted in a deterioration of -2.3, epilepsy one of -5.4, schizophrenia one of -0.8, and arteriosclerosis one of -4.9. Inevitably, the Babcock Test eventually came into criticism since there are mental diseases that affect long term vocabulary knowledge, and which therefore affect what was supposed to have been a stable baseline, but the attempt to create a test aimed at measuring deterioration across time through a test given at one moment in time using factors that decay at different rates was a useful step forward in the evaluation of the severity of different illnesses upon mental functioning.

Naomi Norsworthy (1877–1916):

'What is the secret of her?'

According to Frances Caldwell Higgins, this was just one of the many questions students at Columbia University's Teachers College asked each other in their attempts to plumb the mysteries of Naomi Norsworthy. Deeply averse to anything approaching self-promotion, Norsworthy was notoriously tight-lipped about her past and accomplishments, leaving her one of the campus's most enduring and endearing enigmas. In the years since her passing, however, we have pieced together a good deal of her background. She was the daughter of Eve Modridge and Samuel Norsworthy, the latter an English mechanical engineer who had served for years in the United States Navy, and the former a dynamo of an individual who ran every detail of her family's life and brooked no resistance to her word.

Upon marrying, Samuel gave Eve the choice of Japan or New York as the site of their future home and Eve chose New York to be closer to her beloved England. The couple's first child, a son, died within weeks of birth, but the following three (of which Naomi was the oldest) survived to adulthood, and put a strain on the middling income of Samuel that compelled the family to move to New Jersey, where Eve's minute management of the household allowed the family not only to get by, but to offer shelter and hospitality to neighbors in times of need. Naomi was born on 29 September 1877, and was instantly subject to the over-abundance of concern and management that one would expect of a baby following the loss of a first child. She was as a result a serious, overachieving, self reliant, religiously inclined child who held herself to high standards, particularly after the birth of her younger siblings who looked to her as their example. Unsurprisingly, this constellation of traits and expectations instilled in the young girl a drive to become a teacher, and in school her teachers noted her unusual maturity, resolution, and curiosity.

By 1893, at the age of 15, she had completed all the coursework at the Rutherford School she had been attending, and was skipped ahead to the New Jersey State Normal School, a teaching college in Trenton, where after years of being the top student in her class, she found herself having to work much harder to keep up with the students who were several years older than she. However, her sense of responsibility and preparation were the equal of all obstacles before her, and she was soon receiving top marks once again, and earning the attention of her teachers in spite of her shy demeanor. She graduated in 1896 and received a posting immediately to a public school in Morristown, New Jersey, where she spent the next few years before enrolling in Columbia University's Teaching College in 1899 to fill in some of the gaps in her knowledge base left by

Rutherford and the New Jersey State Normal School. Here, she was singled out by a psychology professor for her outstanding mind, and encouraged to adopt psychology as her field of specialty. That professor made her a student-assistant in the department in 1900. She received her Bachelor's degree in 1901, and her doctorate in 1904 for her thesis 'The Psychology of Mentally Deficient Children', which was the result of months of close observation of patients at different mental institutions.

Norsworthy stayed on at the Teachers College, climbing the academic ladder as she earned a reputation for the crystal clarity of her lecturing style, and her non-antagonising yet inspiring approach to leading students. The popularity of her classes pushed her class sizes into the hundreds, and her sense of responsibility prevented her from delegating the task of grading to assistants, a workload that would all but obliterate the prospect of getting any research done for a modern university professor, but which she seemed to take in easy stride. In addition to her work as a professor, her additional positions as Adviser of Women and chairman of the Advisory Board of the Young Women Christian's Association, Norsworthy was encouraged by her colleagues to use her gift for clear explanation to create a textbook on child psychology, a task that was interrupted by having to care for her mother's cancer (characteristically she did this in addition to her usual teaching responsibilities), and then by her own diagnosis of stomach cancer in 1914. She passed away before completing *The Psychology of Childhood*, but such was her impact on those around her that her colleagues undertook to finish the book for her, and it was published in 1918.

Clara Stern (1877–1945):
Over the course of eighteen years, from 1900 to 1918, Clara Stern and her husband, the psychologist William Stern (known today primarily as the coiner of the term intelligence quotient, or IQ), took thousands of pages of notes on the daily development of their three children which served as the raw basis for a series of books that revolutionised child developmental psychology. *Kindersprache* (*Children's Speech*) was published in 1907 with Clara indicated as first author, and represented a milestone in the minute documentation of children's linguistic development, including the discovery of the fact, since statistically borne out, that elder children refer to themselves via a name in their early years, while subsequent siblings refer to themselves with pronouns.

Following *Kindersprache*, the couple published *Erinnerung, Aussage, und Luge in der ersten Kindheit* (*Recollection, Testimony and Lying in Early Childhood*) in 1909, which has similarly stood the test of time as an important first step in laying out the psychological status of lying and reported events in young children. With the book, they attempted to distinguish at what ages, and in what contexts, a child's departure from the 'truth' of an event could be considered as willful lying with intent to deceive as opposed to a fantastic recreation or an incapacity to perceive aspects of a situation which are honestly meant. The book went on to become an important text in the legal profession for the light it shed on the relative value and reliability of child testimony in court.

The last book which we know Clara had a hand in, though she was not listed as an author, was William Stern's 1914 *Psychologie der fruhen Kindheit bis zum sechsten*

Lebensjahr (*The Psychology of Early Childhood Until the Sixth Year of Life*). After wrapping up the diary project, Clara turned back towards the care of her family, and largely left the world of academic work. The rise of Nazism in Germany compelled the Sterns to emigrate in 1933, ultimately settling in North Carolina, where they lived until William's death in 1938. Thereupon Clara moved to New York to be closer to her children, and remained there until her death in 1945.

Kate Gordon Moore (1878–1963):
Kate Gordon was an intellectually omnivorous psychologist whose interdisciplinary research interests helped change the way that psychology engaged with other academic fields. After her graduation from the University of Chicago and a year spent at the University of Würzburg, she took up faculty positions at Mt. Holyoke (1904–1906), Teachers College, Columbia (1906–1907 – thereby overlapping with Naomi Norsworthy's time there), Bryn Mawr College (1912–1916), Carnegie Institute of Technology (1916–1921), and finally coming to rest at UCLA, where she remained for the rest of her career.

The variety in teaching positions taken up by Gordon was mirrored in the variety of research subjects she undertook over her half-century career, spanning the fields of colour theory, memory, aesthetics, education, and imagination, as well as the links between those fields and literature, music, and history. In her dissertation, *The Psychology of Meaning* (1903), she presents a tour-de-force voyage through metaphysics, economic theory, aesthetic theory, and psychological insight in a grand attempt to get at the core of how a human's attention is grabbed and held:

> Meaning depends upon the possibility of making one thing, an emotion, stand for other things, thoughts, i.e. on the possibility of using symbols. We are accustomed to recognise that progress or improved self-control is conditioned by our ability to use words for experience - by our ability to condense past experiences into meanings which are carried along in consciousness as feelings, and ability to speculate by signs upon future events. The anticipatory excitement which we call volition is constituted by our preconceptions of what our experience is going to be, and this foresight of the 'sort' of experience we are to have is a conceptual readiness for it, largely emotional in nature. (p. 63–4)

She studied attention, perception, volition, and memory in a more experimental manner in subsequent research, including 'Meaning and Memory and Attention' (1903) which investigated how the increasing complexity of a series of nonsense syllables paradoxically made their memorisation easier, 'Some Tests on the Memorization of Musical Themes' (1917) which determined that humans are better able to remember and reproduce random arrays of musical notes than random arrays of syllables, and 'The Recollection of Pleasant and Unpleasant Odors' (1925) which found that a subject's averse or favorable judgment of an odor seemingly played no role in their ability to recall it.

Gordon also carried out research into education, criticising the theory that women's and men's mental faculties differed to such a degree that separate educational institutions were required for each, and experimenting with new tests to measure the ability of children to hold and manipulate mental images. In the 1930s she turned her focus to the subject of imagination and in the 1940s to the perception of imagination across different historical literary cultures, capping a half-century career that continued even past retirement, as she re-entered the classroom in the late 1940s to respond to America's sudden need for qualified professors to meet the demands of the GI Bill expansion of the nation's student body. She died after a brief illness on 4 October 1963.

Grace Maxwell Fernald (1879–1950):

If today multimodal learning techniques are commonplace in classrooms, a good amount of the credit for that can be traced to the decades of research on remedial learning undertaken by Grace Fernald. In 1921, her paper 'The Effect of Kinaesthetic Factors in the Development of Word Recognition in the Case of Non-Readers' appeared in the *Journal of Educational Research* and presented her new 'kinaesthetic' method which incorporated the tactile learning insights of Édouard Séguin and Maria Montessori. The method, still in use for children who struggle with traditional approaches to reading, involves students choosing new words to learn, and then tracing them over and over while repeating the word. Once they feel confident in the individual words, they then are asked to read the target words placed within example paragraphs, working their way up to reading target phrases, until eventually they are able to read sentences at their grade level. A 1977 investigation of the efficacy of the technique by Linda Warren showed a 147 per cent improvement in word recognition performance when compared to Orton-Gillingham phonic techniques.

Fernald joined the State Normal School in Los Angeles (which would become UCLA in 1927) in 1911 and in 1921 established there the Clinical School of the University of California to study the efficacy of different remedial education techniques. Her studies culminated in her 1943 masterpiece *Remedial Techniques in Basic School Subjects*, which was unique among pedagogical texts for its longevity, being consulted for decades after its original printing, and even seeing a reprinting in 1988.

Esther Allen Gaw (1879–1973):

Before her run as Dean of Women at Ohio State University (1927–1944), Esther Allen Gaw was noted primarily for her psychological studies of music, in particular what allows some people to perceive the fine structure of a musical composition more easily and instinctively than others. In 'A Survey of Musical Talent in a Public School' (1920), 'Individual Differences in Musical Sensitivity' (1925) and 'Some Individual Difficulties in the Study of Music' (1922) Gaw, herself a gifted violinist and conductor, used newly available record technology to perform tests on thousands of students and their ability to discriminate between consonant and dissonant intervals, recognise differences in tone intensity, reproduce rhythms of different complexity, relate imagery conveyed by a piece, and reproduce heard tones and melodies. Her findings showed that, contrary to expectations, the amount of musical training a child had did not necessarily correlate to high scores on musical sensitivity, with many students who had never had a musical lesson in their lives ranking in the top 3 per cent for musical discrimination and imagination. Meanwhile, some students who grew up in highly musical households, with music as their companions virtually since birth, showed abysmal pitch and rhythm perception.

These results led Gaw to the question of just what musical education can and cannot be expected to do, and in her papers she urges music teachers to use psychological measures of musical sensitivity to inform their designing of individualised curriculum – the student who is having trouble singing different pitches might not be faulty in his vocal apparatus, but rather in his ability to hear the difference between the tones he is being asked to sing. Another might be underperforming not because they can't hear the differences in pitch, but because their motor coordination keeps them from reliably reproducing what they are hearing in the rhythms they are hearing it. Gaw's tests of the different components of musical sensitivity gave teachers what they needed to see that inability in one area of music could be accompanied by perfectly normal achievement in other areas, and that it was possible to use those gifts to overcome individual hindrances, and bring the joy of performing music to everyone.

Emily Thorp Burr (1880–1945?):

Neuropathologist and industrial psychologist Emily Thorp Burr is one of a number of figures in this book whose candle glows a bit dimmer each year in historical memory. Educated at Cornell and Barnard, and serving as a psychologist and neuropathologist at the New York Department of Public Welfare (1911–1913), Bellevue Hospital (1913–1922), and Vocational Adjustment Bureau (1923–?), her specialty was on developing methods by which those deemed 'mentally inferior' might be tested and vocationally trained to achieve some manner of self sufficiency through industrial work. Her major surviving work is *Psychological Tests Applied to Factory Workers* (1922).

Augusta Fox Bronner (1881–1966):

In the late nineteenth and early twentieth centuries, eugenics-influenced theories of juvenile delinquency held that criminal tendencies were primarily a matter of heredity, a trait passed down through bad blood. A bevy of pseudo-scientific studies bolstered these notions, until it was part of the received social wisdom that methods like sterilisation and permanent incarceration were the only way of stopping the spread of delinquency in civilization. That changed in 1914 with *A Comparative Study of the Intelligence of Delinquent Girls*, the published dissertation of Augusta Fox Bronner which demonstrated definitively that there was no strong correlation between mental deficiency and juvenile delinquency in the population of girls she studied. It was a critically important result that placed environment, family struggles, and mental trauma back at the centre of debates over juvenile crime.

Together with the neurologist and criminologist William Healy, Bronner carried out research at the Psychopathic Clinic of the Juvenile Court in Chicago, and at the Judge

Baker Foundation, which resulted in a string of publications that revolutionised the psychological care of juvenile delinquent cases, including *Delinquents and Criminals, Their Making and Unmaking: Studies in Two American Cities* (1926), *A Manual of Individual Mental Tests and Testing* (1927), *Reconstructing Behavior in Youth: A Study of Problem Children in Foster Families* (1929), *Treatment and What Happened Afterwards* (1939), and *What Makes a Child Delinquent?* (1948). They used the Judge Baker Foundation as a model institution to put their ideas of juvenile reform into effect, including developing team methods whereby case workers, physicians, and psychologists jointly coordinate treatment of patients during weekly meetings and discussions, and the development of new testing methods to identify what educational opportunities might best allow children and adolescents to achieve a level of achievement to foster positive senses of self-worth.

Bronner retired in 1946, after three decades of steady devotion to the cause of improving the treatment of disadvantaged children, and our understanding of their unique mental worlds and challenges.

Jessie Taft (1882–1960):

Jessie Taft, along with her life partner Virginia Robinson, was the leading expositor of the 'functional' school of social work, which was inspired by the psychological theories of Otto Rank and the pragmatic philosophy of George Herbert Mead. Educated at the University of Chicago, her dissertation 'The Woman Movement from the Point of View of Social Consciousness' (1913) was a foundational document in the philosophy of early twentieth century feminism that highlighted the mental toll experienced by women who are prevented from combining what little social power they have been given with enough economic control and force to realise their ambitions.

Taft herself was unable to find a university posting following her doctorate, and spent the succeeding years working as an administrator and social worker in the prison and mental hygiene systems, growing steadily disillusioned with the early twentieth century mania for diagnostic testing, when what was needed was less abstract categorisation, and more on-the-ground practical grappling with the here-and-now problems of her charges. In 1924, Taft met Otto Rank, a former member of Freud's inner circle who had been drummed out, like Melanie Klein, for giving the mother too prominent a place in the mental development of children, as well as for his theory of the centrality of birth trauma. Rank advocated for a 'functional' approach to therapy that was contrasted with the dominant 'diagnostic' approach of most Freudians. The approach, as set out by Taft and elaborated and applied by Taft and Robinson, involved redefining the role of the social worker/therapist. Instead of being the guiding force who leads her subjects through a lengthy process of discovering deep traumas that interfere with their ability to function normally in society, the functional therapist's job is to help the patient access their society's institutional resources to directly deal with the problems identified by the patient as most inhibiting their growth and functionality. Seen by the Freudians as band-aid therapy, Taft, with her background in University of Chicago style pragmatism, saw it as a necessary approach to addressing the short-term limited goals of individuals in crisis, who often did not have the luxury of time required for in-depth traditional Freudian therapy, and who needed more than anything to develop their self-directional skills to put their lives back on track, rather than surrendering their agency to the guiding hand of an omniscient and all-controlling therapist.

For sixteen years, from 1934 to her retirement in 1950, Taft put her theories to the test at the University of Pennsylvania's School of Social Work, which was the first institution in the nation to offer a full curriculum centred around social work, and was the world's primary learning centre expounding the functional approach to a social worker's professional duties. Taft died in 1960, and two years later Virginia Robinson published a biography of her, *Jessie Taft: Therapist and Social Work Educator: A Professional Biography*, before passing away herself in 1977.

Mary Grace Arthur (1883–1967):
Arthur's recognition of the deficiencies and inbuilt biases of the standard Stanford-Binet Intelligence Scale was a defining moment in the history of psychological testing methods. Arthur recognised that Stanford-Binet was, at its essence, a test centred around verbal proficiency that routinely undervalued the intelligence of non-English speakers, children of non-American cultures and non-middle class social backgrounds, and individuals with hearing or speech problems. To replace it, she developed the Arthur Point Scale of Performance, a non-verbal test method that centred on manipulation of shapes and puzzles, as well as a modified form of the Leiter International Performance Scale that requires neither oral inputs or outputs to be administered and evaluated, and which continues today as the Leiter-3 performance scale.

Franziska Baumgarten-Tramer (1883–1970):
A vocational psychologist who studied under Hugo Münsterberg (whom we already met as an important force in the development of Ethel Howes and Grace Kent), Baumgarten was an influential developer of the concept of character tests whose work was translated into fourteen languages, as well as an inventor of new testing apparatuses for potential industrial workers. She was part of the generation, like Mary Grace Arthur and Jessie Taft, who found the verbal and intellectual focus of standard testing methods limiting, particularly in evaluating individuals for their vocational aptitudes. For Baumgarten, what was missing in these tests was any type of consideration for character type, or emotional attributes, which she viewed as equally important to intellectual characteristics, in the determining of an individual's satisfaction in and suitability for different types of work. Her book, *Die Charaktereigenschaften* (*The Personality Traits*) (1933), was a landmark in the study of character traits through lexical analysis, which was an important influence for G.W. Allport and H.S. Odbert's 1936 *Trait-Names: A Psycho-Lexical Study*, and which survives today in the OCEAN personality model. After the Second World War, Baumgarten carried out research on the impact of the war on children (*'Uber die Wirkung des Krieges auf Kinder'* (1946)), and deepened her original studies of industrial psychology (*Zur Psychologie des Maschinenarbeiters* (1947)), continuing her research past her retirement in 1959 at the age of 76.

Margaret Vera Cobb (1884–1963):
Margaret Cobb's father, Nathan August Cobb, was a plant pathologist who drew his entire family into his pursuits, and his elder daughter, Margaret, was not only his assistant at his home laboratory, but was also tasked with the job of teaching her three younger sisters. Attending college at Radcliffe and the University of Illinois, her attention was soon captured by the field of educational psychology, which combined her scientific interests with her pedagogical background. During the First World War, she carried out statistical analysis of army corps intelligence tests that found that army medical officers performed significantly less well than other branches of the army, results which she published in 'Intellectual and Educational Status of the Medical Professions as Represented in the United States Army' (1921). From 1922 to 1928, she worked with gifted children in New York, publishing her conclusions jointly with gifted education pioneer Leta Hollingworth in 1926 in 'Children Clustering at 165 IQ and Children Clustering at 145 IQ Compared for Three Years in Achievement.' Though critical, as so many of her generation were, of the reigning intelligence tests of her time as general measures of children's mental ability, she did herself employ more tightly defined and narrowly focused tests to investigate the heredity of mathematical skills, and found a significant correlation between parental arithmetical aptitude and that of their children.

Deepening her investigation of the links between psychology and mathematics, Cobb co-wrote with Edward Lee Thorndike, *The Psychology of Algebra* (1926), which sought to analyse the impact upon a young child's mind of the various inconsistencies in how mathematics is presented to them from a young age. The variable 'x,' for example, in some contexts is a quantity that a child is asked to find the one and only admissible value of, whereas in other contexts, such as the graphing of lines or parabolas, it is meant to stand as a generic input, related to a generic output value 'y,' which is allowed to take up a number of different values, and in yet another context, that of simultaneous equations, x's and y's are once again values that generally take on a single value, which requires the writing of two different equations that contain them. This method of operation places different streams of learning and problem solving into conflict with each other, creating unconscious tensions which interfere with the child's ability to elegantly move from equation solving to equation graphing. This book is filled with similar intriguing insights about the various ways we shoot ourselves in the foot as math teachers by not considering the psychology behind what we are asking the students to do, and personally as a math teacher I often find myself returning to it when I find myself wanting to give a fundamental re-think to the presentation of math topics.

In 1925, Cobb was elected a Fellow of the American Association for the Advancement of Science.

Nathalie Zylberlast-Zand (1883–1942):
Before her likely execution in a Nazi prison in 1942, Polish neurologist Nathalie Zylberlast-Zand (who often published under the name Zylberlast-Zandowa) was one

of her age's leading authorities on meningitis. Meningitis occurs when the subarachnoid space (the cerebro-spinal fluid filled space between the brain-hugging pia mater membrane and the arachnoid mater) becomes inflamed, often by bacteria that enter the space from the bloodstream through the choroid plexus, where the barrier between brain and blood is most prone to breaching. Zand studied the choroid plexus, which is responsible for the production of cerebro-spinal fluid, laying out her findings in her 1930 book *Les plexus choroides: Anatomie, physiologie, pathologie*, in which she made novel contributions to knowledge about the nature of the blood-cerebrospinal fluid barrier. Prior to *Les plexus choroides*, she published studies of the link between meningitis and mental disorders (1911), and delineated the unique course taken by meningitis in patients also suffering from tuberculosis (1921). In 1940, she was confined to the Warsaw Ghetto with the rest of the Jewish population upon the occupation of Poland by Nazi Germany. She served her community as a physician for two years until late September of 1942, when she was deported to Pawiak prison, presumably for execution.

Susan Sutherland Isaacs (1885–1948):

For three years, from 1924 to 1927, Susan Sutherland was the head of the experimental Malting House School in Cambridge, which aimed to create an environment for children where play was prioritised, and teachers were present not as disseminators of knowledge, but rather as guides for the children in their expressions of creativity and spontaneous imagination. It was one of the first institutions of its kind, and during its short life attracted children's education luminaries Jean Piaget and Melanie Klein to observe the work being done there. In her later work, as head of the Institute of Education's Child Development Department from 1933, and as an advice columnist writing under the pen name of Ursula Wise from 1929 to 1940, she made the importance of children's play in the development of their sense of individuality and independence a central component of her policies and writings. She was not, however, entirely without controversy, as a devoted member of the Kleinian camp of psychoanalysis, which taught complete candidness with children on the subject of sex, which might have played its role in the early demise of the Malting House School.

Tatiana Rosenthal (1885–1921):

The first individual to practice psychoanalysis in St. Petersburg, Tatiana Rosenthal remains today a figure of profound mystery. A devotee of the Russian Revolution of 1905, she saw in the combination of Freudian analysis and socialist theory a force that could remake the world profoundly for the better, and sometime after 1906 she traveled to Vienna where she joined Freud's Wednesday Circle. She returned to St. Petersburg in 1911, and by 1919 was the director of the Neurosis Treatment Clinic of Vladimir Bekhterev's Brain Institute. Bekhterev (1857–1927) was at the time one of Russia's most acclaimed neurologists, who independently discovered the phenomenon of conditioned reflexes more commonly associated with the name of Ivan Pavlov, and was a father of 'objective psychology' which, like behaviourism in the West, held that the only useful psychology was that based in objective observations of outward reflexes and behaviours, rather than on acts of introspection or abstract modeling. As such, he was not a particular supporter of the Freudian system of psychoanalysis, but must have been sufficiently impressed by Rosenthal's education, and particularly by her passion for working with neurotic or mentally impaired children, to promote her as head of his outpatient clinic. The following year, 1920, saw her opening a centre for psychically challenged children under the auspices of the Ministry of Education, which founded principles of nursery school psychoanalysis that would be taken up throughout Russia after her untimely death by suicide in 1921.

Sara Mae Stinchfield Hawk (1885–1977):
Sara Stinchfield Hawk is known today primarily as the co-founder in 1925 of the American Speech and Hearing Association (ASHA), an organisation of speech pathologists, audiologists, and speech/language/hearing researchers that currently boasts more than 200,000 members and affiliates. Educated at the universities of Pittsburgh, Iowa, and Wisconsin (where in 1922 she became the first person in the United States to earn a PhD in speech pathology), she spent some time in Vienna in 1931 studying with Sigmund Freud. For most of her career, she was based in California, where she lectured at UCLA and UC Berkeley, and wrote some of her time's definitive works on speech pathology, including 'The Formulation and Standardization of a Series of Graded Speech Tests' (1923), *Speech Pathology with Methods in Speech Correction* (1928), and *Speech Disorders: A Psychological Study of the Various Defects of Speech* (1933). In 1953, she was awarded ASHA's highest honor.

Marie Gertrude Rand (1886–1970):
The Psychologist of Light, Marie Rand played a central role over the course of a half-century career in the study of human sensitivity to light, the designing of new instrumentation to measure that sensitivity, and the development of lighting protocols to use that knowledge to improve the safety and health of commuters and labourers. Together with her husband, Clarence Errol Ferree, she measured the sensitivity to different wavelengths of light of different parts of the retina, and used those measurements to develop the Ferree-Rand perimeter, a map of the retina which became an important tool in the detection of retinal defects. Employing the results of her research, she designed new equipment for ophthalmologists, glare free lighting for the Holland Tunnel, night vision technology for pilots during the Second World War, and new standards for industrial lighting. In 1935, she became director of the Johns Hopkins Research Laboratory of Physiological Optics, a position she retained until her husband's death in 1943, whereupon she moved to New York to continue her research at Columbia University, where she, Legrand Hardy, and M. Catherine Rittler developed the HRR pseudoisochromatic colour blindness test, which improved on the previous Ishihara test plates by detecting blue-yellow colour blindness, and which avoided the use of the tell-tale Ishihara numerals in order to be used with illiterate adults and young children.

In 1952, in recognition of a lifetime of work in the field of lighting, vision, and optics, Marie Rand was made the first woman fellow of the Illuminating Society of North America.

Gabrielle Charlotte Lévy (1886–1934):
With the early death of Gabrielle Lévy at the age of 48, France lost one of its most gifted, rigorous, and dedicated neuropathologists. She was one of her era's great experts on encephalitis lethargica, or 'sleeping sickness', the brain disease that reached pandemic status in the years between 1915 and 1926, causing over 500,000 deaths, and leaving its survivors with lingering tremors and rigidity. Lévy studied 129 cases at the Salpêtrière Hospital and harnessed her polylingual gifts to scour the medical records of multiple nations for additional references to the disease, ultimately producing a 314 page graduate thesis that in 1925 she published as the book *Les manifestations tardives de l'enchéphalite épidémique*, including her observations of various syndromes experienced by patients who had recovered from encephalitis.

Lévy followed up her work on encephalitis with a case study, co-authored with Gustave Roussy, of seven patients suffering from areflexive dystasia, characterised by a lack of tendon reflexes that affect a patient's ability to stand. The condition described by Roussy and Lévy in their 1926 paper, and now known as Roussy-Lévy syndrome, is the result of a genetic disorder affecting the proteins necessary for myelinating nerves, reducing their efficiency in carrying signals, and causing atrophy in the muscular tissues associated with the affected nerves. In his memorial of Lévy, Roussy declared:

> And I must also say, since her name was often placed alongside mine, that, in collaboration, it was because of her that the first idea of a work was to be undertaken and the very large part in its execution. Also, in all its research, is its role dominant, preponderant. *(Translated from the original French)*

Lévy also worked with Roussy on the study of neuro-oncology, with Pierre Marie on the study of involuntary movements, with Jean Lhermitte on vivid hallucinations (as well as on the post-stroke paralysis syndrome now named Lhermitte-Lévy syndrome), and with Antoine Béclère on the use of radiology to explore the nervous system. She died of a severe neural malady in 1934.

Florence Goodenough (1886–1959):
The question of how to measure the intelligence of children in a way that does not rest on accidental cultural or educational factors was one that preoccupied educational psychologists in the early decades of the twentieth century. We have encountered some of the solutions proposed to that problem already, but one of the most unique was put forward in 1926 by University of Minnesota psychologist Florence Goodenough – the Draw-A-Man Test (now known as the Draw-A-Person Test), which reigned for half a century as one of the most popular diagnostic tools in children's education.

The idea of the test is simple – take a kid, ask them to draw a man, then sit back and see what they produce. By making a test out of something that children naturally enjoy doing, and thus reducing the test error introduced by testing anxiety, Goodenough hoped to produce more accurate, less verbally dependent and invasive, measures of intelligence. She developed a scoring scale based on forty-six features to evaluate the level of each drawing, which was later modified by Dale Harris into a seventy-three feature scoring rubric in his 1963 expansion of the system. The idea was an interesting one, and after the publication of Goodenough's book *Measurement of Intelligence by Drawings* (1926) it grew in popularity until at one point it was the third most employed test in clinical evaluations of children's intelligence.

Goodenough trained at the University of Chicago under gifted children's education pioneer Leta Hollingworth, and received her PhD at Stanford in 1924 with educational psychology pioneer Lewis Terman, where her research was included by Terman in his book *Genetic Studies of Genius*, and where she also played a role in the development of the Stanford Achievement Test, a comprehensive battery of K-12 testing still in wide use today in the United States. (Quick note: though it has the same acronym, the Stanford Achievement Test, which is now owned by Pearson, is a different test than the SAT, owned by the College Board.) From 1925 to her retirement in 1947 she worked at the University of Minnesota's Institute of Child Welfare, where she developed not only the Draw-A-Man Test, but a version of the Stanford-Binet Test (the Minnesota Preschool Scale) that could be administered to preschool children, which she published in 1932.

Perhaps more significantly than either of these new testing methods, however, was her development of a new method of gathering data, called 'event sampling.' As opposed to 'time sampling' which relies on minutely observing all of a child's actions at particular pre-set times, event sampling centres around choosing particular behaviours of interest to the researcher, and training observers to wait for that behaviour, and record all of the causes, features, and consequences of that behaviour whenever it occurs. It was an easier system to teach to non-professionals, which was precisely what Goodenough needed to train forty-one mothers to observe their children during episodes of anger. The results of that study, published in 1931's *Anger in Young Children*, showed that the causes of children's anger evolved over the first five years of their lives, with infants and toddlers reacting primarily to physical discomforts and restrictions, giving way at the age of 4 to socially-centred outbursts.

By 1947, encroaching deafness and blindness pushed Goodenough into retirement, but her will to work proved greater than the betrayal of her body, and by learning braille she was able to continue her researches, and produced three more books in her enforced retirement: *Mental Testing: Its History, Principles, and Applications* (1949), *Exceptional Children* (1954), and *Developmental Psychology* (1959). Though the Draw-A-Man Test fell into disfavor during the 1990s and early 2000s after its results failed to regularly correlate with more modern intelligence methods, her development of event sampling and the Minnesota Preschool Scale, research with gifted children, and insights into the early psychology of anger continue to be widely influential, a half-century after her death.

Marjorie Franklin (1887–1975):
Marjorie Franklin was the founder of the planned environmental therapy movement, which was the predecessor of today's therapeutic community (TC) model of treating mental illness. For two years, she underwent psychoanalysis with Freudian inner circle member Sándor Ferenczi, before moving on to the British Psychoanalytical Society, where she was made a full member in 1931. During the 1930s, she developed a new approach to the treatment of mental illness that focused on the placing of individuals in social environments known as Q-Camps where their positive and well-functioning aspects were harnessed in community social work to develop values of interconnection and senses of self-worth the lack of which Franklin saw as a guiding force behind the criminality or 'maladjustment' of the children she proposed to treat.

The first Q-Camp was set up in 1936 at Hawkspur, with a population made up of children and young adults considered too unruly for placement elsewhere. Problems in financing, outbursts of violence between different groups in the camp, and the onset of the Second World War all played their role in Hawkspur shutting down in 1941. The second Q-Camp, started in 1944 with a population of younger children who had been engaged in criminal activity, fared worse still, with the patients setting fire to and otherwise destroying camp property on a regular basis, leading to the abrupt closure of the facility in 1946.

For many, the failure of the second Q-Camp was a grand moment of, 'What did you expect?' that proved once and for all that bad children could not be treated through the nurturing of their positive features, but rather only through the martial crushing of their bad ones. Franklin, however, never ceased believing in the value of planned environmental therapy, and in 1966 established the Planned Environmental Therapy Trust, the mission of which continues today as The Mulberry Bush, which archives PET materials and carries out research in nurturing school and emotional security projects. Meanwhile, the concept of community-fostered recuperation has seen perhaps its most enduring contribution in the form of the therapeutic community model, which has been shown to help foster recovery from addiction by moving patients through a series of recovery stages, each of which gives them greater and greater degrees of community responsibility until they emerge at the end in possession of useful social skills, and a more solid sense of their own potential, strength of will, and capacity for meaningful work.

Florence Edna Mateer (1887–1961):
In 1897, Ivan Pavlov's *The Work of the Digestive Glands* introduced the world to the concept of classical conditioning, by which new neutral stimuli can be used to produce responses usually only produced by 'natural' stimuli. Dogs, food, bells, drool. You know the story. He won the Nobel Prize for his work in 1904, but it took a solid decade for

classical conditioning experiments to make their way to the United States, where one of the earlier studies undertaken was Florence Edna Mateer's pioneering investigation of conditioned reflexes in young children, contained in her 1915 PhD dissertation, and subsequently published in her book *Child Behavior: A Critical and Experimental Study of Young Children by the Method of Conditioned Reflexes* (1918).

Her work was inspired by that of Nikolai Krasnogorski (1882–1961), who in 1907 applied classical conditioning methods for the first time to human subjects, and who was instrumental in establishing theoretical categories for the evaluation and analysis of conditioning in humans. Her experiments, which she described as an extension of his methods, involved measuring the rapidity of conditioning, un-conditioning, and re-conditioning as a function of gender, age, and mental ability. Her dissertation work included a group of fifty girls and boys under the age of 7, some of which were considered 'deficient' in mental ability, and others 'normal.' The children were placed on a low-lying couch, and then had a bandage placed over their eyes. Five seconds later, the observer would brush their right elbow for ten seconds, and then feed them either a square of chocolate or some sugar water. The idea was to determine how many repetitions of the sequence it would take until the children naturally opened their mouths and began swallowing just at the application of the bandage and touch at the elbow.

Through this study, Mateer found that the number of trials required to activate the conditioned response dropped precipitously until about the age of 4, when the number leveled off at 3–5 trials. She further found that, before 2 years of age, boys tended to acquire the conditioned response more rapidly, but that between 2 and 5 years of age, girls exhibited the response after fewer trials. She also found an inverse relation between the number of trials required to learn the response and the number required to unlearn it, i.e. that the longer it took you to figure out instinctually that bandage+elbow touch = food coming, the less time it took for you to drop that instinctive reaction in trials where the food fails to appear. (Incidentally, if you'd like to read more about her results, the entire text of *Child Behavior* is available at archive.org for free.)

After the publication of *Child Behavior*, Mateer moved to Ohio, where Henry Goddard's Ohio Bureau of Juvenile Research was located, and where she developed new insights into clinical psychology, including a positive reinforcement approach that involved awarding 'OK slips' whenever a child behaved well, and a team centred approach to case work similar to that of Augusta Fox Bronner to ensure that many different opinions and points of view were involved in child evaluations. During these years she also founded the Merryheart Schools for gifted and 'backward' children, which operated from 1927 to 1946, and wrote three texts, *The Unstable Child* (1924), *Just Normal Children* (1929), and *Glands and Efficient Behavior* (1935).

Maud Amanda Merrill (1888–1978):
For four decades, Maud Merrill was a keystone of Stanford University's child psychology program, and the persistent and driving force behind the development and evolution of the Stanford-Binet tests. She received her undergraduate degree from Oberlin in

1911, and then proceeded to work at the Faribault Minnesota State Home for the Feeble Minded until 1919, when she moved to California to begin her PhD work at Stanford University under the supervision of Lewis Terman. Terman had just finished developing his adaptation of the Binet-Simon intelligence tests in 1916, and Maud Merrill would play a central role in the research required for the creation of the second edition, published in 1937. That edition sought to address some of the criticism of intelligence testing that, as we have seen, began emerging in the late 1910s and 1920s, by de-emphasising the role of recall tasks, and introducing more non-verbal ones. The development of the Stanford-Binet second edition involved the testing of over 3,000 individuals from diverse socio-economic backgrounds and geographic regions in an attempt to make the exam more universal in its applicability.

After receiving her PhD in 1923, Merrill continued as a faculty member at Stanford, working with Terman on his Genetic Studies of Genius project, which had begun in 1921, and continues to this day, making it the longest running longitudinal psychological study ever. The study followed the lives of 1,444 gifted individuals to document their educational progress, habits of play, social development, personality, and family life. Merrill worked on the study from 1921 to 1922, and from 1924 until her retirement in 1954, and as such was a contributor to the work summarised in the first three volumes of Terman's *Genetic Studies of Genius* (released in 1925, 1947, and 1959) which combatted the old conception of gifted children as socially awkward, sickly, and isolated individuals.

Merrill continued working at Stanford past her retirement, as the guiding force behind the third edition (or what some call the modified second edition) of the Stanford-Binet tests, published in 1960. This edition attempted to deal with the rising controversy between the ratio and deviation approaches to IQ testing. The former, employed by Terman in his first edition, gave scores based on the ratio between your testing age and your actual age. If you are 5 years old, but are correctly answering questions for 8 year olds, your IQ = 100 * (8/5) = 160. This system was also employed in the 1937 revision, but it was increasingly noted that (1) this method of calculating IQ doesn't make much sense for older children (if you're 16, the difference between having an IQ of 200 and one of 188 is based on whether you're able to succeed in tasks that a 32-year-old could handle, versus those a 30-year-old could, which begs the question, precisely what type of mental tasks *can* a 32-year-old regularly solve, but a 30-year-old can't?), and (2) in practice, this was producing a bulge at the right end of the data, a grouping of unusually high scores that went against people's gut instincts that IQ scores should have a bell-shaped curve that tapers off at the extreme low and extreme high parts of the spectrum. The 1960 edition addressed this by introducing Deviation IQs, which placed the mean score at 100, and introduced a standard deviation of 16. This is still the system used today (though often with a standard deviation of 15), whereby the difficulty and scoring are adjusted until the scores lie along a bell curve, with 68 per cent of individuals having a score between 85 and 115 (or 84 and 116 using the 1960 standard deviation), and only 13 per cent of the population ends up with a score above 145 (or 148 by the 1960 system).

The Stanford-Binet system continues to this day, in the fifth edition introduced in 2003.

Una Lucy Fielding (1888–1969):

If you're a hormone, there's a good chance that major components of your life story are directed by the 'control centre' of the endocrine system, the system made up of the hypothalamus and the pituitary gland. Between them, they either directly create and secrete, or have a major hand in regulating, the majority of the hormones that keep us running on an even keel. In the early 1930s, Australian neuroanatomist Una Lucy Fielding and her colleague, Grigore T. Popa, while studying the brain of the Australian marsupial mole *Notoryctes*, discovered that the hypothalamus and pituitary gland are connected by blood vessels, which we now know are the method by which the hypothalamus secretes hormones to the anterior pituitary gland without also releasing them for general circulation.

Fielding's life outside of her discovery of the hypothalamus-pituitary vascular link, meanwhile, was the stuff of a Hollywood movie. She served as a professor of histology and neurology at the American University of Beirut from 1928 to 1929, and as assistant professor of anatomy at Farouk University in Alexandria, Egypt (today the University of Alexandria) from 1947 to 1951, between which she almost single handedly held the medical department of University College London together during the Second World War, when she assumed the role of head of the anatomy department, and oversaw the evacuation of the medical students to Surrey, rescuing the students while her own research notes, which she was intending to make the basis for a book on comparative anatomy, were destroyed during the bombing of London.

Between the destruction of her notes, the firing of all expatriate staff from Farouk University in 1951, and the extra responsibilities heaped on her by the shortage of manpower during the war, Fielding never had the chance to write her definitive text on comparative anatomy, and the most significant opportunity offered her during her later career was the organisation of fellow Australian Grafton Elliot Smith's neuroanatomical work of 1902. She died in London on 11 August 1969.

Maria Grzegorzewska (1887–1967):

Like Marie Curie a generation before her, Maria Grzegorzewska grew up with the consequences of Russia's occupation of her native Poland, including the need to attend clandestine Polish schools to learn her country's language and history, free of the official program of Russification that sought to leverage the power of the education system to erase and overwrite both. She came from a family that had a strong sense of social responsibility, and most of her career was devoted to the cause of expanding educational opportunities and creating humane care facilities for the mentally impaired in her native Poland. She studied at Kraków's Jagiellonian University beginning in 1909, tutoring in her spare time to save up enough money to study abroad, finally transferring to the University of Brussels in 1913, where she worked with children developmental specialist Józefa Joteyko. When Joteyko moved to the Sorbonne to carry on her research, Grzegorzewska moved with her, compiling her Brussels research into her thesis, 'Study on the Development of Aesthetic Feelings' (1916). That same year, a trip to Bicêtre Hospital, where individuals exhibiting extreme mental disabilities were receiving treatment, convinced her of the need to import such organisations into Poland.

For the next half-century, the realisation of this goal was Grzegorzweska's guiding concern, and in 1919 she returned to Poland and was charged with organising the nation's first coordinated program for the support of the blind, deaf, and intellectually disabled. The result of her efforts was the creation of the State Institute of Special Education in 1922, an organisation that she directed for the next forty-five years of its existence, with a break during the Second World War when the institute was compelled to close. From her position as institute director, Grzegorzewska introduced a number of reforms and support structures for the development of special education in Poland, including the founding of the journal *Special School* (1924), the founding of the State Teacher's Institute in 1930 to train teachers in special education techniques, and the introducing of post-graduate opportunities at the State Institute in 1950. She was also the author of the influential three book series *Letters to a Young Teacher* (1947, 1958, 1961), and a number of articles extolling the virtue of a rehabilitation approach to working with the mentally disadvantaged, and detailing the psychology and best teaching practices for the blind.

Grzegorzewska continued working for special education in Poland until the day of her death in 1967. Nine years later, the State Institute of Special Education (which was renamed the State College of Special Education in 1950) was renamed the Maria Grzegorzewska University, a name it holds to this day.

Elizabeth Caroline Crosby (1888–1983):
Elizabeth Crosby was one of three authors of the landmark 1936 text, *The Comparative Anatomy of the Nervous Systems of Vertebrates*, a two volume, 1,800 page work containing 710 figures detailing the central nervous systems of all known vertebrates from the humble lancelet to the often less than humble human being. (Speaking of which, if you're even now lurching for your social media account to type, 'ACTually, *Dale, Comparative Anatomy* was a 3 volume work…', I'll just say that if you have the 1967 edition, it was indeed reworked to be three volumes, but that the original was two, so hey, we're both right! Isn't that nice?) (Incidentally, and I promise we'll get back to Crosby in a bit, but one of my favorite things about *Comparative Anatomy* is how beautiful and intricate the illustrations are, *except for* this one drawing of a dove on page 211 – how did that happen? Did the publishers say, 'We've only got funding for 709 illustrations,' forcing the authors to outsource their final picture to the students of Mitchell Brook Primary School? In any case, I love that it's there, and if everybody out there writing a comprehensive study of the nervous system could just go ahead and feature a goofy hand drawn picture of a dove stuffed somewhere in the middle, I'd consider it a personal favor.)

For the author of one of her century's definitive texts in neuroanatomy, Crosby came to the subject through an unusual life path. Her 1910 degree from Adrian

College was in mathematics, after receiving which one of her Adrian mathematics instructors introduced her to Charles Judson Herrick (1868–1960), a comparative neurobiologist at the University of Chicago known for his studies in the relation between neurological structures and resulting behaviours. She thereupon switched her field of study to neuroanatomy, receiving her PhD from Chicago in 1915. At that point, however, her mother fell ill, and she returned home to Michigan to care for her until her death in 1918, taking up work as a school superintendent to pay the bills. Finally, in 1920 she felt the call of anatomy again, and took a junior instructorship at the University of Michigan. While on sabbatical in 1923 at Amsterdam's Central Institute for Brain Research, she studied with Cornelius Ubbo Kappers, with whom she would later co-author *Comparative Anatomy*. In 1936, likely in recognition of the mammoth accomplishment that was *Comparative Anatomy*, she became the first woman to receive a full professorship at the University of Michigan Medical School.

In 1962, she showed that, at the age of 74, her encyclopaedic knowledge of the nervous system had not lost a step with the publication of the 581 page *Correlative Anatomy of the Nervous System* (co-authored with Tryphena Humphrey and Edward Lauer) which is an eminently readable account of the nervous system and its structures from the single neuron up through the cerebral cortex aimed at, 'Bridging the gap between the morphology of the nervous system and its applications in physiological, in pharmacological, and in allied clinical fields, particularly in neurosurgery and neurology.' Sixty years on, it's still a great read.

Ada Hart Arlitt (1889–1976):
From being one of the most influential educational psychologists of the 1920s and 1930s, the National Chairman of the National Congress of Parents and Teachers from 1928 to 1943, and the author of a number of books on adolescent and early childhood psychology, Ada Arlitt is today a ghost lacking even a rudimentary Wikipedia stub of an entry whose name persists only in the form of the Arlitt Child Development Centre she founded at the University of Cincinnati in 1925, and which is sustained today primarily through the endowment she provided for it. She received her PhD in 1917 from the University of Chicago for a study on the effects of alcohol on white rat intelligence, and spent most of her career at the University of Cincinnati, where she taught as a professor and directed the Department of Child Care and Training from 1925 to 1951. During this time, she wrote *Psychology: Infancy and Early Childhood* (1928), *The Child From One to Six* (1930), *Adolescent Psychology* (1933), *The Adolescent* (1938), and *Family Relationships* (1942) in addition to her work for the National PTA, and associate editorship of the journal *Child Welfare* (from 1929).

Łucja Frey (1889–1942):
A 25-year-old man gets shot in the jaw by a rifle bullet sometime in late 1920, and a year later, after a surgery at the wound site to correct a persistent swelling of his face and pus leakage from the ear (I'm sorry – I didn't want to type that any more than you wanted to read it), he noticed that, when eating, the left side of his face

would break out in sweat and redden. This combination of symptoms had been noted before in the medical literature, as far back as 1740, but a complete description of *why* this peculiar combination occurs had eluded the medical establishment until Łucja Frey's publication of '*Le syndrome du nerf auriculo-temporal*' in 1923. In this paper, she identified the symptoms as the result of injury to the auriculotemporal nerve, which contains fibers that connect both to the parotid salivary gland, which is responsible for salivating, and the sweat glands of the face. Sometimes, when that nerve is injured, the healing process goes slightly wrong, and the fibers get redirected such that stimuli, like food, which generally trigger the salivation response, instead end up triggering the sweat response, resulting in a dry mouth and moist face. Henryk Higier named the disorder Frey's syndrome in 1936, though sometimes you'll see it in the medical literature as Baillarger's syndrome after the man who described the symptoms in 1853.

The Frey's syndrome paper was only one amongst an astonishing three dozen produced over the space of just five years, from 1923 to 1928, when she worked in the neurological clinic of Kazimierz Orzechowski in Warsaw. These papers cover a broad range of neurological conditions, including the impact of vegetable based poisons on the degeneration of the spinal cord, the topography of the brain stem, Lou Gehrig's disease (amyotrophic lateral sclerosis), the disintegration of weight-bearing joints, brain cysts, and tumors of the frontal lobe, retrosplenial cortex, and cranium. By any measure, it was a breath-taking output of work, and came suddenly to an end with her marriage and relocation to Lwów in 1928. She continued working as a consultant at a Lwów neurological clinic while also raising a daughter, Danuta, but her break-neck publication schedule had come to a firm close, and with the Soviet takeover of Poland in 1939, she lost her husband, arrested and presumed murdered by the NKVD. With the Nazi invasion in 1941, Frey was compelled to live in the Lwów ghetto, where she worked in a polyclinic until her likely relocation to a concentration camp and execution sometime in 1942.

Marion Almira Bills (1890–1970):

You want to hire a worker for Job A. Ideally, you want somebody who not only has the aptitude for that job, but who will be happy doing it, will do it efficiently, and will stay with your company doing it for a long term. How do you find out what one individual amongst fifty applicants is most likely to check all of those boxes? This is the realm of industrial and organizational (or I/O) psychology, which counts Marion Bills as one of its trailblazing figures.

Bills was a student of industrial psychologist and lighting expert Gertrude Rand, whom we met above, at Bryn Mawr, where she earned her PhD in 1917. (Incidentally, if you're playing Five Degrees Of Wilhelm Wundt at home, Bills also studied with Rand's husband, Clarence Ferree, who was a student of Edward Titchener, who was a student of Wilhelm Wundt, so you can put a check next to Marion Bills.) After Bryn Mawr, she held a number of professorial positions until she found her calling as a research assistant at the Carnegie Institute of Technology, where she worked with Walter Bingham on developing new personnel selection and evaluation tools for industry. By 1920, the CIT had become the Life Insurance Sales Research Bureau, and Bills was promoted to the position of associate director. Here, she conducted research on what factors contributed to employee permanence, and found a correlation between higher mental alertness and lower satisfaction in occupations that do not make particularly high demands of them. She published these results in 1923 in the *Journal of Applied Psychology*.

In 1925 or 1926, Bills took up a position as an assistant secretary at Aetna, becoming the first woman officer hired by that company. Here, she continued her research in employee retention, and used her knowledge of psychology to design new programs for the hiring and evaluation of new Aetna employees. One of the results of her research was the development of wage incentive programs for those doing clerical or data manipulation work as a means of compensating talented individuals for the repetitive nature of their work, and thereby increasing the likelihood of their retention, with the cost in bonuses made up for by the savings of not having to train new personnel.

Bills worked for Aetna for thirty years, until her retirement in 1955. She was well-respected for her work and, in something of a welcome novelty for women in early twentieth-century science, was well compensated for it. On her death, she was able to leave 2.3 million dollars to the University of Hartford to endow the Marion A. Bills Scholarship, which continues to this day offering funding for students to travel to the UK to pursue a Master's degree.

Edna Frances Heidbreder (1890–1985):

For decades, *Seven Psychologies* (1933) by Edna Heidbreder ranked as one of the most beloved and influential books on the history of psychology, inspiring generations to take up the study of the mind as their life's work. Its clarity and fairness in telling the story of the development, motivations, findings, and shortcomings of the seven major psychological schools of the early twentieth century stood out in an academic environment often rife with partisan bickering and willful mutual misunderstanding. How did an individual nested firmly in the centre of the intellectual brawl that was early twentieth-century psychology emerge to produce a work of such level-headed objectivity?

Partly, this can be explained by the fact that Heidbreder, like many in the late nineteenth century, but increasingly unlike psychologists of the early and mid-twentieth century, came to psychology through the gates of philosophy. In the age of William James and Wilhelm Wundt, this was a traditional combination, and many professors of psychology were also professors of philosophy who applied their inherited philosophical categories to their introspective analyses, but by the 1910s, when Heidbreder was finishing up her undergraduate work, psychology was coming into its own as a concrete discipline, with measurement, experiment, and neuroanatomical connections replacing the old abstract categories.

Heidbreder was a Latin major at Knox College, where few psychology courses were offered, but philosophy was in ready supply, and received her Master's degree from the

Christine Ladd-Franklin's theory of colour vision used evolutionary principles to reconcile two of her time's rival colour perception theories. (*Source: Wikimedia Commons*)

Augusta Déjerine-Klumpke and her husband Jules were pioneers in the mapping of the human brain, and of the connection of motor and psychological disorders to irregular neural structures. (*Source: Wikimedia Commons*)

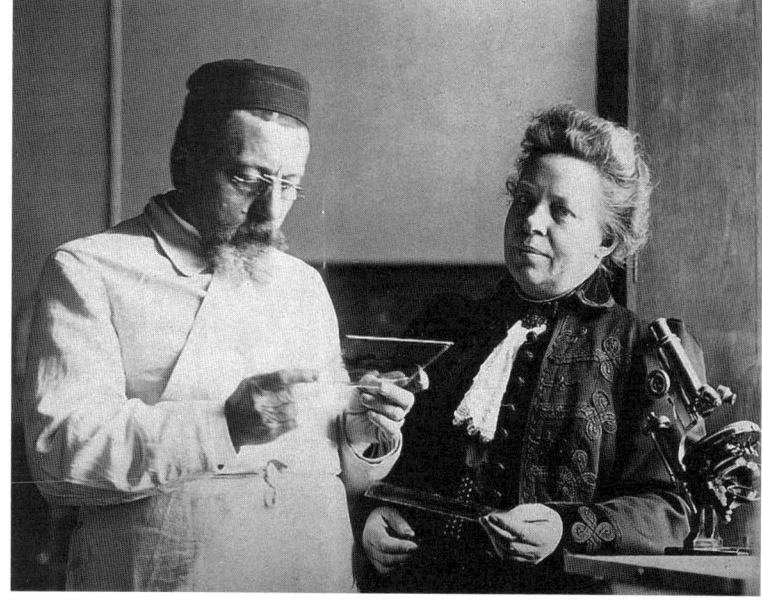

Mary Whiton Calkins developed the paired-association technique for psychological testing which continues to be important to this day, and was the founder of self psychology. (*Source: Wikimedia Commons*)

Maria Montessori was one of the most influential educational and child development theorists of her era, whose approach to early schooling has worldwide impact still. (*Source: Wikimedia Commons*)

Cécile and Oskar Vogt expanded on the research methods of the Déjerines, building up a library of thousands of sectioned brains demonstrating a myriad of different conditions. (*Source: Wikimedia Commons*)

Lillian Gilbreth was, with her husband, a pioneer in the field of industrial psychology, who sought to combine the insights of psychology with workplace efficiency studies to provide safer, mentally healthier procedures for workers. (*Source: Wikimedia Commons*)

Melanie Klein was one of the most controversial psychoanalysts of her era, whose theories about the importance of the mother's breast in early psychological development were criticized by traditional Freudians. (*Source: Wikimedia Commons*)

Tsuruko Haraguchi was Japan's first woman to receive a degree in psychology, and a pioneer in the study of fatigue. (*Source: Wikimedia Commons*)

Leta Hollingworth was a trailblazer in the field of gifted education, pioneering methods for identifying gifted individuals and providing them with stimulating educational programs. (*Source: Wikimedia Commons*)

Inez Prosser conducted early twentieth century studies of the psychological states of children of colour in schools for children of all ethnicities and voluntarily segregated schools that yielded controversial results. (*Source: Wikimedia Commons*)

Anna Freud did foundational work in studying the psychological impact of war trauma and violence on children. (*Source: Wikimedia Commons*)

Rita Levi-Montalcini's research on nerve growth factor was carried out at various temporary labs while attempting to flee persecution from Italian fascists. (*Source: Wikimedia Commons*)

One of the most influential neuroscientists of all time, Brenda Milner's elucidation of the role that the hippocampus plays in memory, and the frontal lobe in allowing us to adapt to changing rules, opened up new fields of research. (*Source: Wikimedia Commons, photo by Eva Blue*)

Marian Cleeves Diamond demonstrated that enriched environments have physical impacts on the size and connectivity of the brain, and that brain growth through enrichment was evident even in elderly rodent subjects. (*Source: Wikimedia Commons, photo by DocumentarianX*)

Ursula Bellugi's studies of sign language changed our perception of how language and the brain relate to each other. (*Source: Wikimedia Commons, photo by Chris Keeney*)

Nobel laureate May-Britt Moser co-discovered the grid cells of the entorhinal cortex which create a hexagonal mental model of the space we find ourselves in, allowing us to navigate it. (*Source: Wikimedia Commons, photo by Bengt Oberger*)

University of Wisconsin in philosophy. Like many students, she was dissatisfied with the gap between what traditional philosophy claimed to know, and its basis for those claims. She thought that the salvation of philosophy could only come at the hands of a greater knowledge of the brain and its functions. To that end, in 1921 she arrived at Columbia University to begin her PhD work. This was an ideal location for somebody with an open mind seeking to answer fundamental philosophical questions, as she not only could attend classes by Columbia legends like John Dewey and the functionalist Robert S. Woodworth, but could also sit on guest lectures from the likes of Bertrand Russell and, perhaps most fatefully, from Harvard professor William McDougall.

McDougall was the founder of what he dubbed 'hormic psychology,' which emphasised the existence of a basic set of principle instincts and emotions which form the universal toolbox of humans' reactions to their world. Heidbreder was interested in the instinctual nature of thinking, which she felt was of a piece with our other instinctual behaviours, and shared the same basis in biological structures and pathways. We are built to think, to perceive and analyse – it's something we can't stop ourselves from doing any more than we can stop ourselves pulling our hands from a hot stove, or seeking food when we are hungry. Thinking isn't some rarefied activity carried out in an ethereal realm of otherness, it is rather part and parcel of how we are built, and in works like her 1924 dissertation *An Experimental Study on Thinking* and her 1926 book *Thinking as an Instinct*, she pleaded for a psychological approach to thinking that incorporated a wider array of sources, including evolutionary/survival, and anatomical considerations.

From 1924 to 1934, Heidbreder was part of the University of Minnesota's staff, where, much like her colleague Florence Goodenough who was developing the Minnesota Preschool Scale, she occupied herself with doing the lion's share of the work creating the Minnesota Mechanical Ability Tests. Following the publication of *Seven Psychologies* in 1933, Wellesley reached out to her with a job offering in 1934, which Heidbreder accepted, and where she remained until her pseudo-retirement in 1955.

Catharine Cox Miles (1890–1984):

Catharine Cox was born to a Quaker family in San Jose, California, and for the first three decades of her life nothing could have been further from her mind than the idea that she would some day be one psychology's most eminent figures. She attended Stanford University, graduating in 1911 with a degree in German language and literature, and proceeded, after some time studying in Germany, to teach German and physical education at the University of the Pacific, until the humanitarian disaster of the First World War, and her Quaker heritage, compelled her to take up a role in delivering relief to the children facing starvation in post-war Germany, culminating in her position as district director to the American Relief Administration for Germany's northeast region.

Following the conclusion of her war relief work, Cox returned to Stanford University in the early 1920s, where, like Maud Merrill and Florence Goodenough, she took an interest in Lewis Terman's longitudinal study of gifted children. As Terman and

Merrill had that project well in hand, however, Cox made the interesting decision to carry out a study of genius not on gifted individuals of the present, but rather via a comprehensive evaluation of the geniuses of the past. Terman had opened the way for this type of research with his attribution of a 200 IQ to the historical figure Francis Galton in 1917. Cox's work consisted of vastly opening the scale of that project, accessing some 3,000 historical sources in the compiling of developmental studies for some 300 individuals of genius and achievement from history. Her results were so impressive that Terman decided to publish all 842 pages of them as the second volume of his *Genetic Studies of Genius* series, the only volume with somebody other than Terman as sole author published in his lifetime.

It was a ground-breaking and promising start, and it seemed an obvious choice at the time for Terman and Cox to continue working together professionally. They decided as their next area of joint effort to attempt a quantification of masculine and feminine mental attributes, with the end goal of creating a scale of masculinity and femininity upon which any given individual could be located. That project went about as well as you might expect, with the two individuals having increasingly heated disagreements about what constitutes a 'masculine' versus a 'feminine' trait. They did publish that scale in their 1936 book *Sex and Personality*, and though it has hardly aged well by today's standards, it was influential for a number of decades. The heat of their disagreements eventually caused Terman to excise a significant amount of material Cox had contributed to their joint work, but the pair did manage to push forward and carry out some tests on differences in word association and interests in men versus women, and determined that, though intellectually there was not a pronounced difference between the genders, there was a significant difference in types of interest, due most likely to cultural differences rather than strictly biological ones.

Cox was also interested in documenting the impact of age on mental ability, a study she undertook with her eventual husband, Walter R. Miles. They found, perhaps not surprisingly, that the largest abilities that suffered decline were in the realm of memory and mental speed, with resultant impacts upon IQ. From 1932 to 1953, Cox held the position of lead psychologist at Yale. Though her publication output diminished drastically after 1936, she was fortunate to live long enough to see something of a renaissance of her work, with the re-publication of some of her historical portraits and results in Margaret and Wayne Dennis's *The Intellectually Gifted* (1976) and in Robert S. Albert's *Genius and Eminence* (1983).

Hazel Martha Stanton (1890–1977?):

Stanton was known in her time as an expert in the field of the psychology of music. Through her position as music psychologist at the Eastman School of Music in Rochester, New York, she was able to carry out experiments and design measurements to determine the impact of musical training on musical capacity and the impact of genetic factors on musical ability, which she published in *The Inheritance of Specific Musical Capacities* (1922), *Psychological Tests of Musical Talent, Eastman School of Music*

(1925), *Musical Capacity Measures of Children Repeated After Musical Training* (1933), and *Measurement of Musical Talent: The Eastman Experiment* (1935).

Louisa Ella Rhine (née Weckesser) (1891–1983):

'The First Lady of Parapsychology' was born the daughter of a gardener and orchardist who passed his enthusiasm for plants onto his daughter, who went on to obtain her PhD in botany from the University of Chicago in 1923. She had married fellow plant enthusiast Joseph Banks Rhine in 1920, and after he obtained his University of Chicago PhD in 1925 the pair taught for a short time at Western Virginia University before developing an interest in the parapsychology movement, which had as one of its chief advocates William McDougall, whom we met above as the founder of hormic psychology.

After some initial disappointments when their joint investigation of the popular psychic medium phenomenon uncovered the largely fraudulent-unto-farcial nature of most mediums, the Rhines decided to turn towards the scientific investigation of the phenomena described in parapsychological texts. To that end, in 1927, they moved to North Carolina so that Joseph could work at Duke University with newly appointed psychology chair McDougall in the creation of a parapsychology department there. Louisa was initially prevented from taking extensive part in her husband's work until their four children were all of school age, but in 1936 she began to study extrasensory perception in children, publishing her results in 1937. Returning to part-time work at Duke, she undertook the sizable task of analysing and categorising the thousands of reports of parapsychological occurrences sent to the department. Sifting through some 15,000 letters, she viewed her job less as an independent investigator of the truth of the claims being sent to her department, and more as a compiler of reported phenomena, with an aim of noting the most commonly documented psychic occurrences in order to then build laboratory-controllable experiments to put them to the test (by 1935 Joseph had established an independent parapsychological lab at Duke with its own dedicated building where experiments were performed including the five symbol card guessing experiment perhaps most famous for its inclusion in the movie *Ghostbusters*).

Using the data from the resulting experiments, Rhine developed a model of spontaneous parapsychological occurrences that centred on the figure of the percipient, the individual receiving psychic messages and intuitions, rather than the agent sending them. She published her results not only in the *Journal of Parapsychology*, for which she was an editor, but in a series of popular books, including *Hidden Channels of the Mind* (1961), *Manual for Introductory Experiments in Parapsychology* (1966), *Mind Over Matter: Psychokinesis* (1970), and *Psi: What Is It?* (1975).

Lois Meek Stolz (1891–1984):

The majority of the psychologists we have met so far in this volume have had as their focus either children, criminals, child criminals, or people possessing some manner of life-altering neurosis or chronic neurological disorder, which calls to the fore the question, was *anybody* working on the psychological nature of regular old, nine to five, two kids and a mortgage, adults?

As it turns out, there were, and one of the most important of those in the mid-twentieth century was Lois Meek Stolz, whose efforts on behalf of adult education, and of providing parents with information about the role their lives, decisions, and behaviours play in the development of their children, made her one of her time's most sought after experts on issues related to the psychology of parenting. She did her graduate work at Columbia University, receiving her Master's in 1922 and PhD in 1925 for her dissertation 'A Study of Learning and Retention in Young Children' (published the following year in book form).

Before receiving her PhD, Stolz was head-hunted by the American Association of University Women to fill the position of education secretary, working under Helen Thompson Woolley (whom we met above) to create adult education programs. In 1929, she was recruited by Columbia University as a full professor of education, and in 1930 was made director of Columbia's Institute of Child Development, a position she held until her 1938 marriage to Herbert Stolz. The couple moved to Chicago shortly thereafter, where Stolz wrote her influential lay work *Your Child's Development and Guidance* (1940). Unlike most psychology books about children, this one was written for parents not about what to expect from their children, but what they should expect about how their own habits, behaviour, household culture, and personality traits have an impact on how their children grow up.

It was a powerful and impactful book, and during the Second World War, Stolz followed it up with powerful and meaningful action. Edgar Kaiser, the owner of the Kaiser Shipyards, needed women industrial workers if he was to meet the wartime demand for ships, and so he recruited Stolz to develop a system of childcare centres for the workers so that they could more easily balance their time at work with the needs of their children. Stolz's first centre in Portland, Oregon was open 24 hours a day and was responsible for watching over the children of 25,000 employees there. Other industries saw the impact of Stolz's model, and adopted it for themselves. Of course, many closed those very same centres after the war, but the concept, and its success, remain as examples of what industry can do when it feels sufficiently moved to try and consider structural reform.

In 1940, the Stolzs moved to Oakland, where for four years Lois worked at the University of California at Berkeley, researching the year-by-year physical development of teenagers, publishing her results in *Somatic Growth of Adolescent Boys* (1951). Soon, however, she was back at the parent-child dynamic, and upon transferring to Stanford in 1944, she took it upon herself to investigate one of the great behavioural puzzles of her time – what was the psychological impact on children of not knowing their fathers during their early years? For those soldiers who left in 1941 and returned in 1945, they came home to a child who was often a stranger to them. Stolz's *Father-Relations of War-Born Children* (1954) used interviews, play-based research, and group observations to evaluate what strain the early absence of fathers placed on the father-child dyad, and how much the fracturing of that bond could be subsequently repaired.

Stolz retired from Stanford in 1957, but continued producing new work, most notably her 1967 book *Influences on Parental Behavior* which, like *Your Child's Development* before it, broke new ground in the psychology of parenting, investigating how different belief structures percolate their way into parenting strategies and philosophies, which in turn have an impact on the children raised under those systems. By focusing on deep questions that probed the fundamentals of parents' moral and existential systems, rather than surface questionnaires, Stolz opened a new method of approaching the great mystery of parenting, what it often is, and what, with a little self-knowledge, it could be.

Charlotte Bertha Bühler (née Malachowski) (1893–1974):

If you ask three different psychologists, 'What was so great about Charlotte Bühler?' you'll likely get three entirely different answers, depending on the part of psychology they specialise in. To some, the Berlin-born Bühler was primarily significant as a founding figure, with Abraham Maslow and Carl Rogers, of humanistic psychology, a branch of psychology that represented a 'Third Path' between the scientifically measurable phenomena of behaviourism and the deep internal processes of psychoanalysis, one that resists explaining human behaviours by reducing them down to their pure biological components, and which focuses on the strong human drive towards self-actualisation along a progressive chain of needs and goals, which it is the job of the therapist to empathetically uncover and support, with as little recourse as possible to labeling the patient's setbacks as the result of deep-seated pathologies.

To child development psychologists, Bühler is the person who began the 'Vienna School' of child psychology, which arrayed itself in stark opposition to the Freudian psychoanalysis so synonymous with the city in the early twentieth century. Bühler and her husband lived and worked in Vienna from 1923 to 1938, when the rise of the Nazis compelled her to flee, first to Norway (where Helga Eng was just finishing her historic run at the University of Oslo), and ultimately the United States. Here she developed new methods of probing the mental worlds of children, including the analysis of diaries kept by children, and the Bühler World Test, sometimes known as the Toy World Test. This was a projective test (a test where you use a patient's response to ambiguous stimuli as a means of gaining deeper information about their psychology) to pull meaning from various forms of play, which discovered, in line with Bühler's humanist approach to psychology, that children actively form goals and place importance on personal achievement from an early age.

Finally, for yet a third branch of psychologists, Bühler is a pioneer in the subject of gerontopsychology, who made the case in her *Der menschliche Lebenslauf als psychologisches Problem* (1933) that the psychological issues and challenges surrounding old age are of a qualitatively different sort than those faced in adulthood, and should be treated as a separate branch of psychology with methods unique to it. Arguably her most famous work, *The Course of Human Life: A Study of Goals in the Humanistic Perspective* (1968), was her culminating statement about humans as goal-directed beings whose sense of well-being during their different stages of life rests upon the striking of different balances between capacities, needs, and accomplishments.

Upon moving to the United States in 1940, Bühler resided first in St. Paul, Minnesota, where both she and her husband had a difficult time finding scientific work that matched their capacities, and ultimately in California, where she worked as chief psychologist at the Los Angeles County Hospital from 1945 to 1958. After retiring from the hospital, she resumed her regular publishing schedule while continuing a private practice in Beverly Hills. She formed personal and professional contacts with Maslow and Rogers, and wrote influential English texts on aging ('Meaningful Life in the Mature Years' (1961)) and humanistic psychology (*Introduction to Humanistic Psychology* (1972)).

In 1971, Bühler moved back to Germany to be closer to her son, and died in West Germany in 1974.

Psyche Cattell (1893–1989):

Psyche Cattell was the daughter of arguably the first professor of psychology in the United States, James McKeen Cattell (thus her too-perfect first name). James had been a student of Wilhelm Wundt (of course) and was highly influenced by the work of Francis Galton. When he became professor of psychology at the University of Pennsylvania, he set about designing a series of what he dubbed 'mental tests' (thereby originating the term) on the student population there, measuring reaction times, sensitivities to weight differences, memory performance, and other factors to get an overall sense of each individual's basic mental attributes.

For somebody so concerned with the minute measurement of mental life, however, James was rather informal about the education of his own seven children, of whom Psyche was the third. Rather than enrolling them in a formal school, he left the task of educating them largely to himself and his graduate students. This somewhat ad-hoc (though high-level) educational program had two important consequences: (1) Psyche did not possess a high school diploma, and thus had a difficult time later at Cornell University, when she completed her course work, but couldn't get a college degree due to that lack, and (2) Psyche's difficulty with reading, likely a form of dyslexia, was never formally diagnosed and supported, a problem which might have detected earlier in a more regular learning institution.

Psyche eventually found her way through the education system, however, earning Master's degrees from Cornell and Harvard, and a doctorate in education from Harvard in 1927. Not surprisingly, her career after receiving her doctorate bore similarities to the work of her father, and centred around testing, where her most lasting contribution was the creation of the Cattell Scale, a system of examination based around common recognisable household objects that took less than half an hour to administer, could be applied to infants, and which was a careful combination of the best aspects of the Stanford-Binet and Gesell systems that still bore enough of a similarity to Stanford-Binet that the results of a Cattell test were meaningful and translatable to a S-B specialist. She published her system, still in use today, in 1940 in *The Measurement of Intelligence in Infants and Young Children*.

Following her work developing the Cattell test at Harvard, Psyche moved to Lancaster, Pennsylvania, where from 1941 to 1974 she ran the West End Nursery School and Kindergarten, which in 1945 changed its name simply to The Cattell School. It was a pioneering institution of pre-Kindergarten education, and served as an important source of material for her 1972 book *Raising Children with Love and Limits* and her regular advice column, 'Children Under Eight,' for the local Lancaster newspaper. From 1939 to 1963 she also served as a psychologist at the Lancaster Guidance Clinic. Finally, in 1974, at the age of 81, she began shutting down her manifold activities, closing the Cattell School, and discontinuing the 'Children Under Eight' column. In 1987, she suffered a stroke that compelled her to move to the Moravian Manor elder care home, where she died in 1989.

Clara Mabel Thompson (1893–1958):
By 1950, psychoanalysis had weathered a steady stream of minor and major revolutions, court scandals, and steady evolutions. At that point, Freud had been dead for eleven years, Adler for thirteen, Ferenczi for seventeen, and Abraham for a solid quarter of a century, and though Carl Jung's black heart would continue to beat for another eleven years, a member of the general public could be forgiven at that mid-century mark for being none too aware of just what psychoanalysis was doing with itself at that moment in time. It was at that instant that Clara Thompson stepped in and, through her work *Psychoanalysis: Evolution and Development*, laid out in a comprehensible and objective way the history of the first sixty-five years of the psychoanalytic movement.

Her goal was to show that the high-profile rifts in the psychoanalytic community ought to be viewed less in the tempting light of the personal drama of the major players involved, but as a grand process of self-development and correction which had allowed psychoanalysis to move from Freud's unique brainchild, with its sexual and internal focuses, to a field that had incorporated new perspectives on the importance of the external environment and culture, and evolved some of its more arcane gender hypotheses into sophisticated views of the interplay of economy, culture, and biology in the formulation of gender perceptions. It was an influential intellectual history, which managed the seemingly impossible task of making Freud, Adler, Jung, Klein, Horney, and Fromm seem like members of a continuum instead of rival dojos.

Thompson's own research focused largely on psychology and women, stressing the importance of economic hierarchies in grafting social expectations onto biological gender differentiators, compelling members of each gender to cultivate different and divergent aspects of themselves in a way that was essentially arbitrary. Women are the women that they are because of the way power has been apportioned – change the latter, and you will widen the personality modalities available to the former to actualise their intellectual potential and biological needs. Though, unlike many (one might even say most) of the psychologists we've encountered so far, Thompson did not live to an advanced old age, her influence continued posthumously with the collection and publication of her works, most notably in *On Women*, published thirteen years after her death.

Ruth Sherman Tolman (1893–1957):

Ruth Tolman's life took her into areas trodden by very few women of her time, opening career paths to her that were all but unthinkable to most. For the wonderful places it went, however, her early life trod much the same ground as many of the figures we have seen so far. She attended UCLA for her psychology degree, where she earned her PhD for a study on adult criminality ('Differences between Two Groups of Adult Criminals' (1937)). She then continued her studies of psychology and the criminal system in *Juvenile Detention in California* (1940), co-written with Ralph G. Wales.

The Second World War, however, would prove the turning point in her life. She was married to physicist Richard Chace Tolman, who during the war was vice-chairman of the National Defense Research Committee and a key figure in the history of the Manhattan Project, who played an important role in bringing his friend Robert Oppenheimer in to direct the effort to create the world's first atomic weapon. Ruth was a great friend of Oppenheimer's (some say they were lovers), and between her connection with Richard and that with Robert, she was plugged into the very heart of the international atomic effort. She soon found herself employing her psychological prowess at the Office of Strategic Services, the organisation that would in 1947 become the CIA. Here, she was tasked with developing criteria to select and regularly psychologically assess OSS field agents, fascinating work which brought her into contact with the unique set of psychological stresses experienced by agents in the field.

This work would prove useful in one of her great post-war endeavours, her position as the head of clinical psychology training at the Veterans Administration, where she specialised in the study of what we know now as post traumatic stress disorder (PTSD), but which at the time was dubbed simply battle fatigue. She also played a role, through her connection with Oppenheimer, in opening up the Institute for Advanced Study at Princeton, which generally concerned itself solely with matters of physics and applied mathematics, to the inclusion of psychology as one of its areas of interest.

Phyllis Blanchard (1895–1986):

A prolific writer on the subject of the psychological development of girls, Phyllis Blanchard received her PhD from Clark University in 1919, where adolescent studies pioneer G. Stanley Hall was president until 1920. Subsequently, Blanchard worked for the New York State Reformatory for Women (1920), Bellevue Hospital (1920–1922), the National Committee for Mental Hygiene (1922–1925), and the Philadelphia Child Guidance Clinic (1925–1956). During these years, she published her observations of girls' different pathways of development in a series of books, including *The Adolescent Girl* (1920), *Taboo and Genetics: A Study of the Biological, Sociological, and Psychological Foundations of the Family* (1921), *The Child and Society: An Introduction to the Social Psychology of the Child* (1928), and *New Girls for Old* (1930). A quote from the section of *Taboo and Genetics* authored by her will perhaps suffice to demonstrate what an advanced and reasonable approach she had in 1921, that many in society still have yet to achieve in 2023:

> In addition to the imposition of these arbitrary standards of masculinity and femininity, society has forced upon its members conformity to a uniform and institutionalized type of sexual relationship. This institutionalized and inflexible type of sexual activity, which is the only expression of the sexual emotion meeting with social approval, not only makes no allowance for biological variations, but takes even less into account the vastly complex and exceedingly different conditions of the emotional reactions of the individual sex life. The resulting conflict between the individual desires and the standards imposed by society has caused a great deal of disharmony in the psychic life of its members. (*Taboo and Genetics*, p. 218)

Julia Elizabeth Heil Heinlein (1895–?):
Heinlein's research centred on two rather disparate fields of psychology: the study of children's manual dexterity, and the controlled study of claimed parapsychological abilities. The former of these fields Heinlein naturally came to as a result of her studies at Johns Hopkins, where she received her Bachelor's (1926), Master's (1928), and PhD (1929). These studies resulted in the publication of two monographs, *Preferential Manipulation in Children*, and *Study of Dexterity in Children*. Most of her career was spent at the University of North Carolina, where she worked from 1952 to 1982. In 1927 she married Christian Paul Heinlein (1889–1977), who was associated with Johns Hopkins from 1930 to 1943 and whose work included evaluations of the Seashore musical abilities tests.

Perhaps as a result of UNC's close proximity to Duke University (a breezy 14.4 mile drive today), where McDougall and the Rhines were developing one of the world's centres of parapsychological research, the Heinleins felt it necessary to devote time to rigorously testing the assertions of the parapsychology community. In 1938 they published the paper 'Are Women Clairvoyant?' in the *Proceedings of the Florida Academy of Sciences*, which rigorously tested 300 women for clairvoyant abilities involving guessing hidden words only to find that, though some reported experiencing definite and clear clairvoyant images of the obscured words, not a single one of them got a single word right. They returned to this field in 1958 with 'What is the Role of ESP in Objective Testing at the College Level?' in the *Journal of Psychology*, this time testing 525 women students to predict the sequence on a True-False exam.

Heinlein retired from UNC in 1962, but continued her association with the university as an emerita professor until 1982.

Helen Lois Koch (1895–1977):
Helen Koch is known today primarily for a series of studies she performed in the 1950s on sibling order, and particularly the differences between regular siblings, fraternal twins (siblings born at the same time, but originating from different eggs), and identical

twins (siblings born at the same time, originating from the same egg, and therefore genetically identical). Her results were contained in her classic 1966 book *Twins and Twin Relations*, which resulted in her winning the first ever G. Stanley Hall Award for Distinguished Contributions to Developmental Psychology in 1967. She also played an important role during the Second World War, developing a training program for nursery school teachers to help meet the explosive growth in nursery schools caused by the demand for women industrial labourers.

Jeanne Lampl-de Groot (1895–1987):
Lampl-de Groot was a co-founder of the Dutch Psychoanalytic Institute, which served as an organisation point for those attempting to flee Nazi persecution. She trained with Sigmund Freud from 1922 to 1925, and was associated with the Berlin Psychoanalytic Institute from 1925 to 1933. One of her primary contributions to psychoanalytic thought was the idea of the 'Negative Oedipal Complex' by which young girls fixate on their mothers as their love object, with fathers as competitive rivals, before they move into the more traditional Freudian castration complex stage. Lampl-de Groot considered the negative Oedipal complex to be an important concept in the explanation of later-life stunted sexual activity in women.

Harriet Eastabrooks O'Shea (1895–1986):
Left to their own devices, scientists sometimes have a tendency to crawl deep into the comforting hole of their particular specialty, and to remain there for decades at a time. We need those figures, but we also need those willing to engage in those grand, frustrating, time-consuming efforts to marry science, organisational practice, and public policy into one cohesive whole, so that the insights of the former have a means of trickling into and having an impact on the public sphere. In the realm of psychology, one of the great unifiers of theory and practice of the mid-twentieth century was Harriet O'Shea, the Purdue University (1931–1964) psychology professor who played a central role in pushing for the merger of the American Psychological Association and American Association for Applied Psychology in 1945, and for the promotion of dedicated divisions of the APA for educational and school psychology. Recognising the need for better organisation and support of women psychologists, she was also a driving force behind the creation of the International Council of Women Psychologists.

Ruth May Strang (1895–1971):
Over the course of *thirty-six* published books and hundreds of articles, Ruth Strang, who rose from the familial expectation that she, as her parents' only daughter, would remain at home to serve the family domestically, established herself as a voice of note in many fields of psychology. A fixture at Columbia University Teachers College from 1923 to 1971, she also co-founded (with Pauline Williamson) the American Association for Gifted Children in 1946, and was a president of the National Association of Remedial Teachers. Chief among her many fields of research was the study of reading and the teaching of reading, of which she was considered a world expert, and about which

she wrote a number of academic and popular works over the course of three decades including *Problems in Improvement of Reading* (1946), *Making Better Readers* (1957), *Helping Your Child Improve His Reading* (1962), *The Improvement of Reading* (1967), *Diagnostic Teaching of Reading* (1969), and *Gateway to Readable Books: An Annotated Graded List of Books in Many Fields for Adolescents Who are Reluctant to Read or Who Find Reading Difficult* (1976).

Beth Lucy Wellman (1895–1952):

Today, we consider it as an indisputable fact that children who attend preschool perform better, and adapt more readily to, elementary school, which itself produces a cascading domino effect of confidence and success that persists through adolescence. It is a piece of common wisdom that informs the decisions of parents the world over, and has been backed by study after study, but where did that idea originate?

Largely speaking, the importance of pre-school for later academic success was laid definitively down by Iowa Child Welfare Research Station psychologist and Iowa State Teachers College professor Beth Wellman in her 1940 book *Iowa Studies on the Effect of Schooling*, and its 1946 successor *Factors Associated with Binet IQ Changes of Preschool Children*. The books were the result of her longitudinal study that tracked the change of performance on mental aptitude tests for children as they made their way through pre-school as compared to a control group of children who did not which found that, over time, while the pre-school children performed better and better on the tests, the non-pre-school children's performance actually gradually decreased, creating an ever-widening gap between the two groups as they entered kindergarten. Subsequent studies of the effect she discovered worked their way into United States educational policy, forming the data basis for the development of the Head Start program in 1965, which has provided early childhood education and resources to literally tens of millions of children in the last half century.

Wellman published *Iowa Studies* the same year that she underwent a mastectomy in an attempt to battle the cancer that would ultimately cut her promising career short, ending her life in 1952 at the far too young age of 57.

Mary Cover Jones (c. 1896/7–1987):

If you have to limit yourself to just one thing to know about Mary Cover Jones, it would be the importance of her 1924 paper 'A Laboratory Study of Fear: The Case of Peter' published in the pages of *Pedagogical Seminary*. It is a landmark document in the history of behaviour therapy, which is the branch of psychology which sought to apply the insights of Pavlovian classical conditioning to therapeutic practice. From a behaviour therapy point of view, psychological problems are less indications of fundamental problems baked into an individual's basic personhood, and more the results of reinforced learning over the course of a lifetime which has pushed people into certain extreme responses to their environment. Behavioural therapists believe that negative and harmful responses to the environment can be unlearned, and positive responses reinforced, through the application of conditioning, and Jones's

Little Peter experiment still ranks as among the most referenced examples of just how that can happen.

Peter was a 2-year-and-10-month-old child who was dreadfully afraid of white rats, white rabbits, and a handful of other objects besides. When he was shown a white rat, he 'screamed and fell flat on his back in a paroxysm of fear' which it took a good half hour for him to work his way back out of after the rat's removal. Jones set about to 'uncondition' this response. Every time she brought a white rabbit into the room with him, she also brought some of Peter's favorite food, or introduced into the room other children who had positive reactions towards the rabbit. Within the space of forty-five sessions, Peter went from abject, paralysing fear of white rabbits to actively cuddling them and letting them nibble on his fingertips.

Today, Jones's approach is referred to as 'systematic desensitisation' or 'habituation' and is a cornerstone of behavioural therapy practice. Jones's next major contribution to psychology came with her move in 1927 to California, where she spent forty-eight years of her life devoted to the sprawling longitudinal study known as the Oakland Growth Study. Beginning in 1932, the idea was to take a group of 200 fifth and sixth graders and do a complete personality tracking of their development through adolescence. As opposed to previous studies, which sought to measure particular limited aspects of each child's growth, Jones sought to encapsulate their growth as full individuals, whose biologies, personalities, social contexts, and larger environment all needed to be accounted for in the summing of their developments. The study was only meant to last some eight years, but due to Jones's close connections with the participants, she was able to continue measuring their progress and life's course for over four decades, using the data she compiled to publish some 100 articles tracing adulthood behavioural traits to measurable antecedent factors, as well as the impact of differing adolescent development rates on the subjects' later lives. She conducted her final interview for the series in 1980, when she was 83 and her subject was in their late fifties.

Tomi Kōra (née Wada) (1896–1993):

Kōra was the second woman to earn a PhD in psychology in Japan, and for the first part of her career stuck close to the inspiration of her predecessor, Tsuruko Haraguchi. In 1914 she matriculated at Japan Women's College, and in the following year attended the funeral of Haraguchi, who had only eight years previously made her pioneering trip to Columbia University Teachers College to study psychology in the United States. Kōra followed suit and, in 1917, also traveled to Columbia, where she studied under Haraguchi's mentor, Edward Lee Thorndike (who also played an important role in the career of Margaret Cobb). After receiving her Master's degree from Barnard in 1920, she was encouraged by Thorndike to remain in the United States and study for a PhD at Johns Hopkins, where ultimately her dissertation was overseen by Carl Richter. That paper, 'An Experimental Study of Hunger in its Relation to Activity,' was published in 1922 and would have been Kōra's entry card to the world of American academia had her parents not urged her to return to employ what she had learned in the service of Japanese girls who did not have the opportunity to travel abroad for education.

Her work in Japan was an extension of her interest in developing and employing instruments to measure physiological changes. Together with Richter, she had developed a salivation-measuring tool before returning to Japan, and during her time as a professor at Japan Women's College (where she began teaching in 1927 after resigning in frustration from Kyushu Imperial University) employed plethysmographs and galvanometers to measure the changes in volume and electrical activity that accompanied different emotional states. With Japan's invasion of China, however, Kōra felt increasingly uncomfortable with the direction her country was taking, and after meeting with notable peace advocates such as Mahatma Gandhi and Rabindranath Tagore, she decided to resign her position at JWC.

During the post-war reconstruction of Japan, women were permitted to serve in government, and from 1947 to 1959, Kōra was a member of the Japanese House of Councillors (the upper house of Japan's legislative system), where she is particularly remembered for defying the Japanese Foreign Ministry in an ultimately successful attempt to reintroduce dialogue between Japan and China, resulting in a 1952 trade agreement between the two former enemy nations.

Lorine Livingston Pruette (1896–1976):

Lorine Pruette worked at the intersection of feminism and psychology, a position she came to naturally after a youth spent witnessing the psychological impact of her father's domineering and potentially physically abusive treatment of her mother. An exceptionally bright but physically frail child, Pruette was located at the intense focal point of her mother's unrealised ambitions. Between the transplanted aspirations from her mother and the repellant example of her father, Pruette grew up with a very definite conception of what she wanted to do in the world, and what she desperately wanted to avoid. She attended college, earning a chemistry degree in 1918, and went on to Clark University to pursue her Master's degree under the supervision of G. Stanley Hall. The example of Hall, whom she much admired, helped to convince her that perhaps not all men had at their cores the same broken and violent nature of her father, and she even went so far as to marry fellow student Douglas Fryer, though keeping her maiden name.

In 1920, she and Fryer moved to Columbia University, where Pruette completed her PhD in 1924 with the dissertation 'Women and Leisure: A Study of Social Waste.' After receiving her PhD, Pruette took up a perpetually revolving door of faculty and consulting positions, most of which only lasted from one to two years. During the Second World War, she carried out some interesting studies of the psychological impact of radio for the Office of War Information, but her greatest impact was in the field of vocational psychology as it related to the feminist project. Her longest lasting job was with the National Bureau of Economic Research as a consulting psychologist (1931–1943), which overlapped with her time as consultant to the National Council of Women (1929–1932), an era during which she authored a number of books and articles on the subject of women's mental health as it relates to their sense of financial independence and personal eminence, including 'The Married Woman and the Part

Time Job' (1929), 'Why Women Fail' (1931), and *Women Workers Through the Depression: A Study of White Collar Employment* (1934).

Rose Butler Browne (1897–1986):
For Rose Browne, the first Black woman to earn a PhD from Harvard University, the virtue of psychology lay in what could be done with it. She was not a grand theorist of new systems or discoverer of new layers of the human mind, but was that often far more important thing – an individual who noticed a great and grave problem, and devoted her knowledge to organising a response to it. Throughout Browne's career, her focus was on the educational opportunities afforded Black children in the United States, and the resources dedicated to supporting those who were teaching them.

Born in Boston, she grew up in Newport, Rhode Island, where she received her Bachelor's and Master's degrees from Rhode Island College. Her first position after receiving her degree was at Virginia State College, in Petersburg, where from 1929 she ran the Extension Service program that organised summer training courses for teachers. One of those attending her classes was the Reverend Emmett T. Browne, a local principal who was to become her husband, and whose church duties determined the family's movements. Over the course of her long career at VSC, and later at North Carolina College, Browne energetically fought not only for equal pay for Black teachers and equal resources for Black schools, but for a new approach to children's reading programs. Noting that Black children were underperforming on reading examinations as compared to national averages, she argued for the importation of Montessori-like tactile techniques to familiarise children with elements of reading even before they start the process of learning to read. She also argued for, and ran, programs to certify teachers in the latest pedagogical theories which they could bring back to their communities and instill in their local educational systems.

Browne continued to operate summer and day programs for children after her official retirement in 1963, and published her autobiography, *Love my Children*, in 1969.

Margaret Schönberger Mahler (1897–1985):
How does the process of building the self work, and what happens when it goes wrong? This was the central question for psychoanalyst Margaret Mahler, whose interest in the origin of psychoses brought her to events that occur within the first few months of life. Her separation-individuation theory of child development, as outlined in 'Observations on Research Regarding the "Symbiotic Syndrome" of Infantile Psychosis' (1960), *On Human Symbiosis and the Vicissitudes of Individuation* (1969) and *The Psychological Birth of the Human Infant: Symbiosis and Individuation* (1974), hypothesised a series of stages through which the mother-child duality progresses on the way to making a child feel like an individual, with a generally positive sense of a supported and valuable self.

In the first, or symbiotic, phase, which lasts some five months, a child becomes aware of its mother, but continues to consider itself as one with her, without any ability to distinguish itself as an individual apart from her. Following the symbiotic phase is the separation-individuation phase, in which the child learns that it is a separate individual from the mother, uses its growing powers of locomotion to learn about the world, and then enters a phase of 'rapprochement' dominated by the conflict between the desire to continue exploring and worry about losing the mother forever that causes the child to seek the mother's company in everything it does. If all goes well, the child as it moves through these phases will gradually internalise the mother and her love as a constant and interior pillar of support for its individual identity. If it doesn't, the result can be feelings of inferiority, or of a deeply fragmented sense of self, which can develop into psychoses.

In 1938, Mahler moved to America, the last of a long series of displacements motivated by European anti-Semitism. It was here that she learned in 1946 of the murder of her mother at Auschwitz, and here, at the Masters Children's Centre which she co-founded with Manuel Furer that she developed the tripartite therapeutic model of child treatment, which instead of seeking to remove the child from the parental setting for therapy, as the Klein approach did, sought as much as possible to include parents as active participants in the therapy process.

In 1970, the Margaret S. Mahler Child Development Foundation was established, and continues to this day to carry out research and educational programs on the mother-child dyad.

Thelma Gwinn Thurstone (1897–1993):

For half a century, Thelma Thurstone was a world authority on the design of intelligence testing, and the development of curriculum. She studied psychology at the University of Chicago, where she received her PhD in 1926. Her subsequent career had to fit itself around the anti-nepotism laws that prevented her from serving in the same psychology department as her husband, L.L. Thurstone (1887–1955), whose work in intelligence testing Thelma assisted with and ultimately made the centre of her life's work. From 1924 to 1948, the pair developed and revised the *Psychological Examinations* test battery, which the Educational Testing Service continued to market until its discontinuation in 1964. They were also engaged in the development of the *Primary Mental Abilities* battery, which Thelma began contributing to in 1941, and which remained in circulation until 1974.

Testing, however, is only useful if there exist curricula to support the competencies measured thereby. Saying, 'This seems low,' is one thing. Saying, 'And here are some materials to help with that,' is another. To this end, Thelma developed the Learning to Think Series, a curriculum of four books published from 1947 to 1972 that taught basic skills in difference perception, vocabulary development, cause and effect identification, and nested categorisation. From 1959 to 1980 she also published the Reading for Understanding curriculum kits, which added a new focus on developing inference abilities while reading that were often lacking in other reading packages.

Josephine Ball (1898–1977):

Josephine Ball was a foundational figure in both behavioural neuroendocrinology (the study of the impact hormones have on behaviour) and sexual behaviour studies. The majority of this groundbreaking work was carried out during the 1920s and 1930s when she had a long spell of job stability as assistant psychobiologist at Johns Hopkins University's Phipps Psychiatric Clinic and as a researcher at the Carnegie Institute of Washington. This work built on the discoveries she made in her landmark PhD paper, 'The Female Sex Cycle as a Factor in Learning in the Rat' (1926), which established links between hormonal activity and learning behaviours, and culminated in two studies on the effects of estrogen (1936) and progesterone (1939) in stimulating and inhibiting sexual behaviour in monkeys, respectively. She also carried out studies on rat sexual behaviour as a function of hormone levels during this era.

All of this work, it should be pointed out, preceded that of the person usually held to be the founder of behavioural neuroendocrinology, Frank A. Beach. While Beach's work was certainly important, his research on the topic was not published until the late 1940s, a good two decades after Ball's PhD dissertation, and a full decade after her sexual hormone studies. Whereas Beach, however, enjoyed stability of employment during his decade at Yale and subsequent two decades at Berkeley, Ball struggled to find a semi-permanent position after leaving Johns Hopkins in 1942. The 1940s and early 1950s were filled with a series of one and two year appointments that took their toll on her academic output, and it wasn't until 1955 that she was able to settle down again, as a psychologist for the Veterans Administration Hospital in Perry Point, Maryland, a posting that allowed her to study the effects of the lobotomies regularly performed there, work which she published as 'The Veterans Administration Study of Prefrontal Lobotomy' (1959).

Elizabeth Bergner Hurlock (1898–1988):

Hurlock was the author of an influential textbook on developmental psychology (*Developmental Psychology: A Life-Span Approach*) that went through eight editions between 1953 and 2012, as well as other texts on child development, but what particularly interests me about her is her 1929 book *The Psychology of Dress: An Analysis of Fashion and its Motive*. Fashion is one of those critical facets of daily life that exerts a constant and profound influence on all of our minds, and yet, because it is a 'feminine' field of interest, it has far too often been consigned to oblivion by 'serious' researchers. Hurlock's work strikes me as an important early attempt to reverse that trend, and think seriously, from the perspective of psychology, about what we wear, and how those choices mentally impact ourselves, and those around us.

Rosalie Rayner (1898–1935):

Before there was Dr. Spock, there was Rayner and Watson. In many ways, the runaway popularity of the Rayner and Watson approach to childcare, as contained in the best seller *Psychological Care of Infant and Child* (1928), necessitated the rise of somebody like Spock to bring balance to the world of popular childcare philosophies. For Rosalie

Rayner and John B. Watson were devoted behaviourists, whose belief in the utility of behaviourist insights for the proper raising of children was cemented by the success of their early 'Little Albert' experiment.

In 1919, Rayner, fresh out of Vassar, began attending Johns Hopkins to pursue graduate studies, where she quickly became the assistant to Watson. Together, the pair conducted what is now known as the Little Albert experiment, which is virtually the photographic negative of the later 'Little Peter' experiment of Rayner's former Vassar classmate Mary Cover Jones. Whereas the Peter experiment involved using conditioning to alleviate a young child's paralysing fear of an animal, the Little Albert experiment was used to *create* a paralysing fear, by sounding a loud noise every time the eleven month old Albert was shown a white rat he had previously liked. Predictably, Albert soon reacted in fear whenever he saw the rat, or indeed (so the pair claimed) anything that shared similar properties with the rat.

To Rayner and Watson, this was a sign that behaviours had as their root causes some manner of conditioning rooted in stimuli, and that therefore the job of a parent was to create a good and reliable environment for their children, to react to their behaviour with a scientific eye towards what factors might be causing the behaviour, rather than with instinctual emotional outbursts, and to at all costs preserve their children's sense of independence by minimising the affection shown to them, and granting them opportunities to self-direct whenever possible. They published this advice in *Psychological Care of Infant and Child*, but not before scandal forced the pair to leave Johns Hopkins.

Watson was married, and at some point during their research together, Rayner had become his lover (he was almost exactly twenty years her senior). Watson's wife found out about the affair, and published Watson's love letters in the press, leading to the pair's retreat in disgrace from Johns Hopkins. They ultimately married on New Year's Eve, 1920, and had two children together, whom they raised according to the principles of their book, and who critics of Rayner and Watson, and behaviourism more generally, never fail to point out both attempted suicide in their later lives.

Rayner died suddenly in 1935 at the age of 36 from dysentery, while Watson would live on another twenty-three years, long enough to see his oldest son by Rayner successfully commit suicide in 1954, and Benjamin Spock's 1946 *Baby and Child Care* supplant *Psychological Care* as the go-to source of parenting advice for American families. And while it's easy to cast Rayner and Watson's book and philosophy as the Black Hat in this drama, as a cold and harsh system that eschewed warm human emotion in favor of rational self-reliance, their research and ideas were also influential in Mary Jones's development of a positive approach to therapy that continues to do lasting good, while their warnings against what we would term today as 'helicopter parenting' seem to ring a little bit more true with each passing year.

Maria Rickers-Ovsiankina (1898–1993):
Rickers-Ovsiankina was born in Chita, the Transbaikal capital that longtime readers of this series might remember also played an important part in the life story of the

physician Anna Bek, who moved there in 1912. It is therefore more than a little likely that Rickers-Ovsiankina knew of Bek, who was a central player in the medical and educational life of the city, and would have been an inspiration as a woman doctor in a Siberian landscape where women professionals were decidedly thin on the ground.

Rickers-Ovsiankina's father was a bank founder and coal mine owner, and as such, with the coming of the Communist Revolution and subsequent Civil War following Russia's withdrawal from the First World War, the family made the wise decision to pull up stakes and relocate to Germany. Rickers-Ovsiankina wanted to study literature, but ended up pursuing psychology at the University of Berlin under applied psychologist Kurt Lewin.

One of her classmates during this time was Bluma Zeigarnik, whose name would come to be inextricably with her own in the annals of psychology. In 1927, Zeigarnik had shown that people seem to remember uncompleted tasks better than they do completed ones, that there is something about finishing a job that allows our memories to flush it more readily than ones we still have to do some work on. This is called The Zeigarnik Effect, and Rickers-Ovsiankina decided to study it more closely for her dissertation. She wanted to know just how strongly we are compelled, as human beings, to finish a task we have been given, outside of any objective value being placed on its completion. In her trials, what she was interested in was less what subjects remembered about their various completed and uncompleted tasks, but rather their compulsions to return to a task after interruption. Her findings suggested that uncompleted tasks create a sort of psychological tension that keeps pushing its way into consciousness until the task is resumed, completed, and then, usually, dumped from the brain in the manner discovered by Zeigarnik. This tendency to finish uncompleted tasks even outside of the task having any inherent value is today called The Ovsiankina Effect, and if I know my audience of science-leaning perfectionists at all, I'm guessing that you are all *intimately* familiar with it.

Rickers-Ovsiankina received her PhD in 1928, and in 1931 emigrated to the United States, where she studied schizophrenia at the Worcester State Hospital, and devoted herself to the study of projective psychology, and particularly Rorschach testing, throughout the 1950s and 1960s.

Theodora Mead Abel (1899–1998):
Up until this moment, we have had but little cause to talk about the impact of culture on the study of psychology. Here and there we have seen individuals trying to improve standardised tests to make their results more reliable across socio-economic and cultural borders, but by and large everything we've seen so far has been well contained within the academic practices and cultural assumptions of Europe and the United States, and the psychological community was fundamentally Fine With That.

That started to change in the 1940s, at least partially thanks to the work of Theodora Mead Abel, whose education was guided by a succession of psychologists and anthropologists who read like a veritable Who's Who of early twentieth century women scientists, including classes taken from Margaret Floy Washburn while at Vassar, and from Leta Hollingworth at Columbia, and collaborations with her contemporary Margaret Mead at that same college. Abel therefore was equipped with everything she needed, upon receiving her PhD in 1925, to combine the best aspects of modern psychological theory with new approaches to anthropology, to create a new psychological style which took into account the importance of cultural differences when formulating psychological theories and creating diagnostic tools.

It took a bit of a while, however, for Abel to find her way to that new psycho-cultural synthesis, and her work of the 1920s and 1930s covered a diverse array of interests, including studies of what psychological factors determine success in learning a new skill set, the nature of electrical skin reflexes (during which she found that changes in skin conductance are *not* reliable indicators of emotional states), the relative contributions of students and teachers to creating a particular classroom environment, the dynamism of behaviour, and 'sub-normality' in adolescent girls.

At some point in the late 1940s, Abel met Zygmunt Piotrowski, a highly regarded expert on the subject of Rorschach testing, and took up an interest in projective testing (which we'll remember from our time with Charlotte Bühler is a test that uses responses to ambiguous stimuli to discover information about mental states). By 1947, she was the president of the Society of Projective Techniques, and co-founder, with Margaret Mead and Rhoda Métraux of the 'Research in Contemporary Cultures' project, and soon began her ground-breaking study with Francis L.K. Hsu on Rorschach tests given to two groups of individuals: (1) People born in the United States, but possessing Chinese parents, and (2) people with Chinese parents who had been born in China, and mostly educated there. The resulting paper, 'Some Aspects of Personality of Chinese as Revealed by the Rorschach Test' (1949) showed the impact that culture had on the results of even the most neutral seeming of tests, and put Abel on the track of further studies of projective test variation across cultures, including studies of Mexican, Native American, and Caribbean cultural responses.

For the next nearly five decades, Abel published a stream of work augmenting the dialogue between culture and psychology, including 'Cultural Patterns as they Affect Psychotherapeutic Procedures' (1956), 'Sex Differences in a Negro Peasant Community' (1959), 'Cultural Components as a Significant Factor in Child Development' (1960), *Psychological Testing in Cultural Contexts* (1973), *Culture and Psychotherapy* (1974), *Childhood and Folklore* (1980), 'Adaptation to the Dominant Society and Identification with the Aggressor as Factors in the Treatment of Chicanos' (1993) and 'Differences in Susceptibility to Influence in Mexican American and Anglo Females', the last written in 1996, when Abel was 97 years old.

Nancy Bayley (1899-1994):
In the table-top role playing game Dungeons and Dragons, you have the opportunity to give your character one numerical value for intelligence, and another for wisdom. You often hear this described as the difference between your innate mental ability, and the knowledge you have picked up along the way. In everyday life, we think of ourselves as growing more wise as we grow older and have more experiences, but that

leaves the question of intelligence – is our basic mental ability, our sharpness of mind and quickness of perception relative to our peers, a constant that remains unchanged over time, or does it wander here and there, waxing and waning at different points in our lives?

One of our most compelling answers to that question came as a result of the Berkeley Growth Study which was overseen by Nancy Bayley from its inception in 1928 until her move to Maryland in 1954. Like the Oakland Growth Study begun four years later by Bayley's colleague Mary Cover Jones, this was a longitudinal study aimed at creating reliable data about the physical, mental, and behavioural development of children with the aim of creating growth tables that physicians and psychologists could use to catch developmental problems in children early, in order to better provide them with support resources. Bayley's group consisted of sixty-one children born in 1928 and 1929, who were measured and tested every month from the age of 4 days old until the age of 15 months, and then once every three months until the age of 3 years, and then every six months thereafter.

The data generated by the study allowed her to develop new measurement criteria for infants, published as *The California First-Year Mental Scale* (1933) and *The California Infant Scale of Motor Development* (1936), and to produce the landmark developmental psychology text, *Mental Growth During the First Three Years* (1933). Equipped with these diagnostic methods, Bayley joined the National Collaborative Perinatal Project from 1954 to 1964, participating in the study of neurological disorders and psychological development in some 50,000 children. This research, combined with her Berkeley Growth Study data, allowed her to publish the comprehensive Bayley Scales of Infant and Toddler Development (or BSID) in 1969. The first edition of the BSID saw use from 1969 to 1993, with its validity as a useful system of early detection for potential developmental problems in cognition, motor function, and physical growth repeatedly validated.

While gathering the data that allowed her to create the BSID, Bayley noticed an interesting thing about mental tests given to children over their first year of life, and that was that the children's intelligence measurements were decidedly not constant. The result intrigued her, and in later studies she found that the basic picture of varying intelligence results holds throughout childhood, thereby casting doubt on the efficacy of intelligence testing generally. Intelligence is a shifting quantity at the mercy of biological and environmental fluctuations over time, and not an innate and fast quality of mind, impervious to the change around it, and so Bayley urged extreme caution in using intelligence testing in any sort of predictive or screening capacity.

BSID-II was released in 1993, BSID-III in 2004, and BSID-IV, the current version of Bayley's great contribution to our understanding of our children's development in their first four years, in 2019.

Grete Lehner Bibring (1899–1977):
Grete Bibring was an important psychoanalyst in the 'Second Generation' of the movement, who played a key part in organising a re-evaluation of the future of

psychoanalysis at a Harvard symposium of 1964 that became the basis of *Teaching of Dynamic Psychiatry: A Reappraisal of the Goals and Techniques in Teaching of Psychoanalytic Psychiatry* (1968). That was an important and highly necessary thing to do by that point, but what I find most interesting about Bibring's career is her 1950s study of the psychological impact of pregnancy and early mothering. As we've seen, so many books released in the early half of the twentieth century went to great lengths to outline the psychological development of children, and the nature of the child-mother or child-father dyad, without as much focus on the mother as an individual facing her own set of unique psychological challenges, susceptibilities, and changes as she moves through pregnancy and assumes the new identity of 'parent.' Bibring's study followed fifteen women through their childbirth odyssey, and in 1962 she published her results in the pages of *The Psychoanalytic Study of the Child* as 'A Study on the Psychological Processes in Pregnancy and of the Earliest Mother-Child Relationship.'

Myrtle Byram McGraw (1899–1988):
And now we come to maturationism. Arising from the research of Arnold Gesell in the early 1920s, as published in his 'Monthly Increments of Development in Infancy' (1925), this was an approach towards child development and learning which held that child maturation proceeds in a series of mostly biologically determined steps, with behaviours and abilities unfolding naturally as the nervous system reaches different stages of complexity. From a maturationist perspective, then, the job of a teacher isn't to try and get children to each milestone as fast as possible, but rather to recognise what stage a child is at, and offer support as their new abilities naturally unfold.

Sounds sensible, but critics of the era noted that with such a concentration on the centrality of biological processes, little room seemed left for the impact of the environment (though Gesell himself explicitly said that environmental factors were still important, if largely secondary). What was needed was somebody to take the best of maturationist theory, and modify it with a closer eye for how maturation cycles are impacted, or perhaps even potentially accelerated, by teaching and outside factors. This was the task that Fred Tilney assigned Myrtle McGraw when he made her associate director of the Normal Child Development Study in 1930.

McGraw turned out to be something of a wizard at observing children's unfolding capacities, and devising new environments to support and extend those capacities in unexpected ways. Thus, while considering infant motor development, did she conceive of the idea of swimming as an ideal training technique, and stunned the world when she discovered that infants as young as 2 months old have an innate swimming instinct that could be harnessed for the improvement of their motor skills. Pushing the idea of maturation enhanced by clever environment manipulation further, McGraw took a pair of twins, James and Johnny Woods, and in order to help Johnny develop stepping motions, trained him to roller skate at the age of 13 months.

She published her results in a series of important articles delineating the interplay between the programmed biological factors favored by maturationists and the demands of the environment favored by behaviourists, including 'Swimming Behavior of the Newborn Infant' (1939), 'Later Development of Children Specially Trained During Infancy. Johnny and Jimmy at School Age' (1939), 'Neural Maturation as Exemplified in the Reaching-Prehensile Behavior of the Human Infant' (1941), 'Neural Maturation as Exemplified in the Righting Reflex' (1941), and *The Neuromuscular Maturation of the Human Infant* (1943).

The onset of the Second World War brought an end to the Normal Child Development Study in 1940, and McGraw didn't hold an academic post again until 1953, when she joined the faculty of Briarcliff college, remaining there until her retirement in 1972. Her impact on the world, however, continued well past her retirement, in the form of infant swim classes, and of classroom structures designed to instruct by supporting developing capacities, instead of imposing desired benchmark competencies.

Ruth Winifred Howard (1900–1997):
One of the first Black women to earn a PhD in psychology, Ruth Howard is remembered today primarily for her study of triplet development. Psychologists and physicians had carried out developmental and comparative studies of triplets before, notably R.A. Fisher's 'Triplet Children in Great Britain and Ireland' (1927), A.E. Clark and D.G. Revell's 'Monozygotic Triplets in Man' (1930), J. Sanders's 'Similarities in Triplets' (1932), and S.E. Torsten Lund's 'A Psycho-Biological Study of a Set of Identical Girl Triplets' (1933), but Ruth Howard's 1934 dissertation, eventually published in 1946 as 'Intellectual and Personality Traits of a Group of Triplets' stands out for the sheer scale of the study. The largest effort previously, that of Fisher, measured the traits of 146 sets of triplets at the age of 6-and-a-half. Howard's paper was based around the measurements of 229 triplet sets, with ages ranging from infancy to 79 years, which allowed her to put to the statistical test her era's common assumptions about triplet frailty and shortened life spans.

Helen Peak (1900–1985):
Helen Peak's formative years as a researcher were spent during the rise of totalitarianism as the hot new thing in governmental theory. Fascism, Nazism, Stalinism, and Maoism each in its turn tried its hand at controlling both the public and private lives of their citizens, with often astonishing results as individuals from the most normal walks of life began earnestly believing a plethora of variously cruel and inhuman things, and acting on those beliefs. For a psychologist in this era, this brought up the question – how hard is it to change a person's attitude about something, and what are the steps by which that is accomplished? How does mild mannered Gerhard the cobbler become Gerhard the anti-Semitic Nazi stalwart?

Answering this question was one of Helen Peak's goals throughout the 1940s and 1950s. In works like 'Observation on the Characteristics and Distribution of German Nazis' (1945), 'The Acceptance of Information into Attitude Structure' (1958),

'Opposite Structures: Defenses and Attitudes' (1960), and 'The Generalization of Attitude Change Within a Serial Structure' (1960) Peak documented not only the specific factors of personal insecurity and hope for profit that motivated the members of the Nazi Party, but began a larger task of probing how motivation impacts attitude, which in turn impacts our ability to be aware of new information that contradicts our current worldview, and to accept it. Changing begins with noticing, but noticing has its own set of pre-conditions rooted deeply in our interests, goals, and perception of threats, all of which can be manipulated by a skilled enough propagandist.

Another important aspect of Peak's work was the thought that she put into how observations are carried out in psychology, as expressed in 'Problems of Objective Observation', a classic chapter of Leon Festinger and Daniel Katz's *Research Methods in the Behavioral Sciences* which served graduate students for decades as an essential guide to the basic assumptions of different attitude scaling techniques, and the problems at the heart of observation quantification, reliability analysis, and correlation practices.

Caroline Tum-Suden (1900–1976):
Known today mainly for the Caroline Tum-Suden Professional Opportunity Award given every year by the American Physiological Society to a pair of graduate students in her name, for which purpose she left the society $100,000 in her will, Tum-Suden was in her time an accomplished neurophysiologist specialising in the study of the adrenal cortex. This is the largest part of the adrenal gland, and is responsible for the synthesis from cholesterol of some of our most important hormones, including the blood-pressure and ion-absorption regulating aldosterone, the stress response/metabolism/inflammation regulating cortisol, and the androgen testosterone. Tum-Suden's work of the 1930s and 1940s studied the adrenal cortex's sensitivity to histamine and potassium levels (1947), the effect of adrenocortical tissue transplantation (1937), and the role that adrenocortical factors play in cardiovascular sensitivity to potassium (1947). The last ten years of her career were spent at the US Army Chemical Research and Development Labs, studying the neural and motor effects of different substances, including the impact of nightshade-derived atropine upon muscular tissues in different species (1958).

Bliuma (Bluma) Zeigarnik (née Vulfowna) (1901–1988):
For all of the challenges faced by psychologists in Western Europe under the expanding shadow of the Third Reich, with careers overturned by dislocation, and facing uncertain futures in England or the United States, those psychological researchers with Western educational backgrounds caught in the Soviet Union after the closing of the Iron Curtain often faced a life that amounted to a decades-long exercise in sustained anxiety, loss, and paranoia that made academic exile in Wisconsin seem positively blissful. Few individuals show just how uncertain life was as a Soviet-era psychologist than Bliuma Zeigarnik, who we met above as the discoverer of the Zeigarnik effect, and who spent the middle third of her life in a state of near perpetual suspicious dread.

Born in Lithuania, the only child of educated, non-practicing Jewish parents, Zeigarnik's youth and education were filled with the same Russian-imposed restrictions on the breadth of her education that Maria Grzegorzewska faced growing up. She was a phenomenally bright child, able to skip several grades at a time, a feat that was all but zeroed out by four years spent at home suffering from meningitis, resulting in her graduating from her gymnasium study at the relatively usual age of 17. In 1919, while studying in the library to take her university placement exams, she met and soon married Albert Zeigarnik, whose family provided the resources for the couple to study in Europe.

Thus did Bliuma arrive at the University of Berlin in 1922, squarely in the middle of that institution's status as a world centre of Gestalt psychology, led by Max Wertheimer and Wolfgang Köhler as a reaction against the structuralist school of Wilhelm Wundt and Edward Titchener. While Wundt sought to break experiences down to their smallest components, and find there the answers to how perception works, the Gestalt psychologists thought that such atomistic analysis lost important aspects of the process of consciousness, that the brain thinks in wholes and combinations that aren't equivalent to the sum of their constituent parts. Zeigarnik attended lectures from both Wertheimer and Köhler, but most of her work during her Berlin years was done with Kurt Lewin, whose 'field theory' held that behaviour could only be understood within the context of the psychological field surrounding an individual at a given moment. During an interaction with a waiter, Lewin was struck by the inspiration that uncompleted tasks represent a kind of psychological tension that is dissipated in the act of completion, and he set Zeigarnik the task of studying the validity of that insight.

Zeigarnik's 1927 paper *'Das Behalten erledigter und unerledigter Handlungen'* ('The Retention of Finished and Unfinished Tasks') was a kind of instant classic in the psychological community, in which she showed that unfinished tasks are remembered some 90 per cent better than finished ones, lending credence to the idea that there is something fundamental in the finishing of a task that causes the brain to unload it from memory. Zeigarnik received her PhD for the work, and remained at the University of Berlin until 1931, when she made the fateful decision to rejoin Albert, who worked at the Soviet Ministry of Foreign Trade. At first, all seemed well – Zeigarnik found a post with the All-Union Institute of Experimental Medicine, and worked closely with Lev Vygotsky, whose psychological approach was also influenced by Lewin's field theory. She worked happily here until 1936, when the 'On Pedagogical Perversions' resolution of the Communist Party's Central Committee sought to root out bourgeois influences in the Soviet Union's academic system. For three years, she went to ground, publishing nothing, and then, in 1940, her husband was arrested, and Zeigarnik was left to raise their 1- and 6-year-old children by herself, only finding steady work again in 1941, performing studies on therapy programs for individuals suffering from head injuries. From 1943, when she returned to Moscow to find her rooms ransacked, to 1948, she published little, and in 1953 was fired in the wake of rising anti-Semitic sentiment following the January publication in Pravda of the Jewish 'Doctor's Plot' to assassinate the leaders of the Communist Party.

The death of Stalin in March of 1953 allowed for a thaw in Soviet anti-Semitism, but Zeigarnik did not truly find her feet again until 1957 when she regained her posting as head of the Institute of Psychiatry in Moscow, and restarted her research career. Much damage had been done by that point – the years from 1931 to 1957 had taught her to be increasingly cautious, suspicious, and guarded in a way noted by her family and friends, and had brought her long stretches of hardship only survived by the dedicated support of a small circle of friends. By 1958, she was prepared to publish again, and did so at a clip, as if making up for lost time, with most of her work in the field of what was then called pathopsychology, but what we would term clinical psychology today.

By the 1980s, Zeigarnik was once again an internationally recognised name in psychological circles, but her health was precipitously declining. She required regular blood transfusions to combat anemia, and lived her last months in what her grandson described as a state of 'exhausting pain' that the strength of her mind somehow still managed to cut through. In 1983 she was nominated for the Lewin Memorial Award, but though she was granted permission to receive it, she was not given permission to travel and pick it up, and in fact died in 1988 before the Soviet bureaucrats could find an acceptable way to get the award, named in the honor of her mentor from so many decades ago, into her hands.

Doris Twitchell-Allen (1901–2002):
Psychodrama. Qu'est-ce que c'est?

In 1910, Jacob Moreno founded in Vienna the 'Theater of Spontaneity' which was originally conceived as a sort of experimental drama company, seeking to burst out of the inhibitions of scripted plays through the use of improvisation, but which by 1913 he had expanded into a new approach to psychotherapy. By being forced to improvise in scenes involving the emotions and situations that were central to their turmoil, Moreno believed, it was easier for individuals to have sudden and new insights into what was truly at the centre of their problems. He brought his techniques to the United States in 1925, publishing *First Book on Group Therapy* in 1932, and since then his ideas about the therapeutic value of mirroring, doubling, role reversal, role playing, and soliloquy in allowing patients to explore their issues as both subjects and observers has become so common that it is hard to find a media depiction of a group therapy session that doesn't employ them at some point or other.

One of the early adopters of psychodrama as a therapeutic method who also had the institutional position to put it to the test was Doris Twitchell-Allen. While writing her dissertation at the University of Maine in the mid-1920s, she became influenced by the psychological ideas of the Gestalt school, and particularly with the field theory of Kurt Lewin that we saw in the tale of Bliuma Zeigarnik. By the time that she took up a position at Cincinnati's Longview State Hospital, she had developed an idea for therapy that combined the ideas of Lewin about the importance of the whole physiological field to the determining of behaviour, with Moreno's pyschodramatic methods. She wrote up the results of her experiments in the new therapy in such works as 'Some

Theoretical and Practical Aspects of Group Psychotherapy' (1951), 'Psychodrama in the Family' (1954), and *Social Learning in the Schools Through Psychodrama* (1978).

In line with psychodrama's stress on the importance of freedom and spontaneity to uncover previously inaccessible layers of truth, Twitchell-Allen created in 1948 the Twitchell-Allen Three Dimensional Personality Test, which consists of relatively unstructured tasks that give the test takers freedom to choose and employ whichever test elements they are drawn to in order to tell stories, or create constructions, with their choices shedding light on the capacities and tensions beneath.

One of the most interesting parts of Twitchell-Allen's career occurred in 1951 after she overheard her son pondering the possibility that some day, when he was grown, he would have to go to war. This sparked a notion – if children destined for leadership from different countries could be brought together now, as children, and experience people from different cultures first hand, might that have an impact on their resistance to developing racial and national prejudices later, and might they then become leaders in world peace efforts as adults? To this end, she started the Children's International Summer Village, with children from 60 countries forming the inaugural 1951 group. CISV International continues today, with over 200 local chapters and 70 National Associations running half a dozen different cultural exchange and friendship programs.

Margaret Ransone Murray (1901–1986):
In 1908, Margaret Reed Lewis (1881–1970) became arguably the first person in the world to successfully perform an in vitro culture of mammalian cells. This was an important step in expanding the toolkit of cell biology, and vastly opened the field of what biological processes could be observed and controlled by scientists. When Margaret Ransone Murray began her graduate career in 1923, it was Lewis's work she turned to for inspiration in culturing methods, and Lewis's example she would follow in expanding culturing methods to ever more delicate biological phenomena.

In the late 1940s and 1950s, Murray applied these techniques to the study of different classes of tumor, particularly those developing from components of the nervous system. In this work, she discovered the origin of nerve sheath tumors, and elucidated the nature and characteristics of neuroblastoma (the most common cancer in babies, usually originating in the adrenal glands). These studies (and work she did culturing and observing other tumor types, such as soft-tissue rhabdomyosarcoma and malignant fibrous histiocytoma) then allowed Murray to begin research into the efficacy of chemotherapeutic agents, while the techniques she developed allowed her, in 1955, to announce the discovery that she had observed myelination (the process by which nerves 'insulate' themselves in sheaths that allow for quick transmission of electrical signals) *in vitro*.

During the mid-1950s, then, Murray lept from triumph to triumph, publishing papers of lasting impact like 'Comparative Effects of Chemotherapeutic Agents' (1951), 'The Classification and Diagnosis of Human Tumors by Tissue Culture Methods' (1954), 'Myelin Sheath Formation in Cultures of Avian Spinal Ganglia' (1955), and 'Some Uses of Tissue Culture in the Classification and Diagnosis of Human Tumors' (1957).

Her later work was an extension of these fundamental contributions, employing her world-class *in vitro* techniques to the study of diseases involving myelin degeneration (1970), and discovering that electrical neural activity could be maintained in nerve cells developed in culture (1965).

Esther Somerfeld-Ziskind (1901–2002):
In the midst of the Great Depression, Esther Somerfeld-Ziskind and her husband, Eugene Ziskind, established a psychiatric practice in Los Angeles that offered treatments for a variety of psychiatric and neurological diseases. As many who most needed treatment at the time couldn't afford it, the Ziskinds often took their payment in trade, accepting anything from oil paintings to musical manuscripts to, in one memorable case, a silk cloak that once belonged to an ex-wife of Groucho Marx.

During these years, Esther and Eugene experimented with new procedures for the treatment of neurological conditions which they then published in the nation's most notable psychological journals. They tested the efficacy of dehydration (1939), phenobarbital (1940), and mesantoin (1950) to treat epilepsy, insulin to treat schizophrenia (1938), vitamin B1 to treat insulin shock (1942), and Metrazol to treat Parkinson's disease (1945) as well as probing the therapeutic value of electroshock convulsive therapy (1945).

In 1953, the pair teamed up with Eugene's brothers to found Gateways Hospital and Mental Health Centre, which continues to this day and in early 2023 announced the development of a new $19.2 million expansion of its behavioural and mental health youth services. Somerfeld-Ziskind was active into her 100th year of life, seeing patients at Los Angeles Children's Hospital and engaging in her favorite activity, getting together with three other musical friends and playing eight-hand piano arrangements on the dual baby grands she kept in her home for just such occasions

Magdalen Dorothea Vernon (1901–1991):
M.D. Vernon was the author of some of the mid-twentieth century's most influential books on the subjects of perception and reading. Her 1965 *Psychology of Perception* was translated into eight languages. Her 1957 *Backwardness in Reading: A Study of its Nature and Origin* went through thirty-one editions from its first publication to 2016. Heck, her 1937 book *Visual Perception* was still in print in 2013! The unusual longevity of her work likely stems from the unusual passion she put into developing the field of experimental psychology. When she and a group of fellow experimentalists felt the British Psychological Society wasn't doing enough to push psychology as a scientific field grounded in experimentation, they went rogue and formed the Experimental Psychology Group in 1946, and published their own journal, the *Quarterly Journal of Experimental Psychology*.

The quality of Vernon's work was such that, even though she served as president of the EPG from 1952 to 1954, she was held in such high regard that she was invited to become president of the rival BPS from 1958 to 1959. Her research centred around issues of visual perception – how do eyes move as they are performing different tasks (including reading) and how do they, and the mind connected to them, deal with changes in illumination level – as well as her famous studies of the visual, auditory, and processing issues that culminate in the ability to read, where those can go wrong, and what therapies work best to treat each type of reading disorder.

One of my favorite facts about Vernon is that she, a member of the Reading Hall of Fame, spent a large segment of her career at the University of Reading, and so her reading research at Reading is something you can currently read at the Reading Hall of Fame.

Alexandra Adler (1901–2001):
In 1942, the Cocoanut Grove was a Boston nightclub with mafia ties steered by the no-nonsense hand of Barnet 'Barney' Welansky, who had inherited it after its first owner, bootlegger 'Boston Charlie,' was assassinated in a restroom by mafia rivals. Welansky's great fear in life, it seemed, was the idea that patrons might leave his bar before paying, to which end he either locked, covered, or in one case just plain bricked-over, all of the establishment's emergency exits. In the meantime, war rationing made freon rare and expensive, so to air condition his club, he relied on a flammable gas substitute, and for decor he made liberal use of palm trees which are, you know, *pretty* flammable. Take all of those factors, add to it front entrances that were revolving doors, and you have all of the makings for a true catastrophe, which occurred in due course on 28 November 1942. Fire broke out at the Cocoanut, and 492 individuals died in the chaos, victims of what is still the deadliest nightclub fire in US history.

After the dead were counted and the wounded were treated, two psychologists arrived in town with the goal of helping the survivors through their grief, and studying the long-term impact of having lived through such a traumatic event. They were Erich Lindemann and Alexandra Adler, and their discoveries were important contributors to the elucidation of post traumatic stress disorder. Adler was, by this point, already a celebrity in psychological circles. She was the daughter of the founder of 'Individual Psychology' and perhaps second most famous rebellious son of the original Freud Circle, Alfred Adler, and had in 1937 gained fame of her own for a post-mortem that she performed on the brain of a multiple sclerosis patient that showed a pattern in the lesioning of periventricular veins with functional repercussions.

In 1943, she published the results of her Cocoanut Grove study in 'Neuropsychiatric Complications in Victims of Boston's Cocoanut Grove Disaster,' which found that 50 per cent of the forty-six survivors she interviewed over the course of eleven months were still grappling with sleep problems, anxiety, and guilt months after the tragedy. Adler carried on studying the lingering personality changes, including excessive anger, lack of interest in life, and profound guilt, caused by surviving a disaster or experiencing profound stress, publishing 'Two Types of Post-Traumatic Neuroses' in 1945, and applying her knowledge to the treatment of returning soldiers from the Second World War.

Of course, Adler's other great role in these years was as the preserver of her father's legacy. Alfred Adler died in 1937, and it was left to Alexandra to carry the legacy of Individual Psychology, with its focus on the importance of an individual's feeling of connection with their family, spiritual community, and larger social group. She assumed the editorship of the *Journal of Individual Psychology*, and the presidency of both the International Association of Individual Psychology and the later American Society of Adlerian Psychology, while also writing influential articles relating advances in medication to the overall program of individual psychology, and authoring the definitive *Alfred Adler's Individual Psychology* (1973).

Angélique Arvanitaki (1901–1983):

In 1963, Alan Lloyd Hodgkin and Andrew Fielding Huxley shared the Nobel Prize for work they had done from 1947 to 1949 measuring action potentials in large squid axons, and determining the role that permeability to sodium ions plays in the firing of a neuron. In the celebration of the recognition of their accomplishment, however, the contribution of the Greek-born, French-raised Angélique Arvanitaki in laying the ground for Hodgkin and Huxley's work was quietly left to lie unmentioned. In the late 1930s, a decade before the work of Hodgkin and Huxley, Arvanitaki developed methods for working with the large nerve fibers in the sea snail *Aplysia*, the land snail *Helix*, and the cuttlefish genus *Sepia* which resulted in a startling succession of discoveries, including the fact that isolated nerve fibers could undergo spontaneous electrical oscillations that built up over time to produce action potentials without being 'plugged in' to a larger neuronal network, and that the firing of a nerve influences the firing of nearby nerves, while her later research investigated the phenomenon of photo-excitation of neurons, which has as of late returned to life in the form of twenty-first century 'optogenetics', which employs light to control neurons and was dubbed a 'Breakthrough of the Decade' by *Science* magazine in 2010.

Tamara Dembo (1902–1993):

One of the greatest trials a human can face is a loss of one of their core physical competencies. When a limb is lost in an accident, or is rendered unusable by disease, how does the brain deal with that loss, and how can therapists best help an individual psychologically adapt to their new state? These questions fall in the realm of rehabilitation psychology, of which Tamara Dembo was an early guiding light. A student of Kurt Lewin at the University of Berlin, her work was an elaboration of unique aspects of his field theory, as seen in her dissertation '*Der Anger als Dynamisches Problem*' (published 1931).

This experiment is perhaps in my top five favorites in this volume, as it centres upon the researcher deliberately making the subjects as frustrated as possible to measure how they react to increasing tensions and contradictions between the goals they are told to accomplish, their ability to do so, and the insistence of the researcher that they are not permitted to leave until they do. As an extension of Lewin's thought, this was a master-class in demonstrating the extremes of an individual trying to navigate both the limitations and demands of the psychological 'field' they find themselves in, and their own innate and internal tensions revolving around the need to complete tasks. As a piece of research entirely on its own, it was a classic study of the nature of frustration, and the components that feed into it.

In 1930, Tamara Dembo travelled to the United States to work with Gestalt pioneer Kurt Koffka at Smith College. With the rise of anti-Semitism in Germany in the subsequent years, Dembo, who was of Jewish descent, made the wise decision to remain in the United States, and upon Kurt Lewin's emigration in 1933, she decided to join him at first Cornell University, then moved with him to the University of Iowa. For the next decade, the pair worked together and supported each other as Dembo produced a steady stream of world-class research, deepening her studies of frustration ('Frustration and Regression: An Experiment with Young Children' (1941)), and opening a new study of aspiration, investigating how children deal both with repeated successes and failures ('Level of Aspiration' (1944)), with half an eye towards informing parents how better to deal, on a psychological level, with underperforming and overachieving children.

Lewin left the University of Iowa in 1943 for MIT, and for a decade thereafter Dembo migrated between different colleges, until finally finding her permanent academic home at Clark University in 1953. It was here that she pulled one more psychological rabbit from her hat, working with Beatrice Wright to design a study of how individuals react to the loss of a limb, or disfigurement. The resulting book, *Adjustment to Misfortune* (1956), was not only a pioneering document in rehabilitation psychology, but directed important attention on the need to actively engage patients in their rehabilitation process, and to pay attention to the therapist-patient dynamic as seen from the point of view of the patient.

The rest of Dembo's career was largely devoted to rehabilitation studies, including her work to establish a new branch of the American Psychological Association, Division 22, for the study of rehab psychology. She received that Division's Distinguished Service Award in 1980.

Helen Flanders Dunbar (1902–1959):

Psychosomatic Medicine. Emotional Thermodynamics. The Clinical Pastoral Education Movement. Helen Flanders Dunbar's omnivorous mind lay at the junction of all these movements and concepts, which shone bright for a dazzling couple of decades before receding back to the fringes of psychological theory. Dunbar was, simply put, one of the most educated people of her time. She earned four graduate degrees in three separate fields over the span of a mere seven years, producing dissertations in philosophy, theology, and psychology. The broadness of Dunbar's expertise was both her great strength as a novel thinker, and the source of the steady diminution of her legacy over the years.

Dunbar's work on symbolism in medieval thought, on the different training systems for religious officials, and on the interface between individual and environment brought her to the conclusion that body and soul/mind are two separate entities, with the latter possessing the ability to influence the former by sending it energy. Disruptions in the flow of this energy, then, can cause what were purely mental problems to manifest themselves as physical problems. For example, prolonged stress can, under P-S theory, result in the development of ulcers or irritable bowel syndrome (IBS). To aid in curing some physical diseases, then, Dunbar advocated the involvement of individuals trained to cater to the needs of the soul, such as pastors and priests, whose education as support staff for clinics lay at the centre of the clinical pastoral education movement, in which Dunbar took a leading role during the 1930s.

During the 1940s, Dunbar took up her position as psychosomatic medicine's most vocal and popular advocate, writing a number of core texts in the field, culminating in the best seller *Mind and Body: Psychosomatic Medicine* (1947). This was likely the high point of her life, as the decade that followed brought a string of trials, both personal and professional, including the suicide of her secretary (1948), the suicide of a patient (1951), a devastating car accident (1954), a sensationalised legal case against her, and friction with the New York Academy of Medicine, all of which drove Dunbar, who had had a melancholic streak to her character ever since her illness-laden youth, increasingly to alcohol as a coping mechanism. She was found dead, floating face down in her pool, on 21 August 1959. In the years since her demise, psychosomatic medicine has come under criticism for placing, in its most extreme forms, too much blame for a patient's physical conditions on their mental state, thereby making the patient feel guilt and self-reproach for the mental mal-adjustment apparently causing their condition on top of the misery they were feeling from the condition itself, while the efficacy of religious figures as central components of psychological and physical therapy programs has been also brought under doubt as a relic of early twentieth century moralism and hierarchical social thinking.

Lois Barclay Murphy (1902–2003):
The 1920s might have been the 'Decade of the Child' with multiple simultaneous longitudinal studies of children's physical and psychological growth, and investigations into the unique traumas and challenges faced during childhood, many of which we've already seen over the course of this book, but it was decidedly a decade that played favorites in terms of what aspects of children it was interested in studying. Loss, aggression, frustration, intellectual achievement, motor development, selfishness, lying, competition, these were the aspects of the child mind that dominated the studies of this formative decade in developmental psychology, and exerted their weighty influence on the research programs in the years that followed.

In 1937, Lois Barclay Murphy opened the Sarah Lawrence Nursery School, a research school dedicated to the idea that children possessed greater depths of positive social behaviours than they were being given credit for, and which developed research programs to bring out those aspects of childhood personality. One of Murphy's most important contributions was in the realm of testing, where she reasoned that most tests given to children to test their intelligence and mental health were fundamentally frustrating to children, and correspondingly produced results which presented children as more self-absorbed and anger-prone than they actually were. Further, these exams were generally taken by the subject children in isolation, without other children to interact with, and thereby also produced what Murphy held to be skewed numbers about children's underdeveloped interpersonal skills.

Murphy sought to correct these trends in child evaluation by designing her measurements to be taken in less structured group situations, noting that children playing together often pull from each other traits and capacities that they do not exhibit when on their own. Children have deep potentialities, Murphy argued in books like *Social Behavior and Child Personality* (1937), which often require the presence of a group, and the shifting dynamics of expectations and roles created by free play situations, to be called forth. Using her methods of observation and measurement, Murphy noted that children have a much greater capacity for sympathy and empathy than they had generally been given credit for, and that those traits begin manifesting themselves as early as preschool.

In 1952, Murphy took up a new position, as co-director of the Coping Project, a longitudinal study that followed a group of children from birth through adolescence, attempting to determine how they coped with the unique pressures and stresses of growing up, publishing her results in *Vulnerability, Coping and Growth from Infancy to Adolescence* (1976). The assassination of John F. Kennedy happened to lie directly in the middle of the study, and provided an opportunity to watch a group of children all at once dealing with a profound loss. Murphy observed that approximately half of the children responded passively to the event – that they felt terrible, but had no desire to *act* in response to the situation, but rather just let the grief pass over and through them. Nine of the children responded in an 'active, helpful' manner, expressing a desire to do something, either to help the First Lady through her grief, or to make the world a place where such things wouldn't happen anymore. The rest fell into the 'vengeful' category, who wanted to see vengeance done and punishment meted out. The helpful group was distinguished by a greater heterogeneity in their responses, and its members, when tested later, showed a greater ability to take an event, evaluate its separate components, and re-assemble them into an explanation for that event, than did the more homogeneously responding and narrowly fixated Vengeful children.

Lois Murphy died of heart failure at the age of 101 and if, by this point, you have the strange sense that the per cent of psychologists who live to their nineties and hundreds seems ridiculously high compared to the global average, you are not wrong.

Georgene Hoffman Seward (1902–1992):
Following Lorine Pruette's trailblazing application of psychology to the unique problems women face as a result of the conflict between their capacities and their opportunities, Georgene Hoffman Seward performed psychological studies on gender differences and roles from the 1940s through the 1970s that sought to tease out which aspects of male/female differences were due to cultural expectations, which to biology, and what the mental impact was of providing women with increasing access

to educational opportunities without giving them correspondingly greater access to employment opportunities and leadership positions. Her most influential works in this effort were 'Psychological Effects of the Menstrual Cycle on Women Workers' (1944) which followed women in the industrial and educational fields and found that their job performance and absenteeism while menstruating were not discernibly different from their non-menstruating colleagues, or indeed from their male co-workers, 'Culture Conflict and the Feminine Role' (1945), which surveyed 147 women psychology students to determine the spectrum of attitudes they had towards women as professionals, and the impact those attitudes had on their expectations of being able to balance a career with the raising of a family, *Psychotherapy and Cultural Conflict* (1956), which investigated the psychological impact of belonging to a minority group within a society, and *Sex Differences: Mental and Temperamental* (1980), which was a grand survey of the existent literature on existing differences between men and women's measured capacities, with an evaluation of which of those were accidents of particular cultures.

Magda Blondau (Blondiau) Arnold (1903–2002):

By the 1960s, the psychological study of emotion was due for a makeover. In the nineteenth century, William James had hypothesised that emotions arise from physiological changes ignited by emotional stimuli. A man points a gun at you, your heart rate and blood pressure spike, and as all of that physical response hits the brain, you experience the accompanying emotion of fear. In the early twentieth century, Walter Cannon and Philip Bard created their own model of emotion, which held that physiological activation was not a necessary precondition for an emotional response, but rather that the arrival of a stimulus causes the firing of neurons in the thalamus, which in turn triggers near simultaneous changes in our physiological and emotional states.

What followed was an emotional tug-of-war that seemed impervious to true and final resolution until Magda Arnold found an entirely new way to frame issues of emotion that would take its study into vast new territories: the realm of information appraisal. One of the great mysteries of emotion is how the same person, presented with the same stimuli, will respond to it in subtly different ways every time. How does a purely physical system of chemical reactions produce diversity and nuance in reaction when presented with similar stimuli? Arnold's appraisal theory reduces emotion to its informational core – before we react to a situation, we appraise it – we put what we are experiencing in the balance against what we know, categorise it, and let that categorisation then determine what reaction pathways get activated.

This way of looking at emotion formation is one that has the benefit of removing the gaps from our understanding, by allowing our diverse gradations in response to be explained by our diverse gradations in analysis capacity, by the subtle balance involved when a set of interlinked chemical systems, shoved by a stimulus, is nudged into an equilibrium which constitutes a 'decision' about the input those systems experience.

With Magda Arnold's two volume *Emotion and Personality* (1960), new vistas into the emotional process were suddenly thrown open, which was a rather remarkable thing for an individual who received her Bachelor's degree at the age of 36. Born in the Czech Republic, most of her early life was dominated by her decision to marry Robert Arnold, who was comfortable in letting her defer her studies and work as a secretary to support him financially on his way to a PhD, and then to let her defer her studies to raise children only to use the father-favoring law code of Canada to pry those children away from her and move away when she was just beginning the graduate studies she had put off for so long.

By the age of 40, Arnold had experienced more than her share of frustrations and setbacks, but instead of letting those dominate her destiny, she somehow found the will to use them as a springboard into a great mid-life intellectual flowering that saw her develop not only the appraisal theory of emotions, but the Thematic Apperception Test that she began developing in 1946 and that presents subjects with cards that contain a picture of an ambiguous situation, and asks the subject to relate what they think happened before the scene on the card, and what the emotions are of the people in the scene. The TAT and its successors continue to be used as a non-score-based evaluation of an individual's overall constructive or destructive state of mind, and as a tool to help individuals explore their own feelings and anxieties.

Lena Levine (1903–1965):
By 1945, the predominantly male leadership of the Western world had marched human civilization squarely into the jaws of two world wars, fascism, and the prospect of nuclear annihilation, leaving many to wonder, not unreasonably, if the blueprint for society laid down by generations of men was, after all, the direction we should all be following. For First Wave Feminism, which had in several of its forms operated under the assumption that the goal for women was to exist and operate in the world as men did, this was a time of deep introspection, from which the tectonic shift into Second Wave Feminism would emanate, with all of that movement's profound insights into the structure of civilizations and the power dynamics baked into them by their builders. What are the unwritten assumptions of the marriage dynamic? What aspects of women-as-property remain in our social codes, sexual mores, and occupational assumptions?

Once feminism began asking these questions, it was only a matter of time before they made their way into considerations of family psychiatric practice, and the psychology of sex. At the forefront of both of those movements in the middle years of the twentieth century was Lena Levine, who received her MD from Bellevue Medical College in 1927 before studying psychiatry at Columbia University. In works such as *The Menopause* (1952), *The Premarital Consultation: A Manual for Physicians* (1956), *The Modern Book of Marriage: A Practical Guide to Marital Happiness* (1957), and *The Frigid Wife: Her Way to Sexual Fulfillment* (1962), as well as in her own practice, she advocated for annual marriage check-ups with a professional to become as regular a part of one's life as annual physical check-ups, for the naturalness and desirability of pre-marital sex for women, for the wider availability of birth control, and for greater attention to be paid in counseling to the dangers of letting one's individuality get subsumed within the job-title and subsequent social expectations of 'wife.' If we live in a world today where sex counseling, and dialogue between partners about the hidden stresses and frustrations of their roles is a regular option available to married couples throughout the world, a fair amount of credit for that lies in the fearless pen and mind of Lena Levine.

Mildred Bessie Mitchell (1903–1983):
When the possibility of sending human beings into space first entered the realm of the possible, one of the earliest questions asked was, 'What *type* of people should we be sending up there?' In particular, how do we determine, ahead of time, who will crack when facing the inky void of isolation, and who will thrive under those circumstances? This entered the realm of extreme occupational psychology, a realm in which Mildred Mitchell had established herself as a leading figure by the 1950s. She served in the US Navy during the Second World War (1942–1945), and remained a member of the navy reserves from 1945–1964 at the rank of lieutenant commander, a time during which she was a clinical psychologist for the US Veterans Administration Mental Hygiene Clinic (1947–1958) and Wright Patterson Air Force Base (1958–1960), and a vocational appraiser for the Veterans Guidance Centre (1945–1946). This experience of matching people with exceptional and unusual skills to new occupations served her well when, in 1959, NASA selected her as its first clinical psychologist, charged with selecting astronauts for the Mercury Program. She describes part of her work in the autobiographical sketch she provided for *Models of Achievement: Reflections of Eminent Women in Psychology*:

> For six weeks in February and March, 1959, I worked long hours six days a week with the thirty-one astronaut candidates from whom the seven Mercury astronauts were chosen. We examined five a week except for the last week we examined six. They came to the base on Sunday mornings and left the following Saturday afternoon. On Saturday afternoons those of us who had been working with them all week met to rank them until we had them ranked one to thirty-one. The candidates were all of superior intelligence, had graduated from military test pilot schools, and most of them had graduated from engineering colleges as well. They had been selected by their services as the best pilots. Although we gave them intelligence tests and various personality tests, we were most interested in their emotional adjustment and their ability to work efficiently under severe stress. For instance, I gave each of them different forms of a battery of six repetitive tests before and after sitting in a heat chamber at 130 degrees for two hours, and before and after being in the high altitude chamber at the equivalent of 65,000 feet for an hour and then at 100,000 feet if they did not pass out. (p. 133–34)

Katharine McBride (1904–1976):
In 1929, Katharine McBride and Theodore Weisenberg set out to perform the largest study of aphasia in the history of psychology. Aphasia is a condition whereby an individual loses significant parts of their language abilities due to some form of

brain damage. Aphasic individuals have all of the mechanical ability to use language, but have lost substantial parts of their ability to understand and employ it, including variously severe inabilities to read, write, form or remember words, or comprehend the words and sentences they hear. In 1926, the neurologist Henry Head published his culminating work on the subject, the two volume *Aphasia and Kindred Disorders of Speech*, a feat made all the more heroic by the fact that Parkinson's disease was wearing away Head's own speech capacities as he was doing the research for it.

McBride and Weisenberg wanted to deepen Head's work, by performing a large scale aphasia study that included control groups and standardised measurements. Their research included 60 aphasic and 170 linguistically normal subjects, who were put through 10 to 19 hours of testing, probing the subjects' ability to read, write, complete sentences, retain words, evaluate analogies, recognise sounds, and hold sequences of digits and letters in memory. They published the results of their study in *Aphasia: A Clinical and Psychological Study* (1936), shortly after which Weisenberg passed away.

In 1938, McBride joined Bryn Mawr as an instructor and in 1942, at the age of just 38, was made president, becoming thereby one of the nation's youngest university presidents. She continued to hold that position until 1970, her thirty-two-year tenure still ranking as the longest any individual has held the Bryn Mawr presidency, during which time she helped push the school through the turbulence of the 1960s, exerting her influence to protect students protesting the Vietnam War.

Nina Ridenour (1904–1996):

Psychological theorists and researchers are great – their work is fundamentally interesting to most people and is easy to appreciate and entertaining to communicate. But what of the psychological administrators, the people who toil in the committees to nail down common practices, build bridges across national boundaries, organise colloquia where ideas can be exchanged, and provide the resources that allow researchers to do their stuff? Without them, psychology as a unified global effort for allowing us to understand ourselves and provide for our mental needs wouldn't be a thing, and yet we tell their stories far too little. To make up for that in some degree, then, let us consider one of the most important administrators in the history of twentieth century psychology, Nina Ridenour.

She received her PhD from New York University in 1941, and after some time gaining administration experience as an assistant executive secretary for the State Charities Aid Association in New York and the State Committees on Mental Hygiene, she found her way to her true calling in 1947 as the director of the division of world affairs for the National Committee for Mental Hygiene, as well as an executive director for the overall National Committee. She discovered quickly that she had a gift for international organisation, and in 1948 was the driving force behind the International Congress of Mental Hygiene, the momentum from which she used to form the investigatory committees that ultimately led to the establishment of the World Federation for Mental Health, which was approved by UNESCO as an advisory body, and continues in operation today, and which took a leading role in outlining desirable mental health policy to meet the particular challenges posed by the worldwide Covid-19 lockdown of 2020.

For the last fifteen years of her career (1952–1967), Ridenour was the executive director of the Ittleson Foundation, an organisation founded in 1932, and which Ridenour steered towards a greater role in national mental health and children's services issues, a commitment it honors to this day in the form of annual grants for innovative mental health programs. During this time she also authored *Mental Health in the United States: A Fifty-Year History* (1961), a work which is well past due for a sequel honoring the work that Ridenour and her successors have done in building the structures upon which our national and global health rest.

Anne Roe (1904–1991):

Across some 115 published books and articles, Anne Roe established herself as one of her time's pre-eminent authorities on the subject of occupational psychology. Like Marion Bills before her, Roe was significant for broadening the concept of what made one Suitable for a job beyond one's mere raw aptitude for it. Roe's work of the 1950s looked into how personality, one's opinion about the nature and importance of interpersonal relationships, and outside interests all affected happiness and performance in different lines of work. Her most well-known book, *The Making of a Scientist* (1953), featured the results of interviews with and tests given to sixty-four prominent scientists spread across three broadly defined fields of research, with the goal of determining what made each turn to the type of science that they did, and what habits of mind and personality tied together the members of different scientific branches.

Extending her investigative procedures further, she published *Psychology of Occupations* in 1956, in which she presented a two-way classification system to help future vocational counselors guide their clients towards suitable careers. Prior to this book, the most popular vocational guidance tool was the Minnesota Occupations Rating Scale, which used competencies and strengths in 7 different abilities to classify workers into one of 214 'ability patterns' which could then be used to refer them to the relevant subgrouping of 432 occupations considered in their study. The system worked, but it was also rather cumbersome to operate and explain, so Roe developed her group-level system to simplify the process, with each occupation fitting into one of eight primary focus groups (service, business contact, organisation, technology, outdoor, science, general cultural, arts and entertainment), and each occupation within that group classified by the level of training, responsibility, and personal autonomy it required, on a scale of 1 to 6. The result was Roe's 6 by 8 profession grid, built around personality type, and interest, rather than on abstract measurements of intellect or inferences made from motor skills, as previous vocational advice charts often relied upon.

In 1959, Roe joined the faculty at the Harvard Graduate School of Education, where she became a full professor in 1963 at a time when sexism was still rampant at the institution, and her work in the 1960s reflected her deepening interest in the continuing and unique difficulties women faced in attempting to reach the highest levels of their professions, as expressed in her articles 'Satisfactions in Work' (1962),

'Women in Science' (1966), 'Women and Work' (1966), and her book *Womanpower: How is it Different?* (1972).

Elizabeth Roboz-Einstein (1904–1995):

Of the many terrifying, awful, not-good diseases that can strike the nervous system, those which attack the myelination of our nerves are among the worst. Our nerves are wrapped in regularly spaced segments of myelin, which is a substance made up of proteins and phospholipids that vastly accelerates the transmission of neural signals. The presence of myelin around our nerves allows them to conduct impulses some twenty times faster, to regenerate to some degree, and to conduct signals without interfering with the action of nearby nerves. Sometimes, however, our body attacks its own myelin, and starts breaking it apart, leaving our nerves unsheathed, with a cascade of consequences depending on which nerves are attacked, the worst being those of our central nervous system, which results in the cacophony of symptoms and impairments characteristic of multiple sclerosis.

Before Roboz-Einstein, we weren't sure just what part of our myelin gets targeted by autoimmune responses like MS, and so building models of the progress of the disease, or thinking about possible cures for it, was a difficult venture, to say the least. If you don't know what aspect of you is being attacked, it's hard to understand the attack, and prepare a defense. After emigrating from Hungary in 1940, Roboz-Einstein spent the next decade primarily as a plant scientist, before becoming interested in neurochemistry and in particular the study of multiple sclerosis in the 1950s. Using electrophoresis, Roboz-Einstein was able to isolate significant amounts of a particular protein she found in abundance in her multiple sclerosis research. She brought the protein to Karian Kies, who identified it as the same protein she had come across while studying a different demyelinating disease, experimental allergic encephalomyelitis.

What Kies and Roboz-Einstein had isolated was myelin basic protein (MBP), which we now know to be a group of seven different related proteins, the most prominent of which makes up some 30 per cent of myelin sheathing, and which is the protein targeted by autoimmune demyelinating diseases like MS (though we are still not entirely sure about the mechanism of this targeting). After the discovery of MBP, Roboz-Einstein devoted herself to further studies of neural proteins, including those involved in the formation and degradation of myelin, and the glycoproteins in cerebrospinal fluid (which she developed methods of measuring), culminating in her 1982 book *Proteins of the Brain and Cerebrospinal Fluid in Health and Disease.*

Liliana Lubinska (1904–1990):

For most cells, the question of how to get something from one end of the cell to the other is not too puzzling: dump whatever you want to transfer into the cytoplasm, and let diffusion do its job moving it from where it is highly concentrated to where it is not. This can happen because most cells are very small. Neurons, however, are made up of cell bodies from which axons stretch that can be yards long, and if you need to get something from one end to the other in a hurry, diffusion just isn't going to cut it.

You need to have some form of dedicated interior transport system to do that, and over the course of a career studying nerve injuries and regeneration, Austrian Paul Albert Weiss (1898–1989) hypothesised the existence of a unidirectional flow of materials from the cell body through the axon to the dendrites.

That idea remained the last word on the subject until 1964, when Liliana Lubinska published 'Axoplasmic Streaming in Regenerating and in Normal Nerve Fibers'. Lubinska was a Polish scientist who got her education in Paris, and then returned to her home country to carry out research, primarily in the field of the peripheral nervous system, until one day when she and Czech physiologist Jiřina Zelená carried out an experiment on a crushed nerve, and noticed chemicals gathering both upstream and downstream from the injury location. Lubinska recognised that this must mean that there existed a backwards, or retrograde, direction for moving substances through a neuron, and designed experiments to test her idea which ultimately established the bidirectionality of transport throughout an axon, with necessary chemicals and organelles being moved from the body down to the dendrites, *as well as* chemicals being hauled from the dendrites back to the cell body, where they can be processed or broken down.

It was a controversial idea, but Lubinska's 1964 masterpiece of a paper soon converted the orthodox Weissians, and inspired a new generation to investigate the suddenly dynamic mysteries of what was once considered to be among our most staid and stable interior companions.

Mary Brazier (1904–1995):

Mary Brazier was a key figure in the harnessing of digital age technology to the improvement of the electroencephalogram (EEG) as a tool in the study of the nervous system. In 1924, Hans Berger recorded the first EEG of a human patient, and waited until 1929 to publish his results, out of fear of being called a crank for claiming to measure the electrical behaviour of the brain using sensors placed along the surface of the head. Indeed, he was greeted with much hostility at first, but upon replication of his results by independent researchers, his techniques were gradually employed by a widening array of scientists wishing to explore the neural activity behind episodes such as seizures and epileptic episodes.

Mary Brazier received her PhD in biochemistry the same year that Berger published his landmark paper, and though her concentration in the coming decade was the field of endocrinology, her development of a method to use changes in the electrical conductivity of the skin to indicate subtle progressions in thyroid diseases brought her to an appreciation of electrical measurement as a diagnostic tool in medicine, and in 1934 she began her studies of nervous system electrical activity.

One of the challenges facing the evolution of EEG methods lay in technical and engineering problems. Berger himself didn't particularly understand *why* his machinery worked the way it did, he just knew the results it produced corresponded with neural activity. It was left to Brazier and her generation, then, to work out the thorny issues involved in, for example, dealing with issues of noise reduction that are the bane of any electrical system – what techniques can you employ to make sure that the phenomenon you are noticing is in the brain you're recording, and not in the machine that's recording it or the atmosphere around it? One way to improve the ratio of signal to noise is to average multiple runs with the idea that the features you want will be more regular than the noise, and stand out more in the averaging. That's okay, but it's not... great, particularly when you don't have a large array of samples of a particular phenomenon to average over. Brazier decided that a better way to go about it would be to take a page from the playbook of electrical and communications engineers, and apply cross-correlation techniques. This is a mathematical method of comparing signals to each other (or a signal to a time delayed version of itself in a technique called auto-correlation), using the features that arise as the two signals are slid across each other to pull out periodic properties and regular features previously obscured by signal noise.

Brazier's 1952 paper 'Crosscorrelation and Autocorrelation Studies of Electroencephalographic Potentials' laid out her new methods of noise reduction and signal boosting that allowed doctors and researchers to pull a much greater depth of information out of EEG signals than previously, and she continued from there to push the boundaries of EEG analysis using the emerging power of computers, writing of the application of computing to EEG signal interpretation in her 1960 paper 'Some Uses of Computers in Experimental Neurology.' From 1974 to 1984, she edited the journal *Electroencephalography and Clinical Neurophysiology* and published two books in the 1980s on the history of neurophysiology, which brought her life's total of articles and books nearly to the 250 mark.

Marcelle Jeanne Beauvallet (1905–1985):
French neuropharmacologist Marcelle Beauvallet is among a handful of figures in this book just teetering on the edge of historical oblivion, though she was for decades an important figure at the Centre Nationale Recherche Scientifique (CNRS), where she attained the position of research director by 1961. Her main research interest was the study of the role catecholamines (such as adrenaline, noradrenaline, and dopamine) play in different physiological processes involving the sympathetic nervous system, and the impact of different drugs on catecholamine levels. Her papers '*L'Amphétamine et les Monoamines du Système Nerveux Centrale*' (1968), 'Action of Reserpine on the Urinary Excretion of Adrenaline and Noradrenaline in the Rat' (1950), 'Noradrenaline Content of the Cerebral Tissue of Animals Under the Influence of Aminodipropionitril' (1961) and 'Norepinephrine and Serotonin in Brown Fat of Cold Acclimated Rats' (1982) all speak to a life-long commitment to understanding the shifting and dynamic chemical-scape of our bodies in action, and our bodies under medication.

Eugenia Hanfmann (1905–1983):
Russian by birth, German by education, and American by profession, Eugenia Hanfmann's diversity of background was echoed in her diversity of research interests. After receiving her PhD from the University of Jena in 1927, she emigrated to the

United States in 1930, taking up an assistant position at Smith College with fellow emigre and Gestalt pioneer Kurt Koffka. The years that followed saw her take up a series of positions, each lasting from three to six years, until she nested into her permanent academic home at Brandeis in 1952. While at Worcester State Hospital in the early to mid-1930s, she carried out studies of schizophrenia patients that led her and fellow researcher Jacob Kasanin to develop the Hanfmann-Kasanin Concept Test, which they outlined in *Conceptual Thinking in Schizophrenia* (1942). Up to that time, researchers were largely of the opinion that schizophrenia affected individuals' capacity for abstract thought, but no adequate test had been developed to investigate that theory. The H-K Test, which was an adaptation of an earlier test by Lev Vygotsky, asked patients to classify a given set of objects, then explain the system that they had developed in their mind for their classification. Administering the test to sixty schizophrenic and ninety control patients, Hanfmann found that, indeed, the ability to form abstract concepts and criteria seemed impaired in the schizophrenic patients.

During the Second World War, Hanfmann brought her psychological expertise to the Office of Strategic Service, helping in the selection of agents undertaking covert propaganda or sabotage work in enemy nations, and then after the war she interviewed 3,000 former Soviet citizens in an attempt to evaluate the psychological impact of living under Stalinist rule, and the long-term effects that had on their personality, publishing her results in 'Modal Personality and Adjustment to the Soviet Socio-Political System' (1958), 'The Mental Health of a Group of Displaced Russian Persons' (1958) and later in *Six Russian Men: Lives in Turmoil* (1976). Concurrent with her work on Soviet psychology, Hanfmann researched the psychological needs of college students, publishing works on how to provide effective college counseling and therapy services throughout the 1960s and 1970s.

Marianne Bellak Frostig (1906–1985):
In 1951, Marianne Frostig founded the Marianne Frostig School of Educational Therapy in southern California with the goal of providing holistic, psychologically-informed education for children with brain injuries that affected particular aspects of their ability to learn and interact. Here, she developed an array of influential special education techniques that continue to resonate today, and which she published in a series of works, from *The Frostig Program for the Development of Visual Perception* (1964) to *Move, Grow, Learn: Movement Education* (1969) to *Learning Problems in the Classroom: Prevention and Remediation* (1973). Seventy-plus years later, the Frostig School continues to operate, offering services to children dealing with dyslexia, dysgraphia, autism spectrum issues, dyscalculia, and attention problems.

Josephine Rohrs Hilgard (1906–1989):
The relationship between hypnosis and psychology has always been fraught, with psychology's attitude towards hypnosis as a useful technique waxing and waning with the reasonableness of hypnotherapists' claims as to the power of their methods. A low point was likely reached in the early twentieth century when hypnosis was lauded

by some in the psychotherapy community as a means of allowing patients to access primal deep memories that would reveal to them the traumas behind their adult neuroses, a claim that psychologists viewed as an unscientific waste of time at best and a dangerously exploitative bundle of charlatanry at worst. The bad blood conjured during that era was some time in abating, and it took the passing of a couple generations before attention was turned back towards hypnosis as a viable tool in psychological research or therapeutic practice. One of the chief voices in that hypnotic renaissance was Josephine Hilgard, who in the 1970s and 1980s performed studies that attempted to answer questions about what type of people tend to be hypnotically suggestible, and what measurable benefits hypnosis can have in the alleviation of pain.

She published her results in 'Imaginative Involvement: Some Characteristics of the Highly Hypnotizable and Non-Hypnotizable' (1974) which reported that those who naturally engage in imaginative activity (day-dreaming, reading, creative acts, drama) have a higher success rate with hypnosis than those who do not, 'The Stanford Hypnotic Clinical Scale for Children' (1978) which established best practices for different age groups, and *Hypnotherapy of Pain in Children with Cancer* (1984), which investigated the use of hypnotherapy in ameliorating children's perception of their pain.

Hilgard was also the discoverer, in 1953, of what she dubbed 'The Anniversary Effect.' What she noticed was that parents who experienced a particularly traumatic event at an early age had an increased likelihood of experiencing breakdowns or psychotic episodes when their own children reached that same age.

Leona Elizabeth Tyler (1906–1993):

'What do you want to be when you grow up?'
'Well, I know I don't want to work in an office, or with customers.'

An occupational psychologist of the early part of the twentieth century hearing this exchange might elbow their way into the conversation at this point and say, 'You weren't asked what you *don't* want to do, tell us what you *do* want to do,' and thereby miss the very important information that the subject is conveying. Early job counseling theory was hyper-focused on competencies and personalities, and matching those with the jobs that best suited them. Then, in the middle of the century, Leona Tyler arrived on the scene with an interest in choice theory, in researching how our brain interacts with our environment to build internal cognitive machinery which pushes us towards certain decision types and away from others. What she found was that, when making big decisions like occupational paths, we are often far more motivated by what we don't want to do, than by what we do want to do. The way that negative experiences in life forcefully encode themselves upon us and our cognitive structures, builds definite and difficult to unmake aversions to certain types of activities, which push our decision making more resolutely than our more hazily encoded experiences of placid satisfaction.

Armed with her knowledge of how our cognitive machinery learns to choose, and how decisive our dislikes are in forming our choices, Tyler advocated for a new approach to counseling which made room for the full range of an individual's makeup, in order to avoid pushing options that the patient has aptitudes for, but which for the moment they are unbridgeably averse to. She developed the Choice Pattern Technique, which gives subjects stacks of occupation cards and asks them to sort them into piles based on their instinctive feelings about those jobs, and then asks them to work logically through the reasons for why they chose the piles they did, in order to build up a better idea of the attracting and repelling forces at work at *that* moment in *that particular* subject, to give them the best advice for what, at this particular stage in their lives, might be best suited for them.

Tyler's ideas about counseling psychology remain influential, and her central works, *The Psychology of Human Differences* (which looked at how cognitive choice structures work across cultures) and *The Work of the Counselor* both went through multiple editions.

Hedwig von Restorff (1906–1962):

Von Restorff's entire psychological output consists of two papers written in the early 1930s, published before the Nazi takeover of the University of Berlin, where she was studying with Wolfgang Köhler, ended her career prematurely in 1935. In one of those two papers, *'Über die Wirkung von Bereichsbildungen im Spurenfeld'* (1933), she described what is still known as the Von Restorff Effect. When given a list of items, one of which has a different dimension than the others, such as a different length, colour, or category type, we tend to remember the distinct item better than the items of homogeneous dimension. Had von Restorff's career lasted longer, she would likely have turned from noting *that* the effect occurred, to studying *why* it occurred – what is the brain doing behind the scenes to freeze that particular item that surprised it into memory? As it was, that work would fall to Leo Postman and William Jenkins, who in their 1948 paper 'Isolation and Spread of Effect in Serial Learning' measured recall performance when placing one meaningful three letter word in a list of twelve nonsense three-letter words that subjects were asked to memorise. Their theory, that distinctiveness and recognition evoke the deployment of extra mental resources that aid in the cementing of memory, formed the basis for future investigations of the Von Restorff Effect over the decades that followed.

Marie Jahoda (1907–2001):

Marie Jahoda grew up in a socialist-leaning secular Jewish family, where a life was not well lived if it was not spent in the service of improving the life prospects for one's fellow humans. Jahoda grew up with the expectation (not hope, mind you, but *expectation*) that she would become Austria's Minister of Education, from which position she could enact reforms that would cause people to think more deeply about the structure of society and what might be done to improve its quality for all. She reasoned that a background in psychology would serve her well in this endeavour, and studied at the Vienna Psychological Institute with Charlotte and Karl Bühler.

In the late 1920s, Jahoda combined her socialist principles with her newly forming psychological insights to carry out a sociological study of unemployment with her husband Paul Lazarsfeld, which she published in 1933 as *Marienthal: The Sociography of an Unemployed Community*.

The rise of the Nazis to prominence in Austria in 1934 meant the effective end of Jahoda's possible career there, and in 1936 she was arrested for her socialist affiliations, and only avoided a three month prison term by agreeing to immediate exile. The first segment of her exile was spent in England, deepening the studies of unemployment that would later culminate in her concept of Positive (or Ideal) Mental Health. In 1945, she moved to the United States, where she carried out research into the psychology of anti-Semitism, racism, and the authoritarian personality for the American Jewish Committee, and worked on a long-term project for the Society for the Psychological Study of Social Issues, of which she became president in 1955, which resulted in the 1964 publication of her two-volume *Research Methods of Social Relations*.

This was also the time when Jahoda formed the concept most associated with her name, that of Positive Mental Health and its foundations. In exploring the psychology of the unemployed, she found that it wasn't so much the lack of *things* that was so psychologically devastating for the individuals she studied, but the inability to live what most humans conceive of as a full life. In particular, she hypothesised that there are five components to having a positive mental life: the ability to manage one's time, the possession of deep social relationships, a sense of collective purpose with one's fellow humans, a reasonable sense of self-esteem, and regular activity. She published her ideas on mental health in her 1955 book *Towards a Social Psychology of Mental Health* and her 1958 volume *Current Concepts of Positive Mental Health*.

Marie Jahoda returned to England in 1958, where she married her second husband, Labour PM Austen Albu, worked at Brunel and Sussex universities developing curricula that prodded students to engage with the psychological problems of actual everyday life, and resided until her death in 2001.

Margaret Jeffrey Rioch (1907–1996):

In the 1960s, as the United States crawled forth from the social expectations of Cold War suburban status quo living, and realised that the road forward for individual happiness lay not in the denial of one's anxieties and psychological issues, but in the exploring and understanding of them, the mental health care industry experienced a sudden and momentous expansion that there were quite simply not enough mental health professionals to service. It was at this moment that the National Institute of Mental Health turned toward a Washington University professor of neuropsychiatry named Margaret Rioch and asked her to develop a training program to meet the emerging mental health need.

The training program Rioch built set a new standard in the efficient and effective production of mental health professionals by tapping into an underutilised segment of the population: women in middle age. Instead of trying to train a batch of teenagers from scratch in how to counsel people during life events that they had yet to experience, she reasoned, why not tap those whose life course had already brought them into contact with many of the issues their patients would be facing? Rioch's program, detailed in a string of papers published in the early to mid-1960s, was a runaway success, and allowed for the staffing of a number of other innovative programs sponsored or created by Rioch, including mental health call centres, and experience-specific support group sessions.

Anne Anastasi (1908–2001):

At this point in the book, you might be wondering to yourself, 'What happened to all the testing experts? At the beginning of this section, it seemed like Every Single Person in psychology was involved in test design, and then they suddenly disappeared. What gives?' There are a number of reasons behind the relative decline in the number of individuals interested in psychometrics over the course of the 1940s and 1950s, from increasing realisation about the limitations of the tests that had been developed over the previous decades, to an increasing interest in social and cognitive psychology that drew away promising young students, but for my money I think a significant factor was the fact that there emerged in this era an individual whose mastery of, and exploration of the weaknesses of, aptitude and achievement testing as they then existed was so complete and compelling that it appeared as a sort of Final Word on the topic beyond which one was unlikely to advance.

Dr. Test. The Test Guru. No matter the nickname, Anne Anastasi was the mid-twentieth century's reigning expert on the subject of testing, its most persistent critic, and its most important innovator. Her book *Psychological Testing* was first published in 1955, and its seventh edition, published in 1997, is a standard text still very much available, a virtual miracle in the shifting trend-prone fields of psychology publishing. Coming from the study of differential psychology, where her 1937 book *Differential Psychology* held as central a role as an authority on cross-cultural psychological studies as *Psychological Testing* would come to in the area of testing, Anastasi brought with her a mind accustomed to unraveling the role that local cultures can play in determining how observations are made, and how the expectations of the observer work their way into and skew their analysis of what they are observing.

She brought this expertise to her evaluation of what tests are and what they can be expected to do. Not only in *Psychological Testing*, but in later works like 'Psychology, Psychologists, and Psychological Testing' (1967), 'Psychological Testing: Basic Concepts and Common Misconceptions' (1985), and 'Evolving Concepts of Test Validation' (1986), Anastasi made the case that tests are made by particular people with particular limitations, and that they are given to individuals at one moment in their life stories, and that to be meaningful at all requires that a test-giver realise the built-in limitations to the tests they select, and work with the test-taker to inform them exactly what the test they just took measured, and what its limitations are as a predictive tool for their future lives. Strict distinctions must be drawn between tests of aptitude, which measure abilities of a limited scope determined by the particular cultural, occupational, and social theoretical concerns of the test-maker, and tests of achievement, which measure what an individual happens to know at a particular moment in time, evaluated against a standard which is, once again, determined according to the educational fads and standards of its time.

Tests are useful, Anastasi emphasised, but they must be well chosen by well trained test-givers with the subject's particular life story in mind, and with a full knowledge of what each test can and can't do, and how differences in culture and class impact the utility of each test's results. They must exercise great restraint in using tests to project an individual's future prospects, and work with the test taker to connect their performance on the test with elements of their lives and experience, to give the number a meaningful context. This was a culturally nuanced view of testing that took into consideration her era's philosophical discoveries about how unspoken assumptions work their way into educational and assessment systems.

For her decades of dedication to the field of testing and cultural awareness in psychology, Anastasi was in late life one of psychology's most decorated individuals, awarded the National Medal of Science in 1987, the American Psychological Foundation's Gold Medal in 1984, and the American Psychological Association's Award for Lifetime Contributions to Psychology in 1994.

Erika Fromm (1909–2003):

If Josephine Hilgard represents a road to rapprochement between hypnotherapy and psychology, Erika Fromm represents a renewed call to battle. Rather than shying away from hypnosis and its association with Freudian theories of the unconscious, Fromm doubled down on that association. She asserted that the basic idea of Freudian psychoanalysis was correct insofar as the road to recovery lay in grappling with the effects of traumas buried in the unconscious mind, but that its benefits were accessible only to a very privileged few. Therapy takes a long time, and costs a lot of money, which is fine if you have an abundance of both, but for most people saying that psychotherapy is the answer to their problems is like saying they should just take a bath in precious gems every night to alleviate their stress. What Fromm liked about hypnosis was its potential to cut straight to the root of patients' problems, by providing a back door to the unconscious mind which could do the work of a year of talking therapy within a couple of sessions to help individuals uncover their defining traumas. She authored a number of books on the clinical value of hypnosis, and the possibility of self-hypnosis, in the 1980s and early 1990s, and served as editor of the *International Journal of Clinical and Experimental Hypnosis*, and served as president of the APA's hypnosis division (which still exists as Division 30) from 1972 to 1973.

Alberta Banner Turner (1909–2008):

When her high school's prom was declared to be 'Whites Only', Alberta Banner went anyway. When her local movie theater refused to sell her a ticket on account of her skin colour, she alerted the press and sued the manager. When Columbus, Ohio had a growing problem with juvenile delinquency, she found the toughest gang leaders and rode in a mobile unit with them to the poor parts of the city to offer mobile services to the youths there. She was in every way a person of action who did the right thing regardless of what imposing power structure stood against the doing. Born in Chicago, she sailed easily through her schooling, and was a member of the first generation of American Black women to earn a PhD in psychology with her 1935 dissertation on the effect of practice on digit memorisation tasks.

In the late 1930s and early 1940s she held a number of home economics and consumer education posts in the American South, which were notable for her outreach efforts to educate the community about the insurance, investment, and banking options available to Black individuals still living very much under the shadow of Jim Crow. In 1942, she began her career at the Ohio Bureau of Juvenile Research, where she would remain for the next quarter-century, devoting herself particularly to the development of programs to meet and offer help to juvenile delinquents where they were, instead of waiting for them to be herded into correctional facilities. During this era, Turner took up a leadership position in the Civil Rights movement, serving as president of the Jack and Jill Foundation, which had been founded in 1938 by a group of twenty-one Black mothers seeking to expand the cultural and educational opportunities of their children and which by Turner's time was a nation-wide force organising regional teen conferences. She was also the first national programs director for Links Inc, a national organisation of Black women centred on community development and service which has counted the likes of Marian Anderson, Dorothy Bell Wright, Rosa Parks, Condoleezza Rice, and Kamala Harris among its membership.

Eleanor Jack Gibson (1910–2002):
A baby placed into an unfamiliar room finds themselves suddenly confronted with a galaxy of objects that they have to decide what to do with. Which are threatening? Which can be shoved in a mouth? Which can be thrown? Which can be crawled on? It's a prospect the full daunting scope of which was outlined with Eleanor Gibson's development of perceptual psychology in the mid-twentieth century. In perceptual psychology, what is important about learning is the gradual build-up of our powers of differentiation. In order to survive, we have to learn better how to perceive ever finer distinctions between similar objects, and 'learning' is the way in which we use new experiences of the world to inform that process of differentiation, and learn what to expect from our environment, and how to interact with it, so that when a tiger comes along, we think 'danger' instead of 'KITTY!'

Gibson's studies centred around probing the boundaries of our perceptual abilities – what distinctions do we come hard-wired with, and which develop over time? In the former category she discovered the 'visual cliff' phenomenon, which ranks among psychology's most famous experiments. Beginning with rats, turtles, goats, dogs, and cats, raised in different environments and placed on a high table with glass sections, she noted which species had innate senses of depth perception from birth, regardless of how they were raised, and which had to learn that ability over time. Confident in her results, she moved on to human children of crawling age, setting up the famous checkerboard floor leading up to a glass-covered drop-off image familiar from every introductory psych textbook ever, and found that 90 per cent of babies avoided crawling onto the glass even if their mother was waiting for them on the opposite side with open arms. The possession of depth perception at the age of walking was an important discovery, and the visuals proved so irresistible that they made their way into the pages of *Life Magazine*, making Eleanor Gibson a scientific celebrity of her time.

Other perceptions, however, take longer to develop, and Gibson's experiments investigating how those develop are equally classic, if less visually iconic. One of the most famous of these was her Random Scribble experiment, in which subjects of different age groups were presented with a random scribble, and then asked to identify which from a stack of variously similar scribbles corresponded to the original they had seen, measuring not only how many the subjects correctly identified, but how many trials it took them to perform the task 100 per cent correctly, thereby determining how age plays into our ability to drag new levels of distinction out of novel situations – presented with a type of object we've never seen before, how long will it take a child versus an adult to break it down, figure out its key features, and determine which are the most salient in the distinctions and identifications they are going to be asked to make?

Eleanor Gibson was an active researcher for a half-century, from her decade and a half (thanks to anti-nepotism laws) as an unpaid assistant in her husband's lab at Cornell, to her long reign as a tenured professor there once her husband found a position at a different institution. Her published works on perceptual psychology span from her early article 'A Systematic Application of the Concepts of Generalization and Differentiation to Verbal Learning' (1940) through her foundational *Principles of Perceptual Learning and Development* (1969) and 'Exploratory Behavior in the Development of Perceiving, Acting, and the Acquiring of Knowledge' (1988) and stretching all the way to *Perceiving the Affordances*, published the year of her death, 2002.

Marie Skodak Crissey (1910–2000):

When do you give up on a child? It's a horrible question to ask, but one which administrators of child services had to confront on a regular basis across the twentieth century – given finite space, financial, and professional resources, how do you apportion them to the children under your care? How do you decide, 'This group can go no farther, no matter what extra care we give them,' and resign them to their fates? For years, hereditarianism, as originating in the works of Francis Galton in the 1870s and 1880s, had a relatively simple answer to that – most aspects of character, intelligence, and ability are hereditarily determined, so if a child has mentally challenged parents, and shows slow early development on the scales worked out by the metric wizards of the 1920s, then the sensible if harsh thing to do is to streamline them towards low-resource environments and save the funding and manpower for the children who have the background and aptitude to make use of them.

Though Marie Crissey studied at Ohio State University with Henry Goddard, whose hereditarianism informed his eugenicist beliefs, she believed that he woefully underestimated the ability of a properly constructed environment to work significant changes in even the most seemingly lost of causes. Some of her first studies tracked the development of children in adopted homes, with half an eye towards determining better psychological evaluation protocols for pairing children with the most suitable parents for their needs, and half towards demonstrating that these children, who were thought to be sub-normal due to the inheritance from their presumably lackluster parents, given a stable home and steady education could test just as well as other children. She provided evidence for her theories with a 1939 study that followed twelve low-testing children as they were placed in a stimulating environment and ultimately found adoptive parents, as compared against a group of twelve normal-testing children who remained behind at an orphanage. She found that the low-testing children, given stimulation, challenge, and attention, steadily increased their performance on tests and became self-sufficient in adulthood, while the normal testing orphans steadily declined in test performance, and had to stay in institutionalised care into adulthood.

Crissey extended this work in the 1950s and 1960s by designing programs to allow even children with severe disabilities to rejoin, and receive the benefits of, a regular classroom education and access to appropriate vocational training. By the early 1970s she was an internationally recognised expert on the complex subject of deinstitutionalisation for the 'mentally retarded,' summarising the progress that had been made in the field over the last two decades in her 1975 paper 'Mental Retardation: Past, Present, and Future' and her 1986 article 'Residential Institutions for Individuals Who Are Mentally Retarded: A Second Look.' In these later pieces, she reflected with some satisfaction that of the 225,000 individuals placed in institutions in 1967, only 100,000 remained by 1985, and many of those in smaller, community-based institutions that avoided the dehumanising warehousing that had characterised the larger facilities of the 1950s, but also wondered if the zeal for deinstitutionalisation that she noted was causing individuals who actually benefited from the structure offered by well-run institutions to be shuffled out of a positive situation before their time, a very real worry in the wake of the dwindling funding for mental health care characteristic of Reagan-Era America.

Keturah Whitehurst (1912–2000):
One often sees the title 'Mother of Black Psychology' bestowed upon Keturah Whitehurst, though most of her work on the subject was published after the foundational work of Mamie Phipps Clark, whose work, as we read earlier, laid out clearly for the psychological community the need for increased attention on the psychological impact of systemic racism and segregationist policies on the Black community. Whitehurst's work was important in accentuating both the need for psychological techniques that take into account the particularity of the Black experience, and in establishing mindful practices for professionals regularly doing work in Black communities, as demonstrated in her influential papers 'Professional Education of Teachers of the Disadvantaged' (1969), 'Techniques and Processes of Socialization of the Black Child' (1972), and 'The Black Child' (1980), the latter of which contains the call to action:

The black child is first a child and secondly a child who must cope with the consequences of being black in America. In this latter sense, the black child is special. Developmental strategies designed to promote the welfare of black children must evolve out of a burning awareness of both truths – his 'childness' and his blackness; these two variables are interfacing at all levels of the black child's development. These two threads are the warp and woof of the black experience. Black children are special because they are treated differently in the distribution of services and resources upon which their growth, development, and self-actualisation depend.

To help see her goals realised, Whitehurst co-founded with Florence Farley the Fisk Children's House, where university students developed structured after-school programs for nearby inner-city youths, providing growth opportunities for the children, and personal experience of the limitations of traditional psychology for the students.

Susan Walton Gray (1913–1992):
During her thirty-three years at Peabody College, Gray not only built the college's psychology department from the ground up, but in 1963 implemented what she dubbed the Early Training Project, which built structured early development programs for disadvantaged children, including at-home visits, integrated parental involvement, health and nutrition services, and summer programs. The children who went through the ETP experienced marked improvement in later schooling as against control groups, and the success of the program was specifically mentioned by Head Start co-founder Sargent Shriver as an inspiration for many aspects of that federal program. During her retirement, she brought emerging technologies to bear on the further development of early education, creating a digital database of information and statistics from twelve different programs that could be accessed by educators looking to develop their own practices or find new and useful patterns and tendencies in the aggregated data.

In 1968, Susan Gray founded the Peabody Experimental School at Vanderbilt University, as an educational research school for children with mental disabilities or who faced potential developmental roadblocks on account of their socio-economic status. It was renamed the Susan Gray School in 1986, and continues to this day.

Marie-Louise von Franz (1915–1998):
A student of Carl Jung, who used her fluency in Greek, Latin, and Arabic to translate ancient alchemical texts for him in order to pay for the training she couldn't otherwise afford, Marie-Louise von Franz was one of the most prolific writers in archetypal psychology of the twentieth century, publishing a three volume analysis of fairy-tales (under the name Hedwig von Beit) that was the culmination of nine years of research. This work had at its basis the Jungian assumption that fairy tales are a gateway to humanity's basic collective unconscious, where the purest form of our shared archetypes lie. Throughout the rest of her career, von Franz wrote steadily about the themes underlying religious visions, alchemical texts, folk traditions, and fairy tales, with her

books and articles translated into multiple languages, and ultimately collected in a twenty-eight-volume definitive edition of her writings, with the first released in 2021.

Bernice Neugarten (1916–2001):
Adult development. We have not had cause to use that term much so far, as most psychologists in the late nineteenth and early twentieth centuries were far more interested in what happens to an individual as they move through childhood and adolescence into their 'completed' form than in what was considered to be the general stasis that followed, broken only by the psychological catastrophe points of menopause and the middle-aged crisis. The work of Bernice Neugarten rewrote All Of That, and presented a view of the aging process far more dynamic and nuanced than anything preceding it. She first began researching the topic in 1952 when she was placed in charge of a new course at the University of Chicago, 'Adult Development and Aging'. Researching her lecture material for the course, she found that the work that had been done on the psychology of aging was spotty at best, and riddled with barely-concealed superstitions and pop psych notions that had never been rigorously investigated.

Over the course of the 1960s and 1970s, Neugarten's published research challenged all our old models of what aging was. In classic articles like 'Age Norms, Age Constraints, and Adult Socialization' (1965), 'Continuities and Discontinuities of Psychological Issues into Adult Life' (1969), 'Age Groups in American Society and the Rise of the Young-Old' (1974) 'Time, Age, and the Life Cycle' (1979) and her book *Middle Age and Aging* (1968) Neugarten argued that the rigid 'stage' conception of aging that had dominated previous discourses was woefully simplistic, and that its accompanying baggage of Empty Nest Syndrome, the Middle Age Crisis, and the psychological impact of menopause were similarly popular fictions. In fact, she found, people from age 50 to 75 often experienced a renaissance in their lives, becoming 'young-olds' who, free from the responsibilities of child-rearing, seek new environments, re-engage with their communities, and try exciting new things. Rather than an era of social disengagement and hand-wringing depression, these are often the best years of individuals' lives.

Further, Neugarten found that much of an individual's fate in old age relied less upon their numerical age, and more upon their economic stability, and their behaviour less upon their biological status, and more upon the social norms for their age that they view around them, and as part of these observations she argued in the 1980s for reform to Social Security systems to allow for higher financial distributions to those in most need of them. She received the American Psychological Association's Gold Medal for Lifetime Achievement in 1996.

Denise Albe-Fessard (1916–2003):
Pain is a useful thing. But what happens when it slides from its natural purpose of telling us what to avoid and what not to move, and becomes a constant presence, conveying no useful information but steadily ruining the quality of our lives? One of the answers of early pain scientists was lesioning – eliminating nerves or neural regions associated with the pain so that we would stop 'feeling' it.

That didn't work so well.

Pain, it turns out, is a more complicated thing than neuroscientists originally reckoned, calling forth a new, and more subtle, analysis of its origins and mechanisms. This task was carried out by a generation of neurophysiologists and pain scientists including Jörgen Boivie, D.D. Price, I.H. Wagman, Derek Denny-Brown, and Denise Albe-Fessard. Albe-Fessard studied to become a physicist, but lacking post-graduation opportunities in her field, put her engineering skills to work building electrophysiological devices for her husband, neurophysiologist Alfred Fessard. During the 1970s she carried out her own research into the subject of pain, attempting to locate the brain regions and relay pathways associated with it, publishing work on the importance of the spinothalamic tract in conveying information about pain and temperature to the brain, the existence of nerve fibers that convey solely pain impulses, as well as those that respond to both painful and regular stimuli, and the multiplicity of inhibitory regions along a pain signal's journey any one of which can dial down our perception of that pain.

Albe-Fessard's papers showed why previous pain inhibition techniques had failed – there are simply too many regions involved in the pain response for a single lesioning to do the trick – and pointed the way forward to new possible treatments – given the many different inhibitory pain mechanisms we have, perhaps the solution to the treatment of chronic pain lies in teaching those mechanisms how to do their job better.

For her efforts to understand and alleviate pain, Denise Albe-Fessard was made a Knight of the Legion of Honor in 1973, and an officer of the *Ordre National de Merite* in 1978.

Beatrice Ann Wright (née Posner) (1917–2018):

Wright's *Physical Disability – A Psychological Approach* (1960) and its major revision *Physical Disability – A Psychosocial Approach* (1983) quite simply set the ground rules for how disability programs and governmental policies functioned in the late twentieth century. Wright began working with individuals possessing physical disabilities during the Second World War, when she was employed by the United States Employment Service to find the most suitable occupations for variously physically disabled workers. In the years that followed, Wright developed her unique approach to the psychology of physical disability, which placed the focus on the individual's environment rather than on their disability. The question ought not to be, according to Wright, what is the individual unable to do and how can they make up for their physical disability, it should rather be, how can the environment in which the physically disabled have to regularly operate be constructed to remove the barriers to their easy access to basic services?

On the patient side, Wright described a four stage process through which a person with a disability could progress to arrive at a more positive sense of self and potential – recognise the abilities one does have, learn that one has an identity outside of one's disability and a personality more important than one's appearance, degeneralise the impact of a disability by recognising what areas of life it directly affects and which are untouched by it, and base your sense of accomplishment on a personal scale, rather than by comparison with those around you.

These ideas lay at the core of the Rehabilitation Act of 1973, which provided more grant funding for vocational programs for the disabled, and established affirmative action in the hiring of disabled persons, disallowing discrimination on the basis of physical disability by any company receiving federal money. Wright's influence was even more profound on the Americans With Disabilities Act of 1990, which laid down federal mandates for accessibility in all public buildings, and minimum requirements for disabled accommodations for companies.

Marian Radke-Yarrow (1918–2007):
During her long career, Marian Radke-Yarrow made fundamental contributions to two central aspects of a child's development: the evolution of prejudice and self-hatred, and the impact of parental depression. During the late 1940s, she studied a group of 250 children spread across 6 Philadelphia public schools to determine when prejudice started, and how it evolved, in young minds. She found that the solution to early prejudice lay less in the dissemination of information, and more in the improvement of interpersonal relations and understanding of group dynamics. Facts and figures can meet internal resistances, be misinterpreted, or simply be shrugged off by a student disinclined to engage with them, in a way that working with others does not allow them to do. Most importantly, in her book publishing her results, *They Learn What They Live* (1952), Radke-Yarrow identified teacher training as a core component in the combatting of prejudice, racism, and self-hatred – rather than scapegoating parents for raising racist children, teachers need to learn how their own structuring of class activities, groupwork, and presentation of information play a role in unconsciously reenforcing existing stereotypes and self-conceptions.

They Learn What They Live and its study of how children learn to hate themselves was, with the work of Mamie Phipps Clark, actively cited by the legal team in *Brown v. Board of Education* (1954) to show the psychological effects of American educational segregation, and the deep untruth behind the phrase 'Separate But Equal.' After completing the prejudice study, Radke-Yarrow joined the National Institute of Mental Health, where she worked from 1953 to 1995, and where she developed The Wilson House, a brick building with two way mirrors through which psychologists could observe families in their regular everyday interactions. Here, she researched child altruism, her results largely echoing the earlier work of Lois Barclay Murphy in the 1930s that children have a capacity for empathy and altruistic behaviour that develops far earlier than commonly thought, and the impact of maternal depression on the development of children.

The twenty-three year longitudinal study on maternal depression was a masterpiece of subtlety that recognised the different shades of depression that might loom over a child's young years, and attempted to evaluate the impact of each. For mothers whose depression tended to result in flashes of anger, the result was often children with stronger fears of rejection, while mothers whose depression was more bipolar in nature tended to have children who performed better than their peers in elementary school, but lost their advantage by adolescence, while across the board young girls were more affected by maternal depression than young boys, who largely seemed to slough it off as not concerning them. Radke-Yarrow published her results in the article 'Patterns of Attachment in Two and Three Year Olds in Normal Families and Families with Parental Depression' (1985) and later in book form in *Children of Depressed Mothers: From Early Childhood to Maturity* (1998).

Dorothea Jameson (1920–1998):

In 1892, German physiologist Ewald Hering proposed the opponent process theory of colour perception, whereby humans perceive a given colour on the basis of a tug-and-pull between three sets of opposing colour receptors: red-green, blue-yellow, and white-black. Our eyes are sensitive to these distinctions, and our brain forms our colour palette from them, which allows us to conceive of a colour as 'bluish green' or 'reddish yellow' but not, as we read in the portrait of Christine Ladd-Franklin, as 'reddish green.' In 1957, Dorothea Jameson, working with Leo Hurvich, performed the experiments that validated Hering's theory, by developing a 'hue cancellation' technique. Jameson would present subjects with a mixed colour, say reddish yellow, then slowly add an antagonistic colour, say green, until the subjects reported seeing only pure yellow as a result of the green eventually 'cancelling' the red component in the colour.

Though her 1957 paper was perhaps her most famous result, Jameson contributed insights into the nature of vision across a hundred papers written over the course of three decades of research, including work on the transition from dark to light vision, the phenomenon of after-images, the measurement of light sensitivity, the variation of perceived brightness under changing illumination conditions, and a special study on the unique properties of the colour white as perceived by the eye and brain.

Frances Ames (1920–2002):

South African neurologist Frances Ames is known for two stances she took in life that expressed a profound gutsiness in the face of power. Firstly, she is remembered for her role in bringing the doctors who colluded to cover up the torture and murder of anti-apartheid figure Steve Biko to justice in spite of the danger to her position. Secondly, as a neurologist she is known for her research in the late 1950s into the therapeutic benefits of cannabis for those suffering from spinal diseases, as contained in her 1958 paper 'A Clinical and Metabolic Study of Acute Intoxication with Cannabis Sativa and its Role in the Model Psychoses' and her 1996 article 'Cannabis and the Brain.'

Carolyn Wood Sherif (1922–1982):

Few institutions seem so ideally suited to psychological study as the summer camp – a group of children, isolated from the world, cut off from their families, tossed amongst strangers, forming bonds and cliques, competing in structured and unstructured activities, engaging with layers of temporary authority – what more could you want as a fruitful ground for psychological observation? The full potential of this tantalising setting was realised in 1961, with the publication of *Intergroup Conflict and Cooperation: The Robbers Cave Experiment* by Carolyn Wood Sherif and her husband, Muzafer Sherif. The Sherifs created an experimental summer camp for twenty-two 11-year-old boys to see how repeated group competition affected the boys' ability to relate with each other, and what types of activities were required to overcome the team bonds that had been formed.

She found that group antagonisms, once formed through competition, were not cast aside through routine workaday proximity – if you were Team Badger, and ate at the same table as a member of Team Snake, that regular exposure wasn't enough to overcome the antipathy you naturally had for them as a result of your loyalty to your Badger comrades. The only way to make a Badger trust a Snake was to put them on the same squad facing off against a new opponent in a larger competition. *Intergroup Conflict* was the book that launched a thousand corporate team-building retreats, and for Wood Sherif it was just the beginning of two decades of phenomenal productivity.

In 1965, Wood Sherif published *Attitude and Attitude Change*, which gave birth to Social Judgment Theory (SJT). According to SJT, given a particular issue, most individuals have a 'judgmental anchor' which represents their position on the issue, and radiating out from it are three 'latitudes' of acceptance, which determine positions they are willing to accept, positions they are neutral about, and positions they reject. The more an individual buries themself in a particular issue, the narrower the latitude of acceptance becomes, and the wider the latitude of rejection. A general fan of science fiction, for example, might be willing to accept most assertions that Star Trek is a good franchise, while a more devout Trekkie might accept only hyper-specific claims centred on the shows and seasons that match their more highly detailed conception of what counts as 'acceptable' Star Trek. When presented with a new idea, we unconsciously relate it in terms of how close it is to our anchor position, and what latitude it falls within, and determine our attitude to the proposition based upon that determination. This results in misperceptions dubbed contrast and assimilation whereby we judge opinions that lie in our latitude of rejection to be further from our position than they actually are, and those within our latitude of acceptance to be closer to our position than they actually are, setting the ground for future reconciliation, and disillusionment, respectively.

Wood Sherif was also a founder of the self-system conception of the individual, which recognised that we conceive of ourselves not as one monolithic thing across all contexts, but are rather a shifting mix of self-valuations that takes on different values as we engage with different groups in our social universe. Who we think we are in church is different than who we think we are in a family setting is different than who we think we are while taking a mono-dimensional psychology test, and to be effective as counselors blithely handing out Standard Evaluation Form Number 7 to every person walking in the door, we need to be aware of that multiplicity and account for it. Only when we do that can we start even beginning to understand the contradictory nature of human action and opinion, how one person, in different settings, can say utterly contradictory things, and absolutely mean all of them.

All of these themes were incorporated into Wood Sherif's great work of the 1970s on women's psychology, highlighting how competition for resources between genders has spawned gender stereotypes, how different social structures have pushed women to internalise conceptions of self that they unconsciously use in evaluating themselves in certain contexts and thereby allow the perpetuation of prejudice in those spaces, and how gender identity is a fluid concept wrapped in social practices within which individuals anchor themselves according to the dimensions imposed upon them by their culture.

Carolyn Wood Sherif died of cancer in 1982 at the age of 60, when by all rights a majestic second wave of creative work ought to have been just opening out before her.

Jeanne Block (née Humphrey) (1923–1981):

Jeanne Block was one half of the mammoth Block and Block Longitudinal Study which tracked 130 children over 20 years of development to investigate the onset of self-esteem issues, the impact of familial stress, the evolution of creativity, and the shifting nature of gender roles in different family settings. As regards gender roles in particular, the study resulted in a plethora of publications in the 1970s that detailed the early coding of gender expectations into children, culminating in the posthumously published *Sex Role Identity and Ego Development* (1983). What Block found was that 'meta-messages' were conveyed to children through *how* they were taught, *when* they were praised versus punished (which falls under the larger category of 'shaping'), and *what* aspects of socialisation are emphasised for each gender. These approaches can convey gender roles in a myriad of non-explicit fashions that sum to substantially different 'learning contexts' that nudge children, by the end of two decades of schooling and family interaction, into the behaviours and self-conceptions accepted by their societies. Unfortunately, as was the case with Carolyn Wood Sherif, Block died of cancer at the too-early age of 58, putting to an early end a career that had already shed much light on the many silent paths through which children are formed into women and men.

Thérèse Gouin-Décarie (1923–2024):

Over the course of her long career, Montreal born and educated Gouin-Décarie has posed and answered a multitude of questions concerning the developmental stages of children, extending and modifying the Piaget theory that she encountered during her graduate school years. Can a 9-month-old tell the difference between a human and an inanimate object, and adjust their behaviour accordingly? (1990); How does the development of a child's sense of perspective relate to their acquisition of personal pronouns? (1999); At what stage is an infant able to concentrate on an object pointed at by a parent? (1995); At what ages are infants and toddlers able to understand and follow the rules of simple games? (1995); How were children affected by thalidomide-related deformities impacted cognitively? (1974); How do children react to strangers? (1974); Could a robot be trained to give a child everything they need to develop normally during the first weeks of their life, or are their interactions between baby and caregiver that are so subtly ingrained biologically that they defy all mechanical reproduction? (1980). Her genius for finding new aspects of early life, gestures and recognitions that we don't often think about but that form the very stuff of how we learn to pay attention to our world and judge our place amongst our surroundings, remained fresh and thought-provoking even as she entered her eighth and ninth decades of life.

Thérèse Gouin-Décarie passed away on 2 April 2024, at the age of 100.

Natalia Bekhtereva (1924–2008):
In 1927, Natalia Bekhtereva's grandfather and one of the greatest neurologists of his era, Vladimir Bekhterev, was summoned to attend Joseph Stalin, who had risen to power upon the death of Lenin three years previously. Returning from the session, Bekhterev pronounced that Stalin was a paranoiac. The next day, *perhaps* not coincidentally, he was dead. Shortly thereafter, Bekhterev's name was expunged from Soviet history books, while Natalia's father was shot, and her mother sent to a work camp in Mordovia, leaving Natalia, at the age of 3, without any family save for her brother, to be raised by the state in an orphanage. In 1941, she began attending First Pavlov State Medical University of St. Petersburg, where she survived the Siege of Leningrad, a fate not shared by the vast majority of her classmates, who died of starvation, disease, or injury during the 872 day operation. She became a doctor of medicine in 1959, and one of the major fields of investigation she launched was a neurophysiological study of how the brain adjusts to prolonged stress, such as she had known all of her life.

A pioneer in the use of long-term electrodes to study and provide deep therapy to the brain, her research uncovered that different zones of the brain operate within a particular range of electrical potentials that they prefer, and that, when experiencing extreme emotions, brain zones tend to get over-excited, shooting beyond those potentials, causing large reactions to small provocations, and effectively shutting down our ability to respond creatively to a given task (the neurophysiological equivalent of 'not being able to think straight'). For some parts of the brain, Bekhtereva found, that is okay, as their part in certain processes can be taken up by other neural zones, but for other parts, the 'rigid' zones, which are central to certain mental chains and have functions not duplicated elsewhere in the brain, the move away from steady potential and subsequent hindrance of function has far-reaching consequences in our ability to respond flexibly to life situations. Fortunately, the brain has its own defenses to move us back into the steady potential zone, but these, Bekhtereva found, come with their own setbacks, as over-activity of the brain's defense mechanisms can downgrade our interaction with the world, leaving us in a state of emotional numbness where, yes, we are not being moved to a state of blind fury at the smallest inconveniences, but where at the same time we are unable to motivate ourselves to do much of much because our brain has also been pushed to a state of non-reactivity, unable to engage realistically with our problems or recognise those of others.

Bekhtereva wrote some 350 articles and 14 books during her long career, including a number which were translated and published in the West, such as *The Neurophysiological*

Aspects of Human Mental Activity (1978) and *The Healthy and Unhealthy Human Brain* (1980).

Jacqueline Goodnow (née Jarrett) (1924–2014):

In 1955, Jacqueline Goodnow's academic career came off the starter blocks with a vengeance with the publication of her Harvard dissertation 'Determinants of Choice-Distribution in Two-Choice Situations.' She was interested in how people change their strategies for choosing between two options based on their perceptions of what the end goal of their activity actually is. Subjects were broken into a 'problem-solving' group and a 'gambling' group, with the former given a series of cards containing elaborate patterns and told that their task was to find a reliable system for predicting which of two face-down cards a provided face-up card is related to, with the goal being finding the system, and no penalty for mistakes made along the way. The gambling group, by contrast, was given a slot machine with two buttons to press, one of which pays out statistically more often than the other, and are told that their goal is to amass as much money as possible after 120 trials.

For both groups, the meat of the problem is the same: on repeated trials, pick one of two choices. However, for gambling type scenarios, once the subject picked up on the fact that one side was rewarded more than the other, they shifted strategy and just pressed that button virtually every time – they don't need to figure out the fine details, they just need to maximise their end result, and so putting all their attention on the more frequently rewarded button makes sense, whereas for the problem-solving group, their goal is comprehending the system, and so probing the two choices more often, even after they figure out that one is more often correct than the other, makes sense to them as a matter of finding enough information to generate the overall system.

It was an interesting experiment digging at some key aspects of how our conception of what counts as long-term success in an endeavour causes us to opt for different general choice strategies, and it kicked off a long career in investigating how the structure of different situations pushes humans along different strategic paths. Her 1956 classic volume *A Study of Thinking* attempted to push beyond the conditioning explanations of behaviourism, and to show how people package the information of the world around them into ordered concepts that they use to guide their choices and strategies within that world, a process that is subject to derailment through familial/societal pressures and time/resource constraints that warp concepts and categories such that they no longer correspond well to reality, resulting in prejudiced or irrational behaviours.

Other central works of Goodnow include *Children Drawing* (1977), which investigated how children of different ages form their rules for how drawing works by presenting them with variously incomplete figures and asking them to add different features to those figures. If you have a circle with two dots drawn too low to be eyes, and are asked to draw a face, what do you do? Do you use them for eyes anyway and draw the face upside down? Do you connect the two points with a line to make a mouth? What if you have a circle with two lines underneath it and are asked to put 'hair' on it? What are the rules for where children allow themselves to draw that hair, and where they do not? In short, how do young minds, presented with a given system that tweaks rules they've already internalised, adapt themselves to carry out new tasks in the presence of previously unencountered restrictions?

Goodnow also carried out research on tactile versus visual learning, probing what types of creative manipulation children are capable of imagining with a tactile object versus a visual one, which properties they are able to keep in memory for longer (1969, 1971), and on the impact of parental opinions and belief systems on the worldview and psychological profile of their children (1990, 1992).

In 1992, Jacqueline Goodnow became a Companion of the Order of Australia, recognised by her native land for her global contributions to our understanding of children's minds and their educational needs.

Carolyn Robertson Payton (1925–2001):
Most outside of psychology know Carolyn Payton as the first woman, and first African American, to direct the Peace Corps (1977–1978), and for her epic struggle against ACTION's Sam Brown who wanted to transform the Peace Corps into an organisation that not only did good work internationally, but brought the lessons learned in that work home to target domestic problems arising from poverty. Within psychological circles, however, she is noted for her work in developing methods to better determine the needs of the Black community in the United States, and to provide services to meet those needs. In the mid-1940s, she evaluated the Wechsler-Bellevue Intelligence Test for its ability to gauge intelligence across different cultural groups, and found it wanting in accuracy as regards Black individuals. From 1970 to 1977 she was director of the Howard University Counseling Service, where she developed a program that gave mental health students clinical experience, and specialised training in the unique needs and challenges of serving both poor communities and ethnically diverse communities, which morphed in 1983 into the Clinical and Counseling Psychology Pre-Doctoral Internship, which serves as a national model for providing psychological services to ethnic minority populations. During the 1980s and early 1990s, she published a series of influential articles on the psychological needs of people of colour, including 'Addressing the Special Needs of Minority Women' (1985), 'Substance Abuse and Mental Health: Special Prevention Strategies for Ethnics of Color' (1981), and 'Implications of the 1992 Ethics Code for Diverse Groups' (1994).

Freda Newcombe (1925–2001):
One of the lingering legacies of the Second World War was a generation of young men whose brains had been scarred by flying shrapnel, and who bore lifelong cognitive performance issues as a result. In 1963, Scottish neurologist William R. Russell hired a clinical psychologist only recently returned from a decade and a half of welfare organisation work in Greece to track down individuals whose brains had been lesioned during the war, and determine how different injuries translated into different function deficits. That hiree was Freda Newcombe, and for the next thirty years, she would expand the frontiers of cognitive neuropsychology through her studies of brian trauma

and its consequences. In 1969, she published the results of her 3 year examination of 153 wounded former servicemen, *Missile Wounds of the Brain: A Study of Psychological Effects*, a grim but fascinating account of her first years of research on the correlation of different brain lesion types to different cognitive malfunctions.

Newcombe's work of the 1960s spawned a profusion of targeted studies on brain damage, including research on injury-related dysphasia (difficulty in understanding spoken language)(1978), impairment to facial recognition (or 'prosopagnosia') (1986, 1993), acquired 'deep' dyslexia (including the substitution of thematically similar but graphically distinct words while reading a passage)(1966, 1973), problems with spatial orientation and visual perception from hemisphere-specific injuries (1969), and cognitive preferences in retaining certain object types in memory (1992).

Newcombe retired in 1990, but continued her efforts to help and guide the injured servicemen she had studied through the challenges of their lives until 1996, including an annual reunion at St. Hugh's College that featured strawberries and cream, a brass band, and the chance to reconnect with stout comrades and old friends.

Elisabeth Kübler-Ross (1926–2004):

We know the cycle so well, we've almost come to believe that it is something that has always been with us, rather than the product of a single human mind:

> Denial. Anger. Bargaining. Depression. Acceptance.

The five stages of grief, as first laid out in Elisabeth Kübler-Ross's best-selling 1969 book *On Death and Dying*, has worked its way into pop culture like few parts of psychology this side of Pavlov's dog or Freud's cigar, but surveying the scope of Kübler-Ross's career, the Five Stages (which were actually something more like a dozen stages in her original writings) are among the less compelling parts of her life's work. Born in Switzerland, she defied the demand of her father that she settle down to life as his personal secretary, and left home at age 16. The story goes that, in 1954, while visiting the remnants of the Majdanek extermination camp in Poland, she saw illustrations of butterflies carved into cave walls by people waiting to die, and was inspired by that sight, of creative beauty in the face of imminent death, to devote herself to the aid of those facing death. She studied medicine at the University of Zürich, graduating in 1957, and soon earned for herself a psychiatric residency at Manhattan State Hospital.

Here, she came face to face with the horrible indifference towards the dying that was baked into the American medical system of the time. She saw that patients with no hope of recovering were stuffed away, ignored, given nothing by way of enrichment or personal attention, and in short, left to die. She set herself against this culture of indifference, and resolved to show that, though these terminal individuals were bound to die, the quality of their remaining life could be improved through individual care and recognition, and made implementing that plan, and conveying the importance of dignity and engagement for terminal patients, a centrepiece of her professional work. In the 1970s, the growth of the hospice care movement attracted her attention for having at its core the proposition that the dying deserved to have their physical, spiritual, and emotional needs attended to by dedicated and informed professionals.

Simultaneously with studying how patients can be best led with care through the final leg of their life's journal, Kübler-Ross investigated how best to help those left behind, not only in the cases of losing an elder loved one to a long illness or simply old age (*On Grief and Grieving*, 2004), but how to cope with the loss of a child through illness or suicide (*On Children and Death*, 1983). She was also on the front lines of those advising sympathetic care and psychological support services for those suffering from AIDS (*AIDS: The Ultimate Challenge*, 1987).

Unfortunately, in addition to her manifestly positive work dragging the West's medical structures to something like a humane treatment of the dying, the later stages of her career were marked by a number of scandals surrounding her support for a sort of neo-Victorian spiritualism encompassing out-of-body experiences and the employment of mediums to act as conduits to the dead. While we can't *not* talk about all of that when reviewing her legacy, I think it safe to say that, in the balance, the damage done by books like *The Tunnel and the Light* (1999) will diminish with each passing year, while the good done by her investigations of what the dying need to feel at peace with their fate, and what the living need to heal from their loss, will only grow in the years to come, as death becomes a less frequent presence and correspondingly more frightening thing, through which we shall all need a helping hand.

Isabelle Juliette Martha Rapin (1927–2017):

Like Elisabeth Kübler-Ross, child neurologist Isabelle Rapin was born in Switzerland, had a profound experience that determined the course of her life in the early 1950s, and emigrated to Manhattan soon thereafter to carry out her life's work. While Kübler-Ross's experience led her to a lifetime of work with the dying, however, Rapin's led her to a lifetime of study of childhood language disorders, and autism. While studying at the University of Lausanne in the late 1940s, she had nothing more definite in mind than a career as a physician, but three months spent first at the Pitié-Salpêtrière Hospital (where the Déjerines had made their mark a half-century before) and then the Hôpital des Enfants Malades, where she came in contact with Stéphane Thierry, who was a leading expert on the development of polio and epilepsy in children, convinced her that child neurology would be her field.

Most of Rapin's career was spent at the Albert Einstein College of Medicine, which she joined as a faculty member in 1958, and remained until her retirement at the age of 85 in 2012. Together with her Einstein colleague, psycholinguist and speech pathologist Doris Allen (1932–2002), she devoted herself to the fight against single explanations and diagnoses for what were, upon closer examination, subtly distinct conditions with unique origins therefore requiring unique treatments. In 1983, she and Allen authored 'Developmental Language Disorders' in which they urged against researchers talking about '*the* dysphasic child' as if all accounts of dysphasia (difficulty comprehending spoken language) had a single origin and explanation, and instead presented evidence for multiple potential conditions which could give rise to dysphasic effects, an effort which was an extension of Rapin's 1975 elucidation of multiple sources for dyslexia.

Her gift for making fine distinctions between different variations of a similar condition served Rapin well in her work of the 1990s through 2000s on autism, including her survey of the different classification systems for autism (1996) and participation in the effort to design best practices for autism screening and diagnosis (1999), her studies of the neurophysiology and genetics of autism (1991, 1998, 2004), her distinctions between different language disorders occurring in autistic children (2009) as well as hearing disorders (1991), and her studies of mobility and gait abnormalities (2014). Her near half a century of work on language disorders and autism culminated in 2006 with the dual release of two books conveying all of what she had learned over the course of her career: *Language: Normal and Pathological Development* and *Autism: A Neurological Disorder of Early Brain Development*.

Natalie Rogers (1928–2015):
Natalie Rogers extended the foundations of humanistic psychology as formulated by Charlotte Bühler, Alfred Maslowe, and most keenly, by her father, Carl Rogers. Here, she played a key role in expanding person-centred expressive therapy beyond its original artistic confines. Carl Rogers had published the book *Client-Centered Therapy* in 1951, where he laid out the ideal therapist as a person who provides positive spaces for others to come into their own realisations about their traumas and talents, rather than as a director who forcefully shoves the patient through a set of predetermined stages by a hyper-focus on the negative aspects of their personality.

Natalie Rogers broadened her father's work by expanding the types of activities that the therapist could provide for patients, including drama, art, poetry, and dance, the differing modalities of expression being offered allowing the patients differing ways of seeing the various sides of their personality, always with the positive support of the observing but not intruding therapist figure.

Natalie was also active in the Encounter Group model of mental health servicing, an idea which Carl had first explored after the war, when tasked with training a large group of individuals to become counselors for the Veterans Administration. Individual training in the facts of counseling and theory of cognitive training didn't seem to be providing the skills that counselors needed to connect with patients, understand their own roles in the patients' lives, and grasp the needs of those patients. So, he developed the Encounter Group model as a means of using a group setting to accomplish more work in personal development and growth than was capable in a one-on-one encounter, using the natural awkwardness, antagonisms, and ultimate healing capacities of large groups as means of allowing individuals to find their own way to what is bothering them and what they can do about it. Natalie played an important role in the expansion of the Encounter Group over the course of the 1970s into something that could be meaningfully done for a group of a hundred or more participants.

Interlude: The Twisting Tale of the World's First Woman Neurosurgeon:
If you stroll through the halls of Internet neuroscience sites long enough, what you are eventually struck by is the plethora of pages that very confidently declare that very different people were 'the world's first woman neurosurgeon.' Partly, this is due to disagreements about what constitutes a neurosurgeon, with some insisting that whatever woman first operated on a brain in the modern era be given the distinction, while others insist on limiting the title to women who underwent a particular course of qualification. While neurosurgery is somewhat out of the way of this book's main focus, I thought a little detour into clearing up some of the controversy, and highlighting some of these pioneering figures might be welcome at this point.

For some, the story begins with Louise Eisenhardt (1891–1967), who in her time was a legend in the diagnosis of brain tumors thanks to her decades spent with Harvey Cushing building up a registry of over 2,000 tumor types. Definitely a pioneering neuropathologist whose work in the field began as early as 1922, who was a key assistant at Cushing's tumor excision surgeries, and was herself an authority on tumors whom other neurosurgeons the world over would send photographs of their patients' tumors to in order to get her insights, as far as I can determine, she wasn't the one performing the brain surgeries that she expertly diagnosed and catalogued, at least not until perhaps the late 1920s, and so, depending on how strict you want to be with the definition, that would place the distinction of world's first woman neurosurgeon elsewhere...

Likely with Alice Rosenstein (1898–1991), who after her licensing as a physician in 1923 began training under Otfrid Foerster at the Wenzel Hancke Hospital in Breslau in 1924, where she was soon performing neurosurgical procedures and experiments in pneumoencephalography (a contrast method that lets you use X-rays to probe aspects of the external structures of the brain). She moved to Frankfurt in 1929, where at the University Hospital Mental Clinic she managed to perform some seventy neurosurgical procedures over four years in spite of the hospital leadership's resistance to neurosurgery as an independent field.

Chronologically, the next claimant to the throne is the American Dorothy Klenke Nash, who received her MD in 1925, and had specialised in neurosurgery by 1928, remaining arguably her country's only woman neurosurgeon until 1960. Neck and neck with Nash, Serafima Bryusova received her medical degree from the Moscow Institute for Medicine in 1923, and after four years of surgical residency was recruited in 1928 by Nikolai Burdenko to work in his clinic as a neurosurgical consultant.

The next two contenders usually thrown into consideration for World's First Woman Neurosurgeon (which I am henceforth designating the WFWN), came along markedly later. The first is England's first woman neurosurgeon, Diana Beck, who received her medical degree in 1925 but didn't start training as a neurosurgeon until she began her time at the Radcliffe Infirmary under the mentorship of Hugh Cairns (himself a student of Harvey Cushing, whom Louise Eisenhardt collaborated with for so many decades) until sometime in the late 1930s (Cairns did not begin his career at Oxford, where the Radcliffe is located, until 1937). She went on to distinguish herself during the Second World War as a neurosurgical consultant at the Royal Free Hospital, and as a neurosurgical advisor to the south-western quadrant of England's emergency medical services. After the war, she established the neurosurgery department at London's Middlesex Hospital.

The second person you hear a lot about is the Romanian Sofia Ionescu, which mostly baffles me in terms of the chronology, but which Ionescu supporters are so sure about that one starts to doubt one's own sense of how time works in the face of their determined advocacy. Don't get me wrong – I am 100 per cent in favor of everybody taking some time out to remember and revere the work of Ionescu, but in terms of being the WFWN, that's going to be a harder sell I think. She applied to the Faculty of Medicine in Bucharest in 1939, a good decade and a half after Alice Rosenstein's first neurosurgical procedures, and as far as I can tell didn't perform her first brain surgery until 1944, a half decade after the first procedures of Diana Beck. Though not the first woman neurosurgeon, however, Ionescu is well worth the recalling, as for forty-seven years she served her country as one of only a few such specialists in the entire nation, and a driving force behind the establishment of a neurosurgical tradition in Romania.

By and large, that's the tale of the first women in the world to become neurosurgeons, but there is still one more story to tell, which is that of the first women to become neurosurgeons from particular cultural traditions and ethnic groups, not only because it's good information to know, but also to give a sense of scale of when neurosurgery became accessible to different women in different parts of the world. Aysima Altinok, who was not only Turkey's first woman neurosurgeon, but one of its first neurosurgeons of any sort, earned her certification as a neurosurgeon in 1959, and was from 1968 to 1992 the chief of neurosurgery at Istanbul's Bakirköy Mental and Psychological Health Hospital, and a co-founder of the Turkish Neurosurgical Society.

The first woman in Asia to become a neurosurgeon was Thanjavur Santhanakrishna Kanaka (1932–2018). Born in Madras, she served as an officer in the Indian Army during the India-China War (1962) and earned her Master of Surgery certification in neurosurgery in 1968, after which she formed part of the vanguard in advocating for functional neurosurgery in India, employing techniques like deep brain stimulation and stereotactic surgery (by which surgical targets are mapped out as locations in a three dimensional Cartesian/cylindrical coordinate model of the brain to allow for minimally invasive procedures) to correct neurological disorders through targeted surgical techniques.

Finally, in 1981 Alexa Irene Canady (b. 1950) became the first Black woman in America to become a neurosurgeon after completing her University of Minnesota residency. She became chief of neurosurgery at the Children's Hospital of Michigan in 1987, and remained in that position until her partial retirement in 2001.

Bernice Grafstein (b. 1929):
As an electrophysiologist and neurophysiologist, Bernice Grafstein has spent her life investigating questions of the mechanisms behind how injured brains and their

sundry neural components repair themselves. Her dissertation, published in 1956 as 'Mechanism of Spreading Cortical Depression' was a tour de force which probed the phenomenon by which a wave of inhibitory neural activity grips sections of the brain. Called cortical spreading depression, or CSD, now, the effect has been linked to neural events as seemingly diverse as migraines, SUDEP (sudden unexpected death in epilepsy), and stroke. Though CSD had been observed by Aristides Leão in 1944, Grafstein's paper provided the mechanism behind the phenomenon, demonstrating that the extended hyperpolarisation characteristic of CSD was likely caused by leakage of potassium ions into the extracellular space around the neurons, and that resetting the system back to its normal state was limited by energy availability. Usually, potassium ions serve the role of regulator for the neuron, flooding into and out of the cell as necessary to maintain the resting state that allows neurons to fire. In CSD, the ion regulation mechanism goes briefly pathological, and the seeping potassium sets up a chain reaction in neighboring neurons, resulting in the characteristic spread of neural inactivity. Though today Grafstein's system has been modified to admit the additional importance of glutamate toxicity as the instigating cause of the potassium surge, her work of some seventy years ago still stands as the basic model for the progression and spread of the cortical depression.

Well established by her 1956 paper, Grafstein took up an interest in the means by which neurons communicate with each other, which led to research in how neurons transport proteins through themselves. Working primarily with goldfish optic nerves, she studied the process of regeneration following the severing of the nerve, and discovered that, while regenerating, neurons vastly pick up the pace of their protein transport chains, with the fast transport system (used mainly for protein particles) doubling in speed, and the slow transport system (used for proteins that are in a mixture of particle and soluble states) tripling in speed, results she published in 1969 in 'Transport of Protein in Goldfish Optic Nerve During Regeneration.' In the 1970s and 1980s, she expanded her interest into nerve regeneration in a more general sense, studying how previous lesions affect subsequent nerve repair efforts (1973), how nerve bodies respond to the severing of their axons (1975), and the effects of nerve growth factor on regeneration (1982).

Today, Grafstein continues her research as the Vincent and Brooke Astor Distinguished Professor in Neuroscience at Weill Cornell Medical College.

Clara Mayo (1931–1981):

Austria-born Clara Mayo ranked during her short life as one of the twentieth century's great investigators into the subtleties of racist and sexist psychology. We have met others so far in this book who looked at how racist and prejudiced sentiments evolve over the course of a child's life – matters of education and upbringing communicated largely through words, but Mayo was interested in something deeper than that – how does the entire structure of non-verbal communication contribute in its silent way to the fostering and maintenance of racist/sexist structures? She performed studies on how members of different races, genders, or social groups held eye contact (1973), defined personal space (1971), employed paralanguage and gesture (1978), and reacted to facial expressivity (1979) to construct the elaborate web of movement and glance by which we tell people, without *telling* them, exactly what we think and expect of them, what the boundaries of their action ought to be, and where we consider ourselves within our mutual power structure system. She summarised her research in her 1978 book (co-authored with Marianne LaFrance) *Moving Bodies: Non-verbal Communication in Social Relationships*. In the last years of her short life, she began investigating the linkages between racism and sexism, co-editing the volume *Gender and Nonverbal Behavior* with Nancy Henley, which was published the year of her death.

Elizabeth Kerr Warrington (b. 1931):
In the late 1960s, after decades of being held back by a retaining wall of behaviourist orthodoxy, cognitive neuropsychology, the study of how brain function relates to psychological phenomena, came roaring back to life thanks to the dramatic results produced by the study of individuals with targeted brain injuries. We have already seen some of this work in the form of Freda Newcombe's ground-breaking studies of individuals with brain lesioning due to wartime shrapnel, and the effects of those injuries on reading, spatial processing, and object memory. Equally important in these formative years of the new cognitive wave, however, was the work of Elizabeth Warrington.

For three decades, Warrington plumbed the depths of memory and object perception disorders caused by brain injuries, allowing her to make minute distinctions in the underlying physical causes behind conditions that, from an outside perspective, seemed indistinguishable. In her 1990 book *Cognitive Neuropsychology: A Clinical Introduction* (1990), Warrington and co-author Rosaleen McCarthy take a triumphant victory lap through the discoveries of the previous quarter-century of cognitive psychological study, linking phenomena as varied as face recognition, voluntary action, word comprehension, sentence processing, speech production, reading, and object recognition to underlying brain regions, outlining how different combinations of damage to those regions produces distinct alterations in the functioning of those fundamental human abilities.

Warrington's work of the 1970s and 1980s showed how much could be accomplished from a cognitive perspective, as she (sometimes working together with other CNP pioneers such as Tim Shallice or Lawrence Weiskrantz) investigated closely how right brain injuries affected patients' abilities to recognise objects rendered from strange perspectives or with unexpected lighting, and to perform figure and letter completion tasks, how some patients suffering from inabilities to generate short term memories can nonetheless continue to develop long term memories, how implicit (things you unconsciously remember) and explicit (things you can actively recall) memory are differently affected in different forms of amnesia, and how our ability to recognise an object after brain damage is dependent upon how we categorise it, revealing important links between brain structure and category structure that further spurred the field of cognitive neuropsychology.

With so much expertise built up in the field of correlating brain injury or disease to disorders in perception and memory, it was only a matter of time before she began running the correlation in the other direction, designing tests of perception and memory which could be used to produce targeted initial diagnoses of underlying neural problems, which could then also be used to develop specific therapy techniques to help patients on the road to rehabilitation. To this end, she co-developed both the Visual Object and Space Perception Battery (1991) and the Verbal and Spatial Reasoning Test (1996), both of which are still in common use.

Florence Denmark (b. 1932):

One of the things that has penetrated the popular imagination in the last decade is the idea that within even the most neutral seeming of questions, experiments, and data collection techniques there are often hidden deep gender biases that silently skew us towards particular and comfortable conceptions of best practice. In this sense, we are just now catching up to the warnings and insights that Florence Denmark had published thirty-five years ago in 'Guidelines for Avoiding Sexism in Psychological Research' (1988), in which she laid out how each step of traditional psychological procedure, from how questions are formed, to how research is designed, to how data is analysed, to how conclusions are presented, is often laden with unspoken gender biases and expectations that have significantly hindered our knowledge of women's particular psychological structures and needs.

'Guidelines' is just one of a hundred articles (and fifteen books) Denmark has authored shining light on the psychology of women, and the relation of the psychological community to women. Denmark began publishing in this field in 1966 with both 'Sex Differences in Attitudes Towards Leaders' Displays of Authoritarian Behavior' and 'The Effect of College Attendance on Mature Women,' a brace of papers which displayed from the start Denmark's range of interest, from measuring how women of her particular time viewed unique political structures, to how they could be personally benefited by expansions in educational opportunities. She went on to investigate how college attendance affects women's conception of themselves (1967), the different approaches to leadership open to, and practiced by, women (1977), the political attitudes of Chinese and Puerto Rican women (1981), the expectations behind the division of household responsibilities (1985), societally ingrained perceptions of step-mothers and mothers-in-law (1989), the role that women play in modern psychology (1998), and the differentiation of the psychological stresses felt by men and women after a miscarriage (1998), not to mention her central text, *Psychology of Women: A Handbook of Issues and Theories* (1993, revised 2007).

In recognition of the importance of her work in detailing the flaws in psychology's past approaches to one half of the human population, and in carrying out new research to make up for those deficits, Denmark has been presented numerous awards and filled important leadership positions, including the presidency of the American Psychological Association, the New York State Psychological Association, the Eastern Psychological Association, the International Council of Psychologists, and Division 1 of the APA.

Every year since 2005, the Association for Women in Psychology has awarded the Florence Denmark Award to women who have displayed a consistent and deep commitment to mentorship and leadership in psychology.

Dame Ingrid Victoria Allen (1932–2020):

As a neuropathologist, Belfast-born Ingrid Allen's work on establishing the role that viruses play in brain disease, on cellular level responses to brain trauma, and particularly on the impact of multiple sclerosis on tissues previously believed to be unaffected by the disease have brought her just international renown. She began her career studying under Ireland's reigning expert on neuropathology, Sir John Henry Biggart, receiving her MD in 1963. One of her early research victories came with the sudden uptick in cases of the brain inflammation disease subacute sclerosing panencephalitis (SSPE) among children in Belfast in the 1960s. Allen, along with John Connolly, L.J. Hurwitz, and J.H.D. Millar, analysed the tissue samples from children who had or died of the disease, and published in their 1967 paper the discovery of measles virus antibodies in the cerebrospinal fluid, suggesting that SSPE was caused by some strain of measles virus, which in turn led to a nationwide push for measles vaccination in Ireland.

Subsequently, Allen turned her attention to another of the great neural pathologies of her time, multiple sclerosis. Macroscopically, the white matter of a brain attacked by MS appears as regions of seemingly healthy and normal tissue interspersed with lesioned demyelinated regions. In 1979, Allen decided to take a closer look at those seemingly healthy regions, and found that, in 72 per cent of the samples she analysed, there were cellular abnormalities, including heightened concentrations of lysosomal enzymes and acidic phosphates. This discovery provided researchers with new chemical tools to probe what happens in the early stages of tissues attacked by MS. Further, as a disease based around how one's immune system attacks one's own myelin (the coating around most nerves that is responsible for quick signal sending), researchers had reasonably tended to confine their studies of MS tissues to myelinated nerves, leaving largely alone nerves that were naturally unmyelinated, such as those in the retina. Allen, however, looked for and found distinct changes in retinal tissue as a result of MS, with retinal ganglion cell loss, neuronal loss, and axonal injury/loss in different regions of the retina (2010).

An author on over 200 papers, Ingrid Allen was made a Commander of the Order of the British Empire in 1993, and a Dame of the Most Excellent Order of the British Empire in 2001.

Doreen Kimura (née Goebel) (1933–2013):

Born in Winnipeg, and raised in the miniscule town of Neudorf (population 272 as of 2021), Kimura started her career as a teacher in a one-room school in tinier-still Cowan, Manitoba, when she saw a magazine ad for McGill University that she replied to, ultimately earning her a three year scholarship to attend the school. Here, she came under the influence of the legendary neuropsychologist Brenda Milner, and earned a Montreal Neurological Institute fellowship that allowed her to carry out her early research, designing dichotic-listening tests to research where auditory signals are processed in the brain. Dichotic type tests have at their centre the simultaneous presentation of different signals to a subject's ears. So, for example, in Kimura's early tests, a subject might hear '1' in their left ear at the same time they hear '5' in their right ear. After a few rounds of presenting different numbers to the subject, they are then asked which numbers they remember.

Kimura extended the basic dichotic testing she had inherited from Donald Broadbent to investigate the 'sidedness' of an array of auditory stimulus types, finding in a series of papers published from 1961 to 1967 that there was a right ear advantage for right-handed subjects when listening to numbers and words, but a left-ear advantage when listening to melodies, a result which suggested to Kimura that the left hemisphere of the brain is specialised for speech functions (a specialisation Kimura saw manifested in subjects as young as 4 years old), while tonal perception and processing appeared to occur primarily in the right hemisphere.

Over the next two decades, Kimura studied the phenomena of apraxia (reduced ability to convert speech intentions into the mechanical production of speech) and aphasia (reduced ability to understand and form spoken language due to brain injury that does not impair the physical mechanisms of hearing or speech), which provided her with insights into the interconnection between motor systems and language systems in the brain that she published in her landmark 1993 volume *Neuromotor Mechanisms in Human Communication*. Kimura found that damage to centres of the brain involved in apraxia not only also impacted gestures that had communicative content, but also impaired the ability of subjects to mimic movements that had no linguistic content, suggesting deep relationships between the neural evolution of our gestural capacities and our linguistic ones which are replicated in our neuroanatomy.

In the 1990s, Kimura controversially studied how sex hormones differentially impact brain development in men and women, detailing important distinctions that ran counter to that era's predilection for cultural explanations for psychological differences between the genders, but that stand more in line with the modern compromise between the importance of early hormonal factors, epigenetic developments, and the weight of society and culture in determining gendered psychological differences.

Reiko True (b. 1933):
In 1989, Reiko True conducted a survey of residents of San Francisco's Chinatown to determine that community's attitudes towards mental health services. She found that, while many reported issues with depression, isolation, and the feeling of emotional extremes, only 5 per cent stated that they had ever sought help to improve their mental health, a number far below that of other West Coast American ethnic groups. This was a disturbing result, but perhaps not a surprising one for True, who has dedicated her career to the unique challenges and failures of the psychological community in building bridges to Asian American populations. Her articles and books of the last four decades have pointed towards unique cultural proscriptions against seeking mental health services, particularly among Asian American women, as well as unique challenges that need to be met in designing psychotherapeutic techniques and methods that take into account the lived reality of the Asian American experience and cultural expectations. One of the dangers, True elaborated in such papers as 'Psychotherapy Issues with Asian American Women' (1990) and 'Asian/Pacific Islander American Women' (1996), of

bringing Asian-American women into psychology as it is currently practiced is that modern evaluations of what count as 'normal' states of psychological health are based on predominantly Western models that will necessarily create potentially hurtful tensions between Asian patients' sense of themselves and their therapists' evaluation of their mental condition.

In 2003, the Asian American Psychological Association awarded True its Lifetime Achievement Award for her persistent efforts in making Western psychology aware of its evaluative assumptions and culturally opaque practices while at the same time suggesting a way forward, allowing psychological aid services to reach at least a key element of the population that has not heretofore had unencumbered access to them.

Anne Treisman (1935–2018):

Opening our eyes to a strange room, we perceive a world of objects all at once, some of which we lock onto and pay attention to, and others of which dance around the periphery of our regard. If we close our eyes again right away and are asked to recite the contents of that room as best we can, our response will be a psychologically tantalising mixture of correct identifications, complete omissions, and intriguing transpositions of qualities. How does that latter happen? How do we momentarily glimpse a square red pillow and a round green one, and later report having seen a round red one? How do our brains only 'kinda' remember objects and what are the mental processes by which that loose 'kinda' memory gets firmed up and locked into accurate whole memory?

Unwrapping those conundrums lies at the heart of Feature Integration Theory (or FIT), as outlined in 1980 by Anne Treisman and Garry Gelade in their seminal paper 'A Feature Integration Theory of Attention.' Though formally laid out in 1980, FIT had its origins a good decade and a half previously, in Treisman's modifications to Donald Broadbent's Filter Model of Selective Attention. Broadbent had been interested in auditory selection – why is it, in a room full of people all talking Very Loudly (I think that's called a 'party' by people with social lives?) your brain can focus attention on the one conversation you theoretically Want to be having, and not pay attention to the content of all the ones around you? Broadbent hypothesised that, though we hear those conversations, our brains filter them out relatively early in the processing phase, dumping all that information before it reaches our conscious awareness, only letting through the auditory information of the person we are actually trying to listen to.

Treisman pushed the rough edges of Broadbent's theory using dichotic listening experiments similar to those we saw being employed by Doreen Kimura above. Instead of hitting a subject's ears with numbers or melodies or single words, however, as Kimura did, Treisman used passages from the novel *Lord Jim*. Subjects were asked to pay attention and recite the words they heard being read by a woman's voice in their left ear, while in their right researchers played either a different passage from the same novel, a translation in a foreign language of that novel, a nonsense passage, or a passage from the novel played backwards, sometimes read by the same woman's voice, sometimes by a man's voice. What she found was that how good the subjects were at ignoring the 'irrelevant' message varied substantially, as measured by the accuracy of the reported text, and the number of 'intrusions' from the irrelevant message that snuck into their reporting. While subjects were able to very well filter out the irrelevant message when the voice was male, producing less than 1 per cent intrusion words and around 73 per cent accuracy in the target text, when the irrelevant message was read by the same voice as the relevant, that result plummeted, with accuracy only at 31 per cent and intrusions up to 20 per cent when the irrelevant message was from the same novel in the same language, and accuracy only improving to 55 per cent when the irrelevant was in the same voice, but a different language.

Something interesting was happening – the brain, apparently, couldn't just 'turn off' the message it didn't want to listen to, with a number of different factors contributing to the degree to which it was successful in hearing what it wanted to hear and ignoring the rest. Attention by this experiment isn't an on-off toggle, but rather some manner of dimmer switch with different weights given to different aspects of an attended signal that allows leakages in what we are supposed to be paying attention to.

Treisman's papers in the mid to late 1960s published the results of her attention experiments, modifying Broadbent's results by showing that some characteristics of a signal allow our brains to reject them before analysing them (such as the difference between a low man's voice and a woman's), but others (such as differences in content read in the same language by the same voice) cross our wires to some degree, and get filtered later down the line *while* analysis is taking place, thus allowing for more misreported messages and cross-contamination. These results, then, paved the way for the experiments in visual attention of the 1970s which culminated in the 1980 development of FIT theory. Treisman's visual experiments, which asked subjects to concentrate on black numbers that were placed next to different coloured shapes in their visual field, found that about one in five times, when asked to identify features of the shapes they weren't told to concentrate on, subjects would mis-assign features, by for example swapping their colours.

What these experiments told Treisman is that we essentially have two stages in the attention process: in the first, we make note of the basic features of objects, but don't bother binding them together in unified wholes. It is when we 'turn our attention' to an object that we head into the second stage, where the different basic features of an object aren't only noted, but are linked together in our conception of that object. This then spurred the important question, 'Just how does that happen?', which became known as the Grandmother Problem – how do the separate parts of our memory of an object, such as our grandmother's voice, face, smile, and hair, get assembled into one unified object, the mental image of our grandmother? This question has launched a thousand research ships onto the waters of cognitive neuroscience, not a few have which were captained by figures we shall soon meet.

And in the meantime, there has been the figure of Treisman, who was made a Fellow of the Royal Society in 1989, inducted into the National Academy of Sciences in 1994, and awarded the National Medal of Science by President Barack Obama in 2013.

Mary Jeanne Kreek (1937–2021):

In an age when people (and particularly Americans) thought of drug addiction as a social problem that could be conquered if people 'just had enough willpower to say No.' Mary Jeanne Kreek was a member of a small group of researchers dedicated to the proposition that addiction was a brain disease that should be researched and treated with the same scientific methods as any other. Early in her career she had the opportunity to work at Rockefeller University with Vincent Dole and Marie Nyswander on the problem of addiction. The team found that opioid addiction seemed to be a primarily metabolic disorder, and that the cravings associated with it could be lessened by methadone, a drug that had been developed in Germany in the 1930s as a substitute for morphine. In 1966, they published 'Narcotic Blockade' (which is, by the by, a *great* band name) which established not only the cadence of addiction as a learned behaviour that morphs over time into a physical dependency with diminishing returns of euphoria, but set out the efficacy of methadone as a treatment option.

In the subsequent decades, Kreek was ever on the front line in pushing the boundaries of the science of addiction, developing a second treatment drug, buprenorphine (which was approved in 1981 as a drug for relieving withdrawal symptoms), and pushing investigations into the epigenetic action of drugs on the brain in the early twenty-first century, detailing dozens of alterations in DNA activity that occur in an addicted brain, some of which plug into feedback loops that help explain the biochemistry of the stress-addiction cycle. For her efforts in drug addiction treatment and in outlining the neurochemistry of addiction, she received the Lifetime Science Award from the National Institute on Drug Abuse in 2014.

Edna Foa (b. 1937):

Born in Haifa, Israel, Edna Foa is today recognised the world over as an expert on the treatment of post traumatic stress disorder (PTSD), Obsessive Compulsive Disorder, and Social Anxiety Disorder. In the late 1970s and early 1980s she developed the prolonged exposure (PE) method for the treatment of PTSD, which features a two-step process of first compelling the subject to repeatedly recount the events that caused their trauma (Imaginal Exposure), followed by the creation of a hierarchy of fears and avoidance behaviours exhibited by the patient, which leads to the stage of 'In Vivo Exposure,' wherein the therapist assigns homework from the hierarchy for the patient to confront over the course of the week, gradually working them from growing comfortable with their minor fears, to full confrontations with the central, dangerous elements of their trauma. Today, PE is a major form of treatment for survivors of rape, abuse, and assault, as well as a growing treatment for substance abuse issues.

Sandra Schwartz Tangri (c. 1936/37–2003):

Tangri was the driving force behind the Women's Life Paths Study, a longitudinal study, which followed 117 women of the 1967 University of Michigan graduating class over the next 25 years of their lives, as they entered different male-dominated professions, performing major status updates in 1970, 1981, and 1992. She broke the class into three ultimate groups, the role-innovators (women who chose work in male-dominated fields), the traditionals (women who chose work in women-dominated fields), and the moderates (women in fields where women made up 30-50 per cent of the work force). She found remarkable stability among the innovators, with 81 per cent remaining in the category after 15 years, while 69 per cent of traditionals remained traditionals. The innovators, she found, were largely women with more education, fewer children, and who were less inclined to marry (and who, when they did marry, did so with the expectation that their husband would support their career).

In addition to her final update on the WLPS, Tangri spent the last decade and a half of her career in a variety of investigations about various elements in the everyday life of women, including a survey of how different women feel and manifest romantic jealousy (1989), a development of different models for workplace sexual harassment (1982, 1997), and a new longitudinal study on the life paths of Black college-educated women who were under-represented in the WLPS (2003).

Eleanor Saffran (1938–2002):
As a cognitive neuropsychologist, Eleanor Saffran spent the better part of three decades developing methodologies to better recognise and categorise cognitive impairments, including work on how to characterise different forms of agramattic and non-agrammatic aphasia (in agrammatic aphasia, patients usually relate sentences without function words like prepositions or conjunctions, and so establishing how agrammatic individuals structure their objects and subjects can be tricky)(1980, 1989), how semantic knowledge (such as correctly categorising objects) breaks down with dementia (1979), how problems with manipulation and function knowledge of objects are rooted in neural architecture (2002), and the creation of strategies for alexic patients (patients who were not born with dyslexia, but developed it later in life)(1989).

Jerre Levy (b. 1938):
Levy was a student of the Nobel laureate Roger Sperry, whose split-brain work investigated how learning can be localised in different hemispheres of the brain when the connection between them is severed. She continued this research in her own work. Sperry's research had provided a tool for treating epilepsy through the severance of the corpus callosum (which links the right and left hemispheres of the brain), which brought with it the added research opportunity of determining which halves of the brain are specialised for which types of functions. Levy's 1969 paper 'Possible Basis for the Evolution of Lateral Specialization in the Human Brain' took these epileptic patients as her starting point for probing hemispherical localisation of cognitive functions. To expand her research in hemispheric specialisation beyond the small sub-population of those who happen to have undergone corpus callostomy (one of the more unfortunate names for a surgery that there is), she began studying differences in perception and performance between right and left handed individuals, resulting in a series of papers in the 1970s and early 1980s that determined the relations between handedness, writing posture, gender, and different linguistic and emotional perception

capacities, and aesthetic preferences. Her work established that, while perhaps some 15 per cent of the population possessed the 'classical' hemispheric division in which linguistic functions are entirely controlled by the left hemisphere of the brain (this group being right handed adult males with a regular writing posture and no family history of left-handedness and no personal history of psychotic behaviour), the rest of the population experiences varying degrees of neural plasticity where those linguistic functions seep over to the right hemisphere of the brain, and that particularly in children, damage to one side of the brain can be compensated for by the other side assuming and undertaking its functions.

As professor emeritus, Jerre Levy today continues her association at the University of Chicago's psychology department, which she joined as an associate professor in 1977.

Sujata Tewari (1938–2000):

Born in West Bengal, Sujata Tewari majored in chemistry at Agra University on her way to earning a PhD from McGill University (in biochemistry) in 1962. She subsequently researched the impact of alcohol on neural development and brain tissues, finding that protein production by free ribosomes (ribosomes not bound to a membrane like in the rough ER) decreased with alcohol consumption (1975, 1978), and that the withholding of alcohol from physically dependent rats (rat alcoholics) similarly resulted in protein synthesis decline (1977), which suggests deep links in how ethanol interferes at the ribosomal level with mRNA trying to do its job of translating DNA commands into proteins, and thereby with normal brain function. In the 1980s, Tewari examined alcohol's impact on fetal development, concentrating on glial cells called astrocytes which do not carry electrical impulses like neurons do, but play important support roles in the nervous system. She found that DNA and protein synthesis in ethanol-exposed astrocytes were reduced by up to 60 per cent from control groups, resulting in slower rates of astrocyte proliferation and ultimately deficient astrocyte functioning. In 1978, she co-founded, with Louis Gottschalk, the Alcohol Research Centre at UC Irvine.

Gudrun Boysen (b. 1939):

Boysen is a world expert on the causes of stroke, and the prevention of recurrent strokes, who served as a member of the executive committee of the International Stroke Society from 1996 to 2000. She has been involved in evaluating the efficacy of a variety of potential stroke treatments, including a three year study of warfarin (1985–1988) that showed a 75 per cent reduction in stroke-like complications after two years compared against aspirin and placebo groups, a 2011 study that found that the blood-pressure lowering drug candesartan did not have any appreciable effect on stroke incidence, a 1994 study that found that very high cholesterol increases susceptibility to cerebrovascular disease, but that moderate to moderate-high levels do not seem to have any effect, a 2003 study that indicated elevated homocysteine (an amino acid that isn't used to make proteins, but that can be changed into one that does) levels were a good indicator of likelihood of recurrent stroke, a 2001 study that, while tinzaparin at high doses does lead to lower levels of thrombosis in patients, it also significantly increases the risk of intracerebral hemorrhage, and everybody's favorite, a 1998 study which found that weekly intake of wine corresponds to significantly lower risk of stroke (daily is also good, but not as good as weekly, so don't go crazy with it).

Harriette Pipes McAdoo (1940–2009):
Prior to Harriette McAdoo, the issue of Black youth and Black families in the United States was generally framed as a problem based on a series of deficits, with the solution structured as the devising of ways to overcome those deficits. This framing, while useful in acknowledging the role that the legacy of slavery and segregation continue to play in the lives and psychologies of Black individuals, also brought with it the unfortunate side effect of creating solutions that had as their underlying thesis that Black families were fundamentally broken, and Black youths fundamentally troubled, which rather tended to double-down on negative self-conceptions. If even the most well-meaning of programs starts from the position of, 'Let's fix what historical societal forces have damaged in you,' you're unlikely to get unambiguously positive results. McAdoo was a pioneer in strengths-based approaches to minority programs, which highlighted the resiliency of the Black population, and the unique strengths, as the foundation from which positive change would come.

To this end, she carried out a number of studies seeking to correct some of the systemic assumptions about Black lives that had persisted since the segregation era. As reported in her 2002 book *Black Children: Social, Educational, and Parental Environments*, though many Black children live with socio-economic disadvantages, their sense of self-worth had come a long way since the Jim Crow era, and indeed in predominantly Black neighborhoods, Black children scored significantly higher on self-esteem evaluations than white children did thanks to the integrated community and family support they felt. Many of the problems come, McAdoo found, with an education system that has been so inundated with studies of Black children as 'underperforming' and 'disadvantaged' that it systematically lowers expectations of them in ways that the children internalise over time. To help challenge this damaging and time-worn narrative of underperforming, prison-prone Black families, she published the anthology *Black Families* (currently on its fourth edition), focusing on stories of familial resilience, kinship-based mutual aid, and community support to begin to turn the tide on the century-old misconceptions and stereotypes that continue to guide well-meaning governmental and educational policies.

Elspeth McLachlan (b. 1942):
When Elspeth McLachlan began her studies of the automatic nervous system (ANS) in the 1970s, it was very much considered the lesser sibling of the somatic nervous system (SNS). While the SNS (which receives sense signals and controls voluntary movements via skeletal muscles) was thought to be an organised structure of tightly controlled communication, the ANS (which controls unconscious body regulatory

processes) was considered as simply a hormonal system that responded in two relatively blunt force ways to the presence or lack of stress. McLachlan was part of a team of researchers determined to reverse that conception, and uncover the levels of organisation and direct control present in the ANS. The existence of deeper functional pathways in the ANS than those simply related to stress responses was established by her in an important 1992 paper, 'Characteristics of Function-Specific Pathways in the Sympathetic Nervous System'. Following hot on the heels of that paper, she and Wilfrid Jänig found evidence for sympathetic nerves and sensory nerves participating together in instances of peripheral nerve injury (1993), suggesting a degree of interconnection that McLachlan ranks as one of the most exciting findings of her career.

Nancy Kopell (b. 1942):

'Computational Neuroscience' as a named discipline with its own specialised university departments has been with us for just a little under four decades now, though its origins as a field attempting to apply mathematical modeling to elucidating the function of the mind lie significantly further back, in the early twentieth century, with the first attempts to understand and model the electrical firing of neurons by Louis Lapicque (1907), the first explanations of learning as a matter of increasing the connectedness of neurons by Donald Hebb (1949), and the publication of the Hodgkin-Huxley model of the shape of a neural action potential and the role of ions in determining it (1952). The field was developed in heavy mathematical earnest by a generation of researchers born in the 1940s, including David Marr (b. 1945), Eric Schwartz (b. 1947), and Nancy Kopell, who came to prominence in the 1970s and 1980s.

One of the concerns of computational neuroscience is how to model the characteristic behaviour of the brainwaves produced by different mental states and activities, which would then allow closer investigation of the physical connections underlying the model. Kopell worked with mathematician George Bard Ermentrout (b. 1954) to develop the Ermentrout-Kopell model for bursting neurons. Bursting neurons are those that have a regular schedule of intense firing alternating with long quiescent periods, and their particular rhythms often lie at the heart of repetitive processes, like walking or breathing. There are many ways for bursting behaviour to manifest itself, with the E-K model tackling the 'parabolic' bursting behaviour, wherein the frequency of the spikes changes over the course of the burst in a pattern resembling a parabola. It's a neat model that takes a slow periodic function and grafts on top of it a function consisting of two cosines which come into effect whenever the neuron is ready to burst, but remain under wraps during the quiescent stage. They published their model in the classic 1986 paper 'Parabolic Bursting in an Excitable System with a Slow Oscillation', a paper which was itself situated in the midst of a burst (pause for laughter) of activity by Kopell in describing the activity of coupled oscillators (oscillating systems that are chained to one another, like a spring connected to a block which has under it another spring connected to another block) and how those models shed light on the neural networks that run certain large scale repetitive actions, like the back-and-forth motions of simple life forms.

Sandra Rees (b. 1942):

Australian neuroscientist Sandra Rees has been asking questions about the factors that impact fetal and neonatal brain development since the late 1970s which have guided our sense of the waxing and waning vulnerabilities of fetuses in their growth cycle to the various chemicals and traumas we shove at them. Her investigations have included an analysis of the impact of cortisol on fetal neural development (i.e. what happens when a mother is under persistent stress while pregnant)(2002), the effect of different types of fetal brain injury (2008), the developmental problems produced by exposure to inflammatory chemicals (2008), and the role that a less advanced fetal growth schedule can play in the decreased development of myelinated nerve fibers (1990). In 2019 she was inducted as a Fellow of the Royal Society of Victoria.

Vivien Casagrande (1942–2017):

You see a tiger, hear its roar, and within a moment, you are ready to turn tail and head for the hills, without much bothering to notice what else is happening visually in the vicinity of the tiger. This is a great trait to have, which begs the question, well, how did we get here? To solve this problem, Vanderbilt professor Vivien Casagrande developed a three-stage research attack plan that sought to get to the bottom of how we primates have developed some clever short-cuts to up our response time when different senses are giving us mutually corroborating information about important features of our environment. Her theory was that the visual thalamus, far from being a simple gateway through which retinal data flowed on its way to the visual cortex, where all the 'real' analysing happens, is itself an important organising centre that pre-packages salient auditory and visual information before it hits the cortex, to speed up our reaction time. Her research into the active roles the visual thalamus might play not only with regard to the visual cortex, but other regions of visual processing, earned her fellow status in the American Association for the Advancement of Science (2006), and her dedication to teaching and mentoring won her the Vanderbilt Brain Institute's Outstanding Teacher of the Year Award (2015).

Ann Graybiel (b. 1942):

In the mid-1970s, two women, using two different methods, discovered striosomes, which are chemical 'compartments' of neurons located in the striatum, a subcortical region of the brain involved in motivation, reward calculation, and planning. The first of those women, Candace Pert, we shall meet later, while the second, Ann Graybiel, we shall meet... right now! Her 1978 discovery of the histological distinctness of striosome regions of the striatum was followed by a whirl of papers investigating the properties of striosomes, including the role they play in addiction and habit formation (1990, 2018), the impact that their function as limbic controllers of the striatal matrix has on the progression of diseases like Huntington's disease and dystonia (2007, 2011), and the part they play in decision-making (2015, 2017). Graybiel also studied how information from the primary somatosensory cortex (the part of your brain that has regions dedicated for receiving sensory data form each part of your body) is remapped when received by the striatum, which sends the signal outwards on multiple branches that terminate in the putamen (which plays a role in motor planning, execution, and learning), the multitude of connections providing a possible explanation for our ability to crisply modulate our motion (1991).

Marsha Linehan (b. 1943):

Marsha Linehan is the creator of Dialectical Behaviour Therapy (DBT), which she first laid out in 1987 as a treatment for borderline personality disorder. Prior to Linehan's work, individuals with Borderline Personality Disorder (BPD) were primarily treated either through medication or psychoanalysis, with most behaviour therapists (excepting Ralph Turner) happily ceding the playing field. Not so Marsha Linehan, who developed DBT as a biosocial approach to helping chronically suicidal individuals. DBT is grounded in the notion of radical acceptance through mindfulness – of recognising the magnitude of the feelings one has, and experiencing them fully, but with a cultivated sense of perspective that disrupts the cycle of heightened emotion leading to distress about the existence of heightened emotions, leading to even further heightened emotions. Patients are aided in understanding and identifying their emotions, led in developing interpersonal techniques that validate their own self-esteem while paying heed to the needs of others, and taught methods for how to employ mindfulness to navigate extreme life situations and how to lay out coping strategies in advance of probable stressor situations. In a 2012 study of BPD treatments, DBT proved significantly more effective than standard courses in reducing patient anger, parasuicidality (attempted suicide), and improving mental health.

Sandra Bem (1944–2014):

Androgyny is not something we've had much cause to talk about so far in this book, and that's because we have not until now reached Sandra Bem, who was a founding figure in androgynous psychology. Her 1974 paper, 'The Measurement of Psychological Androgyny,' established gender identity as a spectrum, rather than a binary choice. In it, she outlined the Bem Sex-Role Inventory, a sixty question survey that allowed individuals to self-test what aspects of the reigning 'masculine' and 'feminine' traits they identified with. Some individuals naturally felt relatively equal pulls to both trait types, and were designated by Bem as Androgynous, a distinction she looked upon as an entirely healthy condition. In 1981, she posited her Gender Schema Theory by way of explaining how dominant gender conceptions percolate into and affect how individuals process their own identity, as well as how efficiently they process the often gendered information thrown at them by their cultures and societies. In a world where heterosexuality is the norm, for example, gendered 'traits' get charted out in a certain way that children learn to orient their gender in relation to as either sex-typed (their gender generally matching their biological sex), cross-sex-typed (their gender matching the opposite of their biological sex), androgynous (a balanced mixture of traits from both genders), or undifferentiated (a lack of identification with any particular gender traits). In a predominantly homosexual society, Bem points out, gender traits would be defined in potentially radically different ways, with corresponding consequences for gender typing. In subsequent research, Bem studied best practices for supporting and raising children whose gender identity was not sex-typed, what assumptions about gender constancy should be investigated in the early educational system, and on how gender schema undergird gender roles and workplace expectations.

Jean Lau Chin (1944–2020):
Jean Lau Chin died on 13 May 2020, from Covid-19, within days of the death of her husband of fifty-two years from the same disease. It was a tragic end that was overlooked by most in the dire hellscape of misery that surrounded it, which brought to a close a remarkable life. Chin was a clinical psychologist who took an active interest in diversity leadership and 'culturally competent healthcare' as ways to help improve the mental well-being and access to mental health services for populations caught in the cross-fire of globalisation. Like Reiko True, she saw that the health services of the United States were ill adapted to meeting the needs of their non-Caucasian populations, with pay-walls that restrict the access of poor immigrant families, and unresolved linguistic and cultural issues that prevent Hispanic and Asian families from receiving the same level of care as their English speaking neighbors and co-workers. For Chin, change had to start at the top, with institutional leadership that reflected the multicultural nature of modern global life, and health services that do not erect arbitrary barriers to care. She expressed these ideas in articles such as 'Culturally Competent Healthcare' (2000), 'Making Way for Paradigms of Diversity Leadership' (2010), and 'The Leadership Experiences of Asian Americans' (2013), and in her books *Diversity and Leadership* (2014) and *Community Health Psychology: Empowerment for Diverse Communities* (1998).

Elizabeth Loftus (b. 1944):
For some, Elizabeth Loftus is an inspiring psychological rebel who followed her scientific instincts to rewrite what we know about the permanence and reliability of memory so fundamentally that it kicked off slow but tectonic changes in our base assumptions of how the legal system can and should work. For others, she is an irresponsible maverick whose theories about repressed memories and the malleability of our recall capacities have consistently put her on the wrong side of some of feminism's most iconic moments. Reviled by the #MeToo movement for her testimony and consultation for the defense in the trials of Harvey Weinstein, Ghislaine Maxwell, and Bill Cosby, but revered by some emergent strands of feminism that are troubled by #MeToo's basic tenet of automatic belief, Loftus is not a figure it is easy to come to a unilateral conclusion about.

Luckily, we don't have to. We can put away all of the complicated nuance of what Loftus *means* for a future and presumably wiser generation to wrangle with, and concentrate instead on what she has *done*. In the mid-1970s, she began probing the edges of memory's elasticity. She showed subjects videos of car crashes, and noted how she could manipulate their reported memories of the videos by the way she posed her questions. Respondents asked how fast the cars were going when they *collided* reported an average of 32 mph, while those asked how fast they were going when they *smashed each other* reported an average of 41 mph. Memory, rather than a stable object etched in stone in the mind, seemed to be actively participating with its current environment, reshaping itself from hints and suggestions contained in pertinent questions and comments. You could not only get different responses by asking different questions, but that change would also impact how subjects would remember that event going forward.

This had drastic implications for the entire legal system, and its basis in witness testimony. Psychologists like Hugo Münsterberg had questioned the reliability of witnesses as far back as the late nineteenth century, but Loftus's experiments revealed new, almost unbelievable, levels of manipulability. One of her students, Jim Coan, when tasked by her with creating a method to implant false memories in individuals, developed the famous 'Lost in the Mall' experiment in which he presented his family with four events that they were to write down all of their memories of, the trick being that one of them, featuring his brother getting lost in a mall as a child, was entirely fabricated. Not only did the brother write down a number of further details surrounding the time he was supposedly lost, but when told that one of the four events was false, he could not identify that it was the mall event that never happened, and was shocked by the revelation. Loftus adopted the technique, and in a group of twenty-four participants, a full quarter of them expressed memories of the event, and when it was revealed that one of the memories was fake, a fifth of them could not identify the mall event as the intruder memory.

To Loftus, what Lost in the Mall and the Car Crash experiments revealed was how even the most well-meaning of people could unconsciously redirect memories towards desired conclusions, while ill-intentioned people could easily exploit these memory-nudging techniques to get the results they wanted. This caused her to write extensively on issues of hypnotic regression therapy, expressing skepticism about the neutrality of the hypnotherapist in the production of repressed memories, and on the reliability of reports obtained from police questioning. So far, so good, and few would argue that the way law enforcement often asks questions isn't due for a massive rethink, though critics do claim that the malleability Loftus found in her experiments is not necessarily extendable to the more durable and resilient memories of actual lived traumatic experiences. Where people tend to jump ship on Loftus is on her willingness to testify for the defense in cases where victims are recalling instances of rape or abuse. From her point of view, she is merely presenting the results of her life's research on the malleability of memory for the benefit of engendering a more just legal system. From the perspective of her detractors, her testimony, and the weight of her fame, muddy the waters in a way that tips the scales too harshly against plaintiffs already facing an uphill battle.

As the author of over 600 articles, 24 books, and an expert witness in over 300 trials, Loftus has been one of the most polarising and influential figures in modern psychology for half a century, but as to what history will remember about her, who can say? Memory is, after all, a highly malleable thing.

Eva Syková (b. 1944):
A former director of the Institute of Experimental Medicine of the Czech Republic's Academy of Sciences, neuroscientist Eva Syková is a leading researcher on techniques to deliver stem cells to damaged areas of the central nervous system, including the use of stem-cell seeded nanofibers and macroporous polymer hydrogels to bridge any nerve gashes created during injury to the brain or spinal cord, and act as templates for new neural growth. In experiments carried out in the early 2000s and 2010s, she reported significant recovery improvement using her hydrogel methods, and is currently engaged in clinical studies of ALS patients, and patients with spinal cord injuries.

Eve Johnstone (b. 1944):
Eve Johnstone is recognised for her pioneering studies of schizophrenia, beginning in the mid-1970s. In 1976, she employed computed tomography (CT) scanning technology, a form of X-ray based imaging that produces cross-sections of internal organs and which had only been available since 1972, to compare the brains of schizophrenic patients against a control group, finding that the schizophrenic patients had increased cerebral ventricle sizes (these are the cavities in your brain that hold cushioning cerebrospinal fluid), and that increases in size tended to map to increases in cognitive impairment. This was an important result not only for giving schizophrenia a new biologically grounded dimension (though we are still unsure as to whether the enlargement is a cause of schizophrenia or is caused by it), but for proving the efficacy of imaging technology for studying mental disorders. From this important early result, Johnstone went on to develop the Edinburgh High Risk Study, which followed teenagers who had strong familial histories of schizophrenia, in order to develop better early detection and treatment procedures, and which has resulted in the publication of some five dozen papers analysing its findings. She has also consistently been at the forefront of adopting new imaging techniques in the study of schizophrenia as they become available, including the use of MRI (1986), diffusion tensor imaging (2001), voxel based morphometry (2005), and functional magnetic resonance imaging (2006).

Gail Wyatt (b. 1944):
The first licensed Black woman psychologist in the state of California, Gail Wyatt has been for decades at the centre of the study of how the long shadow of slavery continues to cast a pall over the sexual experiences and expectations of Black women. As a sex therapist and psychologist, Wyatt has carried out multiple studies of Black women's sexual experiences, educations, and beliefs, and has discovered that, while different generations hold a number of different beliefs about political and social issues, that nevertheless there is a fundamental continuity to their sexual lives that speaks of large trans-generational forces insistently exerting their pressures on women who are often not aware of what is being done to them. In her book *Stolen Women:*

Reclaiming our Sexuality, Taking Back our Lives (1997), Wyatt details the effects of centuries of slavery in the United States, during which Black women were subject not only to regular rape from their owners, but to those owners' characterisations of Black women as fundamentally lascivious which served as their justification for their actions. Told for centuries that their nature was uncritically sexual, and that their responsibility was to just stoically accept whatever sexual assaults happened to them, Black women even after emancipation carried with them a sense of powerlessness to direct their own sexual destinies that make them still, in Wyatt's memorable phrase, stolen women, physically free of slavery but still mentally confined by the categories created during that era.

Wyatt locates early sexual education as the best way to emancipate Black women from their hyper-sexualised self conceptions. In her studies, she found that Black families traditionally avoided conversations about sex with their children, which she feels put Black girls at a disadvantage by placing them at the mercy of the reigning stereotypes about what Black women ought to be (which, at the time of writing of Wyatt's book, was centred around the hip hop music video paradigm which taught that Black women are, and ought hope for nothing more than to be, constantly sexually available members of a charismatic male's harem) without giving them the biological, historical, and sociological facts about themselves and their bodies to make educated decisions about their futures.

Patricia Arredondo (b. 1945):
Just as Reiko True and Jean Lau Chin took the lead in pointing out the shortcomings of the mental health community's accessibility to people of Asian heritage, Patricia Arredondo has been a steady advocate for more targeted counseling services and psychological resources for Latinx individuals, writing articles on developing standards for multicultural counseling competencies, running longitudinal studies of Latinx students' college experiences, and co-authoring the book *Culturally Responsive Counseling with Latinos/as* (2014). She is currently a professor at Arizona State University.

Chapter Twenty-Seven

Signs: Ursula Bellugi and the Neuroscience of Language

Sign Language is a grammarless series of bluntly defined iconic hand gestures.

Until William Stokoe (1919–2000) published his groundbreaking *Sign Language Structure* in 1960, this was overwhelmingly the opinion of linguists and psychologists regarding the complexity and potential for expressive nuance of American Sign Language. They saw it as a slow and cumbersome poor replacement for spoken English, and some even doubted its status as a language, based on theories that brains were hard-wired to understand language in a linear acoustic manner, and that anything lacking that modality couldn't even be considered a language.

Stokoe was among the first to attempt to study the basic structural elements of sign language, but its representation as a fully grammatical and self-consistent language independent of English had to wait until 1979, when Ursula Bellugi (1931–2022) and her husband William Klima published *The Signs of Language*, which presented the results of years of close study with native signers and pulled back the veil on the stunning richness and potential of non-verbal communication.

How Bellugi managed to find grammar where other researchers had only seen a blockish succession of iconic signifiers can perhaps be traced back to the circumstances of her youth. Born Ursula Herzberger in Jena, Germany, in 1931, her father was a mathematician and physicist, and her mother an artist, a combination of influences which, as we'll soon see, one can't help but think came into play in her later ability to discern fine gradations of motion and rigorously categorise them.

Being of Jewish ancestry, her father could not stay long in his position as a professor at the University of Jena with Hitler's rise to power and subsequent expulsion of Jews from academic positions. Fortunately, his former teacher Albert Einstein managed to find a job for him as an optical researcher at Kodak, in Rochester, New York, allowing the family to emigrate in 1934 and escape the fate that befell so many less well-connected Jewish individuals under Hitler's Final Solution. Of her youth, there is virtually nothing in the public record until her matriculation at Antioch College in Ohio in the late 1940s. Here, she met Piero Bellugi (1924–2012), an Italian conducting student who had recently travelled to the United States to study with Rafael Kubelik and Leonard Bernstein. The couple married in 1954, two years after Ursula received her Bachelor's degree.

At this point, Bellugi's activity again falls into something of a black hole. Over the course of the next five years, we know that she had two children, and that in 1959 she divorced Piero and maintained custody of the children. Her graduate studies in the 1960s, then, were carried out while also attempting to navigate life as a single mother, and in the face of institutions that were still not accustomed to the presence of women researchers. She pursued her doctorate in education at Harvard University, writing her dissertation under Roger Brown on negation in children's speech. While studying at Harvard, she was also taking linguistics classes at the Massachusetts Institute of Technology, where Noam Chomsky (b. 1928) was the rising superstar, and where she met Edward Klima (1931–2008), who had been teaching there since 1957 and was heavily influenced by Chomsky's biologically-centred approach to linguistics.

By 1968, the pair were married and had moved to La Jolla, California, where Klima taught at the University of California, San Diego (which had just enrolled its first students in 1964), and Bellugi worked at the Salk Institute (which was likewise less than a decade old at the time, having just been founded in 1960). She founded the Laboratory for Language and Cognitive Science in 1970, and soon set out as her first research question the problem of how language works for people who cannot participate in its acoustic and verbal components. Recalling this time later, she related how she knew so little of visual linguistic systems at the time that her first step was simply to look in the Yellow Pages under the 'deaf' category and dial the numbers she found there. Through these first rudimentary and tentative connections with the world of American Sign Language (ASL), and the research of Stokoe, she began laying out her program, which was to categorise ASL as a language, rather than as a brute system of one-to-one signs.

An initial hindrance in the study was the fact that she and her colleagues simply didn't know the depths of what they didn't know. In *The Signs of Language* she cites an experiment where different native signers were asked to relate in sign language the sentence 'His face became red in the wind.' Some signers seemingly reduced this sentence to the far less rich FACE RED, or WIND RED, as if confirming all the earlier prejudices that ASL is a shambling Frankenstein-like mode of just-barely-communicating. Bellugi, however, looked more closely, slowing down the footage of the signers to notice that signers changed how they delivered different signs to convey different meanings, that just as regular English grammar adds endings and explanatory phrases to words to convey extra grammatical meaning, so do signers modulate their delivery of the signs to give them simultaneous extra layers of grammatical signification. For example, a sign that means SICK when held stationary will, when rotated in elliptical motions, take on the meaning of 'gets sick often,' whereas, if delivered with tensed hands and more rapid than usual motions, it will take on the meaning of 'very or intensely sick.' A performer's modulation of base signs, then, conveys extra information that is totally opaque to outsiders, who are only reading the direct translation of each sign, but which can be elegantly picked up on and interpreted by native visual speakers.

Because producing words with hands takes physically longer than producing words through vocalisation, ASL has evolved to a hyper-efficient state whereby a great deal of grammatical structure and meaning are all offered simultaneously to the eye, so that while a speaker is still droning on with the sentence, 'The child was susceptible to repeated sickness,' the signer disposes of the same content in a couple of densely meaningful motions, resulting in rates of information transmission essentially no different from that of spoken language. Most importantly, there are rules. The modulation that turns SICK into HABITUALLY SICK, for example, cannot be used for the adjective UGLY, because it is only usable for states that one can fall intermittently into. You can be habitually WRONG, you can be habitually QUIET, but you can't be habitually UGLY – you are either are, or are not. These rules for which modifications can be used with which types of words, as well as other rules for how compound words work, and where location references can be used, add up to the grammar of ASL, a grammar which is based on maximising the efficiency of its mode of communication, and which is much more than just an importation of English rules, and which as such had to wait for discovery until someone came along who was willing to engage with ASL on its own terms, instead of repeatedly looking for English equivalents and, failing to find them, declaring that ASL lacked structure.

The Signs of Language was published in 1979, and was such a cornerstone of not only signed language studies, but of a re-evaluation of what languages can be, that it was reprinted in 1995, which is a rather rare thing in a field where technological developments and theoretical debates tend to bury their elders with grim rapidity. In the meantime, Bellugi was attracted by a new linguistic puzzle in the form of Williams syndrome (WMS), a condition that affects some 30,000 individuals in the US currently. Individuals with Williams syndrome have a deletion in chromosome 7 of a couple dozen genes that have a profound, characteristic, and unusual effect on development. Williams syndrome children often suffer from a narrowing of blood vessels in the heart that leads to a chain of other severe health conditions requiring vigilant medical supervision. They also have a reduction of IQ similar to that of Down's syndrome patients, but crucially have uncharacteristically sharp linguistic and musical competencies, and a hypersociality and openness to strangers that has caused Williams to sometimes be referred to as reverse autism.

Bellugi studied the performance of individuals with Williams syndrome as against those with Down's syndrome on a variety of tests and neural measurements to determine just how they can operate so well linguistically, when so many other features of their cognitive life were so drastically impaired. In one test, for example, children were given a storybook with no words and asked to create a story to fit the pictures they saw. WMS subjects evinced a rich array of prosodic story-telling devices to capture and hold the attention of their listeners, including relation of the imagined mental states of the characters, sound effects, and crafted dialogue between the characters. This affective richness of description, combined with WMS children's good performance on tests of facial recognition and memory, created something of a conundrum.

Whereas neuroscience had ready-to-hand models of individuals with good linguistic abilities but poor spatial cognition in the form of people with right-hemisphere brain damage, that damage also tended to go arm in arm with poor performance on affective narration tasks and facial recognition. WMS individuals didn't *quite* fit the right hemisphere damage model, and analysis of their brains, while showing a general shrinkage in brain mass that they shared with Down's syndrome individuals, did not show any right hemisphere lesioning to speak of, leading Bellugi to conclude that WMS represents a 'consistent but new and different biologically-determined neurobehavioural pattern.' Just as ASL users found their own way to language without the benefit of supposedly critical verbal and acoustic elements, so have the brains of individuals with WMS found their own unique way to uncharacteristic proficiency with language that has found a successful workaround for how language is 'supposed' to work on the neural level.

Throughout the 1990s and into the early 2000s, Bellugi published her findings on Williams syndrome and what WMS individuals could teach us about the development of language in articles and books which, when combined with her work on deafness and sign language in the 1970s and 1980s, summed to a remarkable half-century of linguistic research that earned her more awards than can be listed here, as well as a fellowship at the American Association for the Advancement of Science, the American Psychological Society, and the American Psychological Association, and membership of the National Academy of Sciences (2008). She continued working at Salk, watching the institute grow to a world-class centre of biological research, until her retirement in 2018 at the age of 87. She died peacefully at her La Jolla home in 2022, at the age of 91, leaving behind her vast new vistas for what language can be, if we step outside of our own assumptions long enough to observe, understand, and reflect.

FURTHER READING: *The Signs of Language* is a fascinating and eye-opening book that is pretty easy to find copies of, and will make you fundamentally rethink all of the aspects of language that you consider central to communication, but which are in actuality more or less artifacts of spoken language's limitation to linear expression. Information about Bellugi's life is thinner on the ground, and largely limited to the various obituaries published in 2022, of which the best is probably that of the Proceedings of the National Academy of Sciences.

Chapter Twenty-Eight

When Memory Has Gone: Suzanne Corkin's Journeys through the Hippocampus

Forgetting is the horrible, beautiful necessity that keeps the past from swallowing the present but that, given too free a hand, picks apart the very strands of selfhood. As recently as a century ago, since we didn't truly understand memory we had no idea of how to account for its loss. But then an irresponsible craze for lobotomising patients in the early to mid-twentieth century produced a new wave of humans with hacked brains whose various degrees of amnesia pointed the way towards a new model of memory and its afflictions.

No case taught us as much as that of H.M., whom we already met in the story of Brenda Milner, whose declarative memory was all but eliminated by an anti-epilepsy surgery when he was 27, and whose neural story was told over the span of multiple decades by MIT researcher Dr. Suzanne Corkin (1937–2016). She began her work with H.M. in 1962 and continued probing the intricacies of his mental world until his death in 2008, and in between shed vitally important nuance on our brain's remarkable variety of memory mechanisms.

Prior to Milner, Corkin, and H.M., there were numerous theories about how memory might work, dividing the various phenomena of mental storage into theoretical kingdoms that were difficult to prove or reject. Is episodic memory (memory about the succession of sights, sounds, and impressions that make up a moment in life) stored the same way as semantic memory (memory for facts and dates)? How does memory for the performance of a physical task (like soldering a circuit or juggling a ball) differ mentally from memory for people and places?

What was needed to start teasing apart the fibers of our capacity for information retrieval was a person with highly localised brain trauma whose memory could be tested to determine what parts of the brain were needed to remember different kinds of data. Unfortunately for humanity, but fortunately for the study of memory, the fad for lobotomies that swept America and Europe from the 1930s through 1950s provided just such people. The most infamous practitioner, Walter Freeman, could knock out dozens of lobotomy operations a day using an ice pick thrust under the eye socket.

I'll give you a couple of moments to process that.

For all the initial recklessness of its application, controlled use of lobotomy procedures showed some hope as a cure for people suffering from extreme epilepsy. Patients, cut off from a normal life by the omnipresent threat of grand mal seizure, and suffering the side-effects of prolonged exposure to anti-seizure medication, grabbed at the lifeline of experimental brain surgery and some found relief there.

Henry Molaison did not. Submitting to surgery at the age of 27, both hippocampi and a large portion of his amygdala were removed in an attempt to cure his epilepsy. While the severity of that epilepsy did decrease, it came at a terrible cost. Henry was unable to form new memories. He was trapped within a thirty second span of available recall, a 'permanent present tense' as Dr. Corkin described it in the title of her book detailing her decades of study of Molaison's case.

Let's stop and appreciate that for a moment. Think what it must be like to have no memory of what you did this morning, or five minutes ago, of seeing everybody around you always as strangers, of having no ability to plan for the future because you have no notion of what you've done. There is freedom there – freedom from the reliving of painful mistakes and loss, but also endless frustration, knowing that so much of you is lost to yourself. Henry Molaison, through that surgery, ceased being a promising and diligent skilled worker, and became H.M., neuroscience's most famous case study.

The examination of his capacities fell to Suzanne Corkin, who had begun her studies at Montreal's McGill University under Brenda Milner, investigating the role that the sense of touch plays in forming memories. Over the next fifty years, Corkin's studies of Molaison provided the crucial data to prove that the hippocampi are the areas of the brain responsible for the formation of episodic memory and that even episodes dating before the loss of the hippocampi can only be extracted as hazy outlines afterwards, while previously learned semantic memory stays largely intact. Further, she found that, though totally unable to recall newly learned information, H.M. did still exhibit certain unexpected abilities.

When asked to complete the names of famous people, he would sometimes come up with answers he ought not have known because they did not become famous until after his operation. Shown a picture of Billie Jean King, he wouldn't be able to bring her name to mind, but asked to complete the name 'Billie Jean' with any last name of his choice, he picked 'King'. How was somebody with no ability to recall even the last two minutes of his life able to unconsciously complete a famous name he had no business knowing? He was also able to, for certain experiments, perform them better each time he did them, regardless of not remembering ever having done them before. Using fMRI techniques to examine the processes of his brain, and cleverly designed experiments to isolate and probe rigorously defined types of memory tasks, Corkin detailed what information is and is not consolidated by the hippocampus, providing thereby a new appreciation of the realm of non-declarative memory and the richness of our cortex's ability to create its own forms of recall in the form of working memory, face perception, and task familiarity.

Though Corkin is primarily known through her studies of H.M., analysis of his abilities made up only about 22 per cent of her lab's output at MIT. The rest included equally vital if less narratively dramatic work that sought to define precisely how Parkinson's and Huntington's disease, which attack different sections of the brain's striatum, impair sequence and motor learning, thereby elaborating the striatum's role in the creation of forms of non-declarative memory. She also sought insight into the neurochemistry of aging by harnessing the newest imaging methods to document changes in white matter over the course of time, and how those changes have various impacts on the subject's ability to retrieve episodic, semantic, and task-centred memories.

Like her mentor, Brenda Milner, who was still actively involved in research into her nineties, Corkin continued her research past the age of retirement, only stopping when liver cancer claimed her life at the age of 79 on 27 May 2016. She thought of herself and Henry Molaison as partners in a great project to open before the world at last the physical mysteries of memory, and from their work together there have sprung thousands of researchers and tens of thousands of papers which have mapped the persistence, fragility, and most importantly, variety, of our strategies for storing the past.

FURTHER READING: *Permanent Present Tense* (2013) is Corkin's history not only of her time studying H.M., but also of the development of memory science from its early theoretical days right through to the modern age of 7 Tesla brain imaging technology, and contains more twists and turns of the sneakiness of information storage and retrieval than I could have ever included here. Just when you think you've got figured out what Henry should and shouldn't be able to do, she develops a new test which reveals unexpected abilities that themselves call for a new complexification of our sense of what stores what in the brain, and riding that roller coaster is a unique experience you should treat yourself to. For a decidedly different perspective, in 2016 Luke Dittrich published *Patient H.M.*, a book that sharply rebukes Corkin for her handling of H.M.'s life and legacy in a tone so harsh (especially considering the condition that she was in when he was interviewing her as a result of her cancer treatments) that 200 neuroscientists wrote in protest after the publication of an excerpt of the book, accusing him of grossly mischaracterising her personality and work.

Chapter Twenty-Nine

Filling in the Gaps: Naomi Weisstein's Active Brains and Activist Life

For the last thirty-two years of her life, Naomi Weisstein (1939–2015), the mercurial spirit who was simultaneously a cognitive scientist, rock star, psychological critic, and women's liberation pioneer, whose mind leapt so easily between disparate disciplines, and whose body seemed to contain boundless reserves of energy and will, lay bed-bound, a victim of a disease that, when she contracted it, had neither diagnosis nor prognosis, but that in one stroke severed her from the majority of the work by which she had defined herself for two decades.

Her parents were New York socialists who placed themselves somewhere on the Menshevik to Bolshevik spectrum of leftist thought, and Weisstein absorbed their political commitments, setting herself on a scientific trajectory from an early age, invigorated by Paul de Kruif's *Microbe Hunters*, the 1926 book about the first generations of germ researchers that also inspired both Marie Maynard Daly (the first Black woman chemist in the United States) and Gertrude Elion (the Nobel laureate who redefined how medicinal research is done) into scientific careers. She was the star pupil in all of her classes throughout elementary school and junior high, and attended the prestigious Bronx High School of Science, which had only begun admitting girls in the 1940s, and which has the distinction of being the world's leading secondary school in the production of Nobel laureates.

In these years, she leaned into her studies, and looked forward to a college environment that would allow her to focus on her work and not on performing to the expectations of campus fraternity culture, ultimately selecting Wellesley, which only just last year began accepting male applicants. She throve in the all-woman atmosphere, and earned admission to Harvard for her graduate studies, where she encountered the overt sexism of her male classmates, and characteristically met it head on, gathering friends to provocatively dance outside the windows of a university library that refused to let her study in it because her presence would be too distracting, and organising a blacklist of campus males whose behaviour had placed them beyond the pall of dating consideration for Naomi and her circle.

In 1964, Weisstein received her PhD (though she had to do much of the research for her dissertation at Yale because she was banned from using necessary equipment by the department's male professors for fear that she would break it), and the following year married the historian Jesse Lemisch, who was to be a supportive companion of her superstar years, and a steadfast caretaker through her years of struggle. Moving to Chicago, where Lemisch was a professor of history, Weisstein initially was led to believe that she would receive an instructorship at the University of Chicago, but ultimately, due to nepotism laws, had to settle for a non-faculty position as a lecturer.

Weisstein, though popular with the students at the University of Chicago, was something of a target to the psychology department, which tried to fire her, and when the college prevented them from doing so, kept her on at a low salary, such that when Loyola University offered to take her on, she was more than willing to make the switch, thereby beginning a period of almost unfathomable creativity across multiple fields. In 1968, she published 'Psychology Constructs the Female', which still ranks as a central text of Second Wave Feminism and of feminist psychology. The article doesn't so much say new things – her expressed skepticism about psychological standardised testing, as we have seen, was already present in the work of Jessie Taft and Nancy Bayley half a century previously, and her culminating statement of the importance of socio-economic expectations over biological factors in the development of gendered behaviour patterns had been laid out by Helen Thompson Woolley, Margaret Mead, and Karen Horney in the early twentieth century, and in fact stretches back through the ideas of John Stuart Mill and Mary Wollstonecraft – but in how well tuned this constellation of ideas was to the spirit of the time. The idea that the unconscious expectations of male psychologists as filtered through their clinical observations and test designs, when combined with the overt pressures for women to conform to certain social roles, have created a picture of women's behaviour and psychological capacities that aspires to monolithic universality while in actuality only being an expression of one historically-determined possibility, was an empowering one to the emerging Women's Liberation Movement as it charted the course of feminism's great second wave.

Chicago was one of the epicentres of women's liberation in the late 1960s and early 1970s, and here Weisstein took a pioneering and often punishing role as an organiser of feminist institutions, including the establishment in 1969 of the Chicago Women's Liberation Union, and the creation of the Chicago Women's Liberation Rock Band, which Weisstein led from 1970 to its dissolution in 1973. She was a popular lecturer on feminist topics, so popular in fact that she was asked to stop lecturing in compliance with the egalitarian spirit of the movement that all should have equal access to opportunities (an egalitarian spirit Weisstein had also to directly grapple with when deciding whether to keep spirited members of the band who couldn't sing or play an instrument, and thereby sacrifice the quality of the music and therefore the dissemination of its message, or drum them out to produce better records and live performances).

These were *also* the years of Weisstein's great studies in visual processing, which is what I want most to talk about, because it's the thing most biographical accounts of Weisstein gloss over with a couple of sentences vaguely saying that the work was 'important' without saying what it actually was. One of her first great results was contained in the paper 'Neural Symbolic Activity: A Psychophysical Measure' (1970), which described what happens when subjects look at a field of vertical black and white stripes which has the figure of a cube placed in the foreground, obscuring part of the field. Previously, studies of visual perception had focused on the initial stimulus and how it is received. Weisstein's study moved that process a step ahead, looking at what higher-order activities the brain might be engaging in while it is recognising the pattern in front of it. She found that subjects reported a change in contrast of the figure that they were looking at that was precisely consistent with what would happen if their brains had, behind the scenes as it were, recreated the parts of the stripe pattern that were obscured behind the image of the cube. It was an interesting result which suggested that the brain doesn't just passively accept the objects it sees, but actively engages with them.

She continued this line of thinking in her landmark 'Visual Detection of Line Segments: An Object-Superiority Effect' (1974) which found that individuals were much better at identifying which of four slanted lines they had been shown for a fraction of a second when that line completed an object that was recognisably a three-dimensional shape, and that our ability for line identification plummets precipitously as the object it is a part of loses its three dimensional appearing structure. In short, we have something in us that seizes upon objects as a whole, and devours them that way, which we can harness to identify specific parts in a way that is more effective than simply being presented by those parts on their own. Visual data isn't a single packet of information passed on down the line, but is a matter of different layers of processing all moving along at once, making this paper an important part of the visual theory of 'parallel processing' which holds that the brain has separate tracks for analysing the shape, colour, motion, and depth of the visual data it is receiving, which combine to form our understanding of the object we are observing.

One of the last scientific papers Weisstein published was 'A New Perceptual Context-Superiority Effect: Line Segments Are More Visible Against a Figure than Against a Ground' (1982) which continued her visual processing research, employing the famous face-vase optical illusion ('Do you see a vase or two faces?'). Subjects were asked to stare at a fixed point, and to alternately consider either the faces as the main figure or the vase as the main figure, and were then shown a flash of a diagonal line segment and asked to identify which segment they saw. Performance, as it turned out, was dramatically better when the segment was part of the illusion that the subjects were considering as the main figure, than when it was part of the 'background,' again suggesting that there is some extra level of processing to our visual experience which assigns extra processing to unitary figures than it does to whatever has been consigned to background status.

When this article was published, Weisstein had already been struggling for two years with a precipitous downturn in her health. In March of 1980, she experienced an episode of sudden and total weakness while trying to climb a set of stairs, and within a year she was compelled to use a wheelchair as her primary means of locomotion, and by 1983 she was almost completely bed-ridden, even the slightest action or stimulus utterly exhausting her. The disease that had so suddenly and completely changed the direction of her life was myalgic encephalomyelitis, or chronic fatigue syndrome, which had only first been described in the 1930s, and wouldn't receive its full diagnostic description until 1987, after outbreaks in Nevada and New York directed the attention of the medical community towards what had hitherto been considered a condition too mysterious and rare to devote too many resources towards (in the 1970s a rash of cases in London were deemed the result of 'mass hysteria' so little was the condition believed to be grounded in biological reality).

For the remaining thirty-two years of her life, Weisstein was carefully attended to by her husband, who would read to her the day's news every morning, and oversaw the limited contact with old friends and colleagues that her condition would allow her, while taking down notes for articles and memoirs that she dictated when she had the strength to do so. As if fate had not dealt her a cruel enough hand already, Weisstein was also diagnosed with ovarian cancer some two and a half decades into her bout with CFS, which ultimately claimed her life in 2015.

Her husband survived her by three years.

FURTHER READING: In 2020, Martin Duberman edited a collection of Naomi Weisstein's essays that is readily available, and contains a nice selection of her articles from the early days of Women's Liberation, and some of her reflections on her career, and the direction of modern feminism, but not so much on her scientific findings. If you are a member of the American Association for the Advancement of Science, you can access the articles I mentioned above for free, but if not they're something on the order of $30 a piece to look at a digital copy, because nothing advances AAAS's goal of progressing general scientific literacy like restrictive pay walls on half-century old papers. *Autobiographical Writing Across the Disciplines* (2004) is a bit harder to find but contains a good fifteen pages or so of autobiographical content written by Weisstein. The Chicago Women's Liberation Rock Band's album, *Papa Don't Lay that Shit On Me*, was released in CD format in 2005, and is available for download on whatever your favorite streaming service might be.

Chapter Thirty

Making Working Memory Work: The Multidisciplinary Neuroscience of Patricia Goldman-Rakic

You're a monkey, and somebody in a white lab coat has shown you a location where a delicious, ever-so-nummy, bit of banana has been placed, and then obscured your visual field. It is up to you, my good monkey, to hold the location of the food in your monkey mind as second by agonising second drags on so that, when the obstruction is at last removed, you can identify which location has the banana and claim your reward. To pull this feat off, you are going to have to access the capacities of your 'working memory' – the system that allows you to hold things in your mind for a few seconds, without necessarily storing them in your brain forever by placing them into 'long-term memory.'

Working memory represents a tantalising middle ground of semantic knowledge. There are things that we know because they are right in front of us, and things that we know because our hippocampi have made the effort to package them away in our permanent memory system, and then there are those things that we can't see, and don't remember forever, but DO remember for a short while, which begs the question, how does memory like that, which clearly must be using its own unique system, *work*?

Ask any neuroscientist who, more than anyone else, gave us the keys to unlocking working memory, and they will say, without hesitation, Patricia Goldman-Rakic (née Shoer, 1937–2003), the multidisciplinary researcher whose many-angled investigations of the prefrontal cortex beginning in the 1970s peeled back its mysteries down to the molecular level, and gave us our sense not only of where the capacity for working memory is located in the brain, but how the neuronal structures that make up that location allow us to hold onto knowledge for short but crucial spans of time.

Her mother was a Russian immigrant and her father was the son of Latvian immigrants, and when she was born with her twin sister Ruth on 22 April 1937, in Salem, Massachusetts, her parents could hardly have realised the heights awaiting her. With Ruth she attended Peabody High School, and the pair moved on together to Vassar, where Ruth began her path towards a career as an esteemed virologist and vaccine researcher, and Patricia graduated cum laude in neurobiology in 1959. She went on to receive her PhD from UCLA in 1963, and eventually found her way to the lab of Haldor Rosvold at the National Institute of Mental Health in 1965, a posting that would determine the course of her career.

Rosvold was experimenting with discovering the purpose and functioning of the prefrontal cortex through the application of targeted lesions on the brains of monkeys (and now is as good a time as any to once again step back and appreciate how much sacrifice of non-human life has been made in the quest for better understanding our brains, and how fortunate we are to be living in an age now where better and better imaging techniques and computer models are beginning to lead the way out of the absolute charnel house of neurobiology's past). Put somewhat brutally, this work involved damaging a portion of a monkey's brain, and then testing what deficits in function resulted. Goldman-Rakic, as it turned out, was a positive maestro of surgical technique, able to tightly target regions of the prefrontal cortex for lesioning (sometimes even in utero), and thereby to localise with great accuracy what parts of the PFC are responsible for what functions.

Her work determined that a caudal portion of the principal sulcal cortex is the region most responsible for the spatial aspect of working memory, which was an important result, but Goldman-Rakic wanted to know more, and with improvements in neural visualisation that were occuring in the 1970s, and which she went to MIT to study in 1974, she was able to use amino acids that had been tagged with tritium to find columns of neurons in the PFC which she recognised at once as being similar to those found earlier in the primary visual cortex (V1), which meant that the methods that had been developed for studying V1 could be brought to bear on studying the PFC. In 1975 she was made chief of developmental neurobiology, and began assembling a team of research all-stars (including Roger Brown and Tom Brozoski) with complementary but very different skill sets, her neurobiological Ocean's 11, to focus in at the cellular level on how the PFC does what it does.

The result was a chain of some of neurobiology's most cited papers, including 'Columnar Distribution of Cortico-Cortical Fibers in the Frontal Association, Limbic, and Motor Cortex of the Developing Rhesus Monkey' (1977), 'Cognitive Deficit Caused by Regional Depletion of Dopamine in Prefrontal Cortex of Rhesus Monkeys' (1979), and 'Development of Cortical Circuitry and Cognitive Function' (1987). What Goldman-Rakic (who added the Rakic to her name upon her 1979 marriage to neuroscientist Pasko Rakic) and her colleagues discovered was that the PFC had different regions for attending to the spatial features of a scene, than for attending to the *nature* of the objects within it, with the spatial neurons interconnected in a way that allows continued short-term firing of the neurons associated with one region which then trigger inhibition of neurons associated with other regions. If, for example, I know that the bucket with my beloved banana is located at an angle of 40 degrees to the left of me, then during the obscuring phase of my test, the neurons responsible for that region of my vision will keep firing on a loop, and sending out signals to other neurons responsible for monitoring other angles that inhibit them from firing, so that I can keep the one desired region active and accessible for when I need it during the reveal phase. The architecture and interconnectedness of the PFC, then, gave rise to its ability to temporarily hold information in our minds, and then to let it disappear without a trace.

These results would have been more than enough to retire on, but of course Goldman-Rakic did not stop there, and pushed her investigations from the cellular level down to the molecular one, investigating the chemical components of PFC function, and in particular on the crucial role that dopamine receptors seemed to play for the carrying out of working memory tasks. These studies on the neurotransmitters of the PFC, then, led to important insights into the nature of schizophrenia, which had until the 1960s been most commonly treated through lobotomisation, when it wasn't shrugged away as the result of bad parenting. Goldman-Rakic saw that some of the symptoms associated with schizophrenia were similar to those she had witnessed in her lesioned rhesus monkeys, and when pursuing the matter saw that the schizophrenic patients had marked decreases in dopamine receptors in their PFC, meaning that the intricate ballet of neuronal firing required to support the working memory system which she had discovered was essentially knee-capped for these individuals, preventing them from effectively holding things in short term memory, with resulting deficits in ability to communicate and function. Even something like forming a meaningful sentence is made much more difficult without the aid of working memory, which allows us to hold the different parts of a proposed sentence in place while we are linearly communicating them, and as to the ability to remember a phone number, the name of a person you just met, or even getting an object from a room that you need, deficits in working memory can add up to a much more challenging life.

Uncovering the neurochemical nature of at least part of the symptoms of schizophrenia went far to promoting a wider sympathy for those living with the disease, and certainly promoted a more targeted investigation into it as a complicated biological, rather than unfortunate social, condition. By the early 2000s, she was pushing forward the boundaries of our understanding of schizophrenia, and of the interconnectedness of the PFC with other regions of the brain, while the scientific community honored her with a slew of awards including the Merit Award of the National Institute of Mental Health (1990), the Karl Lashley Award of the American Philosophical Society (1996), the Ralph Gerard Prize (2002), and the Gold Medal for Distinguished Scientific Contributions of the American Psychological Association (2002). At the peak of her abilities, with her multidisciplinary approach to team-building and problem solving allowing her to attack the problems that interested her on multiple simultaneous fronts, yielding groundbreaking strings of scientific papers, she ought to have had another two decades to act as a wise, rigorous, and guiding force in neuroscientific research, but fate intervened on 29 July 2003, when a car struck her while she was crossing the street in Hamden, Connecticut.

Patricia Goldman-Rakic died of her injuries two days later, on 31 July and the neuroscientific community rose as one in a chorus of shock at the loss of such a talented colleague and good friend, and respect for all that she had done in taming the wilderness of the prefrontal cortex. When it comes to the great mass brain of civilization, most of us are destined for its working memory – present for a few lingering moments after our passing and then allowed to slip quietly into oblivion – but Patricia Goldman-Rakic, for all that she accomplished, and all that she taught us, will never have that fate. The individual who showed us how working memory works has ironically slipped its grasp, and nestled her way firmly into the long-term memory of our species, there to reside for as long as the human race continues living, and remembering, and occasionally hiding bananas from monkeys.

FURTHER READING: There were many obituaries and remembrances released upon the death of Goldman-Rakic in 2003, but I think the most complete and characteristic by far was that by Kavli Institute neuroscientist Amy Arnsten, who worked with Goldman-Rakic in the late 1990s, in the pages of *Neuron*. Goldman-Rakic wrote a number of papers summarising some of her most important work in an accessible form, and if I had to recommend just two which are available free from pay-walls, I'd say 'Cellular Basis of Working Memory' (1995) and 'Working Memory Dysfunction in Schizophrenia' (1994) are two good starting points.

Chapter Thirty-One

Life on the Grid: Nobel Laureate May-Britt Moser and the Fine Art of Knowing Where You Are

Bees know what they're doing. When they set out to build a hive, and have to decide what structure to use for the stashing of their honey, they adhere with instinctive diligence to a hexagonal pattern. Not only is this a very strong structure (hexagon junctions are three 120 degree lines, which is as mechanically resistant to pressure as you can get), but also happens to use the least amount of wall-making (and biologically expensive to produce) wax to store the most amount of honey. The other regular polygons that are able to seamlessly form a grid, like triangles and squares, don't come anywhere close in terms of efficiency to the stalwart hexagon. If you want to fill a space with a pattern, and want to do it seamlessly, there's no doubt about it, hexagons are the way to go.

It was with general delight bordering upon euphoria (as well as not a little sense of vindication for their life choices) that math nerds everywhere hailed the publication in 2005 of a paper by the Norwegian neuroscientific duo of May-Britt and Edvard Moser announcing the existence of grid cells in the entorhinal cortex which are responsible for producing a virtual map of the space that we are in which is organised in a hexagonal pattern. Just like a bee using hexagons to save wax for storing its honey, so does our brain employ hexagons to save energy for virtually storing its representations of the space in which we exist.

It is a magnificent result with the Mosers deservedly won the Nobel Prize for in 2014 (which made May-Britt just the 11th woman to win a Nobel Prize in Physiology or Medicine – the following year Tu Youyou (see the first volume in this series) became the 12th and most recent – out of 225 so far awarded), and to most onlookers it came seemingly out of nowhere. May-Britt was born in 1963 in the small town of Fosnavåg (population 3,621) on the Norwegian island of Bergsøya, where she was known as 'The Professor' by the townsfolk because of her seemingly inexhaustible curiosity about nature and its workings. Her father worked as a carpenter, and her mother studied medical texts in her free time, and so, though May-Britt did not grow up in a family of professional academics, she certainly grew up in an atmosphere of intellectual striving, and when she entered school she described it as an 'El Dorado' where she could finally ask all the questions she wanted.

She met Edvard when they both attended the elite *Ulstein Vidaregåande Skule*, a high school where they found themselves taking the same science classes together. Later, in 1983 while studying at the University of Oslo, they were able to reconnect, and married in 1985. At the time, May-Britt was unsure whether to follow her interest in mathematics, or psychology, or neuroscience, but with a chance to work at Terje Sagvolden's lab while an undergraduate on the neurochemistry of rat attention deficit disorder, she found that she had a gift for experimentation (when the Mosers had their own lab, it would be May-Britt who tended to work on the experimental aspect of problems while Edvard tackled the theoretical) and a fundamental interest in analysing behaviour through the lens of neural structure and chemistry. Soon, the pair were off to the office of Per Andersen (1930–2020), who had played an important role in the discovery of long-term potentiation in 1966, and was engaged at the time in research on the hippocampus. They asked/insisted to be allowed to work in his lab on the connections between hippocampal structure and function, a sizable undertaking that Andersen was skeptical of as an undergraduate project, but ultimately permitted. Their method was essentially to slice different parts of the hippocampus away and measure what happened, and they soon found, contrary to expectations, that the dorsal side has much sharper reactions to spatial activity than the ventral side, a fact which would play an important part in their later research.

During her graduate work, May-Britt had the opportunity of working at University College, London, with John O'Keefe, who would share the Nobel Prize with her and Edvard in 2014, and who in 1971 had made the momentous discovery of place cells in the hippocampus – neurons that fired when rats moved to particular recognisable locations in their environments. From O'Keefe, the Mosers learned key skills in measuring the activity of single neurons, which they would not be long in applying. In 1995 they received their PhDs in neurophysiology and were almost instantly offered joint positions at the Norwegian Institute of Science and Technology, where they had to build their lab essentially from the ground up, while also attempting to do right by their two young children.

The Mosers knew about the existence of place cells in the hippocampus which reacted to spatial features, but had a feeling that they were receiving their information from somewhere else, most likely the entorhinal cortex (EC) to which it is strongly connected. Previous studies of this connection had yielded only fuzzy results, but that was because they had been carried out on the EC's ventral side, which was connected to the hippocampus's spatially less reactive ventral side. They elected to use O'Keefe's single neuron monitoring methods, on the dorsal side of the EC, to watch what neurons light up as a rat explores new surroundings. Their setup consisted of a computer which displayed the locations in the test area where EC neurons fired, and interesting results were not long in coming. It seemed that there were locations in the room where, every time the rat passed through them, EC neurons would fire. The Mosers decided to move their rats to a larger test area to see if there might be a pattern to those firings and when they did so, sure enough, what they found was that the neurons were firing in a hexagonal grid spread over the test space.

It was too beautiful a result to be real, and the Mosers spent some time trying to determine if it might be a result of their testing equipment, but at the end of the day beauty won out. Rats navigate their world by the creation of a virtual representation of it, efficiently laid out in a hexagonal grid system, created by specialised cells the

Mosers dubbed 'grid cells.' They published their results in 2005 in *Nature* and went on to plumb the depths of this discovery, which attracted researchers from the world over to study with them in Trondheim. They found that, as you travel along the EC, different sections are responsible for the creation of different size grids, with each step along the EC producing a grid approximately 1.4 times the size of the grid before it.

To push the boundaries of their experimentation, the Mosers also became ace instrumentation designers, adapting and creating technology to simultaneously record the activity of hundreds and even thousands of neurons at once, including the development of miniaturised microscopes to allow the use of two-photon microscopy in free-moving animals. In 2015, the year after their Nobel win, the Mosers announced the discovery of 'speed cells' which, in combination with grid cells and the 'head direction' cells discovered by James Ranck in 1984, provide the suite of virtual spatial tools an animal requires to know its location in, and track its movement through, a given space.

The Mosers divorced in 2016, but continue working together, and in 2020 took up leadership positions at the K.G. Jebsen Centre for Alzheimer's disease, a logical next step as the EC is one of the first targets of Alzheimer's, a fact which explains why Alzheimer's patients tend to struggle with knowing where they are, and get easily lost, as their internal grids of their surroundings start breaking apart under them. By learning more about how the EC's structure allows it to create our sense of space, we can learn more about how Alzheimer's does what it does in the early phases of breaking apart that sense, and armed with that knowledge, we can begin to design practical defenses against it, so that all of us can keep track of where we are, where we have been and might just have a glimmer of insight into where we are going.

Chapter Thirty-Two

Dr. Tania Singer and the Neuroscience of Empathy

The year is 1990 and a man is sitting across from a monkey.

Between them is an object that will, in mere moments, become the Raisin Heard Round The World. This is the lab of Giacomo Rizzolatti, and the monkey is part of an experiment to determine what pre-motor cortex neurons fire in the performing of an action. By hooking an electrode up to a neuron and a loudspeaker and listening for activity, they can determine whether that neuron fires or not when a particular action is performed.

They have found one such which fires whenever the monkey reaches to pick up a raisin. Well and good. One down, several million to go. But this time things go a little differently. By chance, the scientist happens to pick up one of the raisins while the monkey is still plugged into the system, and the loudspeaker crackles with activity. Thinking it a technical malfunction, he does it again, and again the speaker lurches to life. A realisation slowly works its way through the team – the monkey neuron that fired when performing an event also fires when *watching* the event performed. The first 'mirror neuron' had announced its existence in rather spectacular style, and opened the floodgates for a radical new understanding of the interactivity of our mental life.

Subsequent research revealed conclusively that, when we watch actions being performed, groups of pre-motor neurons fire in our brain that we ordinarily use to perform those actions ourselves. When I see you do something, in a way, I am doing it as well, and that lets me understand your action intentions much better than if I worked through just abstract reasoning alone.

The natural next question, then, was how deep these mirroring processes went. Could they be used to gain more insight, for example, into our shared emotional lives? Dr. Tania Singer, now director of the Social Neuroscience Lab of the Max Planck Society, saw empathy as a particularly promising emotion to investigate. Subjectively, it feels very much like the mirroring we've been talking about so far. When I see you in pain, it somehow causes me pain as well. How, precisely, does that work?

The challenge was to find a neural explanation for the phenomenon by creating situations fraught with empathetic meaning. Singer's famous experiment, published in 2004, involved that ever-promising combination of romantic love and electric current. She took couples and attached electrodes to their hands. These electrodes delivered either non-painful or rather painful shocks to one or the other members of the couple. On a screen, each participant was able to see three seconds in advance how bad the upcoming shock was going to be, and who would be receiving it, all while an fMRI scanner recorded their brain activity.

It turned out that many of the same areas responsible for the pain response when the scanned participant felt a substantial shock also lit up when her partner did, that the brain relives to some degree the pain of others. More than that, the activation was more intense for those who, in questionnaires given after the experiment, described themselves as generally more empathetic.

It was a landmark result, but Singer was far from done. For her next investigation, she decided to see how issues of fairness and gender might play a role in neural empathetic responses. This time, she hired actors to run through an elaborate priming scenario with the participants. A participant would show up and play a distribution game with actors who were introduced as other participants. The game involved money sharing, and one of the actors was instructed to always be generous, and the other to always be greedy and unfair, in their distribution of the given money. By the end of the experiment, participants generally had a very positive conception of the fair actor, and a generally distrustful one of the unfair actor. Next, Singer replicated her earlier experiment, except instead of the participant being linked with her loved one, she was now fed information about the two players.

Then something really rather amazing happened. For female participants, the empathy-related brain responses showed up regardless of whether the fair or unfair player was receiving a painful shock. Her distrust or personal dislike of the person getting shocked played no role in the intensity of her empathetic response. Male participants, on the other hand, glowed with empathy whenever the Fair actor was shocked, but registered no response at all when the Unfair actor was and, in fact, showed marked activation in their pleasure-associated reward centres when they knew that the guy behaving unfairly was getting a nasty jolt.

It was a fascinating result that has since spawned a flood of interesting questions about when our Empathy Engines are engaged, and when they are left dormant, and evolutionary questions about why the difference between men and women was so substantial in this and subsequent experiments.

What I love most about Dr. Singer, however, is that it would have been the easiest thing to disappear down the hole of academic research on these topics, but instead she chose a very different path. She took an interest in the neural, psychological and philosophical differences between empathy and compassion, with an eye to using her biological insights to help people craft programs and environments that are more nurturing of compassion as our default response to each other's misery.

As Dr. Singer put it in an interview with *Psychologie Heute*, 'Empathy is quite generally the ability to share feelings with others: when you, for example, are hurt, or worried, or afraid, and I am standing there, then I as an empathetic human experience negative feelings as well. Such affective resonance is practically universal: pretty much everybody does it... Compassion, on the other hand, is a reaction to another's suffering from an entirely different world. We can verify this through brain physiology: when somebody is in pain, a compassionate reaction does not replicate the painful state itself, but rather produces feelings of concern and warmth as well as a motivation to help the sufferer.'

Empathy can be 'painful', and felt by the brain as such. Compassion, Singer claims, leaves a far less destructive toll on us, and if we could teach people how to attain compassionate neural states, the potential for easing the burden on doctors, caretakers, and those who have to regularly enter into the misery of others, would be of incalculable benefit.

In order to get a discussion going about how compassion might be taught, she has organised collaborative workshops between artists, sociologists, neuroscientists, and economists, one of the results of which was a book detailing the evidence for the teachability of compassion, and ideas for how to go about making a society that considers the inculcation of a compassionate standpoint a primary objective.

In 2008, Singer undertook a one-year longitudinal mental training investigation called the ReSource Project, which evaluated 90 different measurements on 300 different subjects in order to determine the neural, subjective, health-related and behavioural changes that meditative and mindfulness practice may have on us, while more recently, she has headed the CovSocial project, which has sought to evaluate the extent of the psychological damage created by the Covid-19 pandemic on the residents of Berlin.

It is the rare scientist who not only discovers new and profound things about human nature, but who also has the desire and ability to turn those discoveries into practice for the realisation of a better human community. Dr. Singer's science has inspired a wave of study about the limits and development of our interconnectedness on a neural level. What will the world look like once that wave of applied, compassion-oriented neuroscience has worked its wisdom upon the world? I don't know – but I'm looking forward to living in it.

FURTHER READING: Dr. Singer's classic 2004 paper, 'Empathy for Pain Involves the Affective but not Sensory Components of Pain' is the place to start, and if you're a reader of German, her 2014 interview with *Psychologie Heute* which I quoted above is available online. For a general history of mirror neurons and their role in empathy, I love *The Empathic Brain* by Christian Keysers, and if that inspires you to look more closely at the structure of neurons and the diversity of what they can pull off in concert, then Joseph LeDoux's *The Synaptic Self* is one of my favorite books ever, a sweeping account of the nitty gritty of brain function that is still accessible to an enterprising layman.

Chapter Thirty-Three

Dealing: Dr. Iris Mauss and the Science of Emotion Regulation

'Well Dale, we, the universe, hate to break it to you, but your desk is on fire, your copy of Thor 337 was lost in the mail, you've been assigned to teach Summer School, you have five articles due yesterday, and your cat ran away to be with a family she likes better than you.'

We have all had days like this. Life is an unavoidably stress-wrought venture, but some of us seem to deal with it rather better than others. Some will crumple in the face of a treacherous cat, and others will shrug it off with relative equanimity. The interesting question is, why? Is it a genetic ability, a cultural side-effect, or the result of a better approach to stress that is accessible to us all?

Though we've been dealing with stress as a species from day one, the science of its regulation is relatively new, but as both the speed and demands of society increase under the inexorable press of a globally organised urban planet, few developing sciences will be as critical to our sustained success and mental health.

And when it comes to emotion regulation, you can't go far without running into the work of Dr. Iris Mauss, currently at UC Berkeley's Emotion and Emotion Regulation Lab. Together with her students, she researches how humans interface with and direct their emotional lives, and how cultures can invisibly curtail our range of emotional coping mechanisms. For those working in high burnout, high stress professions (like emergency ward nurses or police officers), or even those facing the day to day grind of poverty or family care, it is research that could prove crucial in allowing them to function in spite of potentially overwhelming emotional demands.

Mauss was born in a small town in Western Germany, and grew up a literature junkie of the first order, devouring the psychologically complex novels of the French and German traditions. They inspired her with a love of psychology and how it asks big questions about human nature that are kept firmly grounded in the methodology of scientific investigation. She chose psychology as her university specialisation but was ready to fall back to the study of literature if she couldn't get accepted to one of the coveted psychology positions of her day.

It is our good fortune that she was accepted, and after some initial work investigating how environmental stress like noise affects health, she came to San Francisco to intern at a half-way house that gave psychiatric support to patients facing a return to the Real World after time as an inpatient. But clinic work requires a very particular skill set and personality type, and Mauss realised that her strengths and nature made research, not clinical work, the best choice for her.

When it came time to do her doctoral work, she came back to the Bay to study with James Gross of Stanford University. 'I was lucky to have a graduate advisor who not only mentored me in the science but also the intangibles that make or break a researcher.' Properly ensconced in the Bay Area, Mauss began her research into some of the big questions about humans and their mechanisms of emotion regulation, or the 'altering of one's emotions according to one's goals.'

'One of the things we have found is that emotion regulation seems to play a crucial role in determining people's well-being and health. In other words, people who are able to use emotion regulation well tend to be happier and healthier than people who don't.'

Interestingly, two seemingly contradictory strategies for emotion regulation appear to result in significantly higher resilience in the face of stressful events: Reappraisal and Acceptance. Reappraisal involves a reframing of the emotions surrounding an event. This is a conscious decision to choose 'positive' interpretations of events over negative ones, and those who mindfully employ it tend to rate their lives as happier, and their stresses less monumental, than those who don't.

The second strategy, Acceptance, is at first glance entirely antithetical to Reappraisal. Acceptance involves, as the name implies, accepting negative emotions in the short term, letting them wash over you – 'I liked that cat, and am sad I won't get to scratch his chin anymore, and all in all it sucks.' While prolonged negative emotions can lead to diminished health, an immediate and temporary acceptance of them can, it is theorised, mobilise 'meta-emotional' structures that diffuse the impact of the negative emotions and work to a person's long term good and psychological stability.

'We've found that accepting one's negative emotions is, somewhat paradoxically, associated with less negative emotions in response to daily stressors. In other words, the more you accept your negative emotions, the less negative emotions you tend to experience. Over time, reduced levels of negative emotions appear to add up to better well-being, almost like a healthy diet leads to better health over time. To people from a background in mindfulness this will not be surprising, but to many people used to thinking of negative emotions as "the enemy" and who have been socialised to aggressively pursue happiness, this may be a new consideration.'

Reappraisal seeks to recast emotions tactically, and Acceptance to accept them as they are. And both, it turns out, are right. By employing acceptance and then reappraisal, people can gain the benefits of both and significantly boost their resilience to stressors. The key lies in a re-evaluation of what 'negative emotions' are and mean. As the inheritors of a work-mad Victorico-Protestant ethic on one hand and the consumers of an advertising culture that blares constant happiness as both norm and ideal on the other, we have been taught from all directions that negative emotion is a thing to be denied and sublimated at every turn. To let it have its say for even a moment is considered a personal failure and social embarrassment. But there is harm in that, Mauss writes, not just in the manic pleasure-sating it idealises, but in how it lessens our ability to cope effectively with stress.

Culture, she maintains, plays a significant role in how we allow ourselves to deal with the slings and arrows of daily existence. 'Emotion regulation doesn't happen "out of the blue" or for no reason. The beliefs that people have about their emotions – how they should feel, what they should do about it, and what they can do about it – underlie when and how they will use emotion regulation. These beliefs are shaped by what we learn in childhood and through our culture. If we want to understand emotion regulation, we need to understand what beliefs people hold about their emotions.'

For that army of people who experience unavoidable stress on a regular basis, learning about the options they have to regulate their emotions may mean the difference between stability and breakdown. If we could teach the habit of emotional self-analysis and the psychology of resilience to doctors working with terminal patients, prison guards, social workers, and so many more besides, not only will it help them, but as a result our public spaces and services will be that much less grim, and the small tragedies of normal people broken by the demands placed on them by profession or circumstance that much less common.

Just as the ideas of Dr. Tania Singer about the light mirror neurons shed on empathy and compassion have in them the seed of a brand new, neurologically informed approach to a myriad of social services and professional structures, so does Mauss's work in emotion regulation have the potential to give us moderns a shield against the persistent thrum of our own manifold stressors. She and her colleagues have gifted us with a bit more insight into our best tricks and worst habits as emotional actors, interpreters, and manipulators.

Chapter Thirty-Four

Brains In Love and Brains Alone: The Social Neuroscience of Stephanie Cacioppo

One would think that there is no aspect of the brain's multitudinous biochemical majesty that lies outside of the interest of neuroscientists. There is an embarrassment of research riches packed into that 1,300 grams of neural matter, each topic surely affording a unique avenue of insight into the nature of humanity that any neuroscientist would be glad to call their own. And that is true, nearly. A budding young scientist proposing to research the topics of memory, spatial perception, or task planning, or the neurochemical roots of addiction, will find the doors of funding flung wide open for their approach. However, until quite recently, this was decidedly not the case for the biochemical origins and aspects of what is for many of us the most important part of our biological lives: love.

To propose research into what love is, neurochemically, what biological systems foster it, and which systems its processes in turn support, was for much of the twentieth century to be met with derision. Love was deemed either too complex a phenomenon to admit of meaningful targeted research, or too soft and nebulous a topic to attract sufficient funding for long-term study. And so, while we spent decades honing our knowledge of how experiences get enshrined in long term memory, or why our sense of directional hearing is so woefully inadequate when compared with other animals, we spent shockingly little time (with some bold exceptions, such as 1969's *Interpersonal Attraction* by Berscheid and Hatfield) answering the simple question, 'What is Love, what does it do for us, and what happens when it is gone?'

In the late 1980s a movement was founded which had as its goal the identification of the neuroscience underlying human interactions. Social neuroscience, as that movement is known today, made its way into the larger consciousness as a result of a 1992 article by John Cacioppo and Gary Berntson, published in *American Psychologist*. The article, and the research that gave rise to it, inspired a rising generation of students with the radical idea that one of humanity's defining characteristics, our intricate social lives and the flood of joys and anxieties they bring with them, was a fit subject for research, including the individual whose studies into the neuroscience of affection attained such heights of fame that she was dubbed Dr. Love by those covering her work, Dr. Stephanie Cacioppo.

Cacioppo was born Stephanie Ortigue in a ski resort in the French Alps, the only child of two parents who were madly in love with each other. Her mother was a professor of economics, and her father was the manager of a frozen-food business, and together they demonstrated daily a romantic ideal of mutual dedication and support. Young

Ortigue, for her part, was a person of solitude, who enjoyed her studies and challenging herself physically, but who never felt the social drive that seemed to characterise the adolescents around her. She was, as many only children are, something of a solitary dreamer, given to flights of imagination.

Perhaps because of her distinctly thoughtful character, she was the particular object of affection for her grandmother, who was Ortigue's most reliable source of social comfort growing up. When Ortigue was 9 years old, that grandmother died of a stroke, which inspired in the young girl a need to understand what had happened to the most important person in her life, in order to find a way to overcome it, so that nobody else would have to lose somebody as central to their lives as her grandmother had been in hers.

Ortigue entered the university system at an exciting time. The early 1990s saw not only the birth of the social neuroscience movement, but also the discovery of mirror neurons at the lab of Giacomo Rizzolatti. Between them, these two developments provided both the departmental motivation to look at the neuroscience of interpersonal relations, and a powerful explanatory tool for exploring the uniquely powerful empathy of primates.

Ortigue made the decision to do her PhD work at the University Medical School of Geneva, studying brain injuries and disorders, developing the all–consuming devotion to her research that would become the hallmark of the next two decades of her life. Sleeping too little and forming few social connections, all in the name of maximising her time at work, she compiled over the early 2000s a portfolio of work covering a broad range of neural phenomena, including alloesthesia (in which sensations experienced on one limb are 'felt' by the patient on the opposite limb), hemispatial neglect (in which an entire half of one's visual field is ignored by the brain), colour neglect, auditory hallucinations, induced speech arrest, and the neural origin of out-of-body experiences.

Over the course of her work in Geneva, she found that one thing which allowed patients with brain injuries to commit to therapy and succeed in rewiring their brain to compensate for lost functions was to tie that therapy into something that the patient felt passionate about. There seemed to be something about love and passion that engaged neural machinery which somehow augmented the body's self-repair processes. It was an intriguing phenomenon which she wanted to study further, and after surviving a barrage of professional advice to avoid the subject of love entirely in order to avoid committing 'career suicide,' she eventually transferred to Dartmouth College in 2006 to do just that.

The mid to late 2000s, then, saw Ortigue producing the work which we most associate with her name today, and which earned her the title of 'Dr. Love' in the popular press. She collaborated with Francesco Bianchi-Demicheli on a series of papers that sought to map the neural regions associated with love, affection, and sexuality, including early papers in 2006 on the neurophysiology of the female orgasm, and on 'The Power of Love on the Human Brain,' followed by papers on the cerebral networks surrounding sexual pleasure in women, the role that mirror neurons play in allowing the individual members of a couple to expand their sense of selfhood, the factors that allow humans to understand the intentions of others, and the neuropsychiatry of sexual desire.

From 2005 to 2012, Ortigue (who became Stephanie Cacioppo upon marrying the afore-mentioned John Cacioppo in 2011) turned the power of fMRI imaging onto the question of how the brain experiences love, and how the neural machinery of love marshals resources that the body can use to improve performance on a wide range of tasks. As opposed to the old adage that Love Makes People Dumb, Ortigue found that love can make us cleverer and heartier. Individuals primed with the name of a loved one performed better on cognitive tasks than those primed with the name of celebrities or acquaintances. Looking closer at what regions of the brain are involved in love, Ortigue found not only those regions one might expect, ancient seats of emotion we have long known are involved in our more primal reactions to our surroundings, but regions of higher-order thought as well, including the angular gyrus, a relative newcomer on the evolutionary scene which has ties to language, autobiographical memory, and imaginative thinking.

Love, it turns out, is smarter than we think, or at least stimulates parts of our brain which might cause us to think quicker and act more imaginatively than we could when motivated by lower order attachments like friendship. During this roughly decade long period of study, Ortigue showed us not only how our body is augmented by love, but broke down the components involved in the process of falling in love – how do lust and desire shade into love, and how does our distribution of neural stimulation change in that process? What does your brain pay attention to when looking for a sexual mate, versus a lifelong partner? (Some of the answers to that question might *not* surprise you.) How do love and mirror neurons combine to allow us to know another person's intentions, and to experience a loved one's reactions to situations on an almost instinctual level, as if they were our own?

After marrying Cacioppo in 2011 and moving to the University of Chicago to work with him at the Brain Dynamics Laboratory, we see a definite shift in Ortigue's research focus, towards topics more aligned with his research interests. Over the course of the 1980s, while Ortigue lived out her life as the dreamy apple of her grandmother's eye, Cacioppo was developing the field of social neuroscience, culminating in his work on how loneliness is experienced by the body and mind. This was important work for a world in which people grow daily more isolated from each other, and Ortigue's papers of the late 2010s increasingly focused on his chosen field of the neuroscience of loneliness.

John Cacioppo passed away in 2018, and in 2022 Stephanie published *Wired for Love: A Neuroscientist's Journey Through Romance, Loss, and the Essence of Human Connection* which tells the story of her early research into brain injuries and the neuroscience of love, interspersed with autobiographical material centred on her relationship with John. She continues her work at the University of Chicago Brain Dynamics Laboratory, directing her group in studying the impact of emotions on human performance, and further delineating the physical and psychological effects of loneliness on individuals.

Chapter Thirty-Five

Achtung, Brainy: Grace Lindsay and the Mathematical Modeling of the Human Brain

You are placed in front of a screen that is black save for one spot of red in the centre, and are told to focus strictly and solely upon that red dot. You think, 'This will be no problem. As a college educated, long time subscriber to, and very occasional reader of, the most sophisticated periodicals, I can surely count on myself to focus on a single red dot.' So, you're looking at the dot when suddenly a bright yellow triangle appears at the bottom left of the screen and zips to the bottom right. Your eyes betray you, and dart to the enticing new event, a betrayal that lasts a fraction of a second, but is perhaps all the more profound for its brevity.

What happened? Why did your focus waver, against your will? It was a question that, a century-and-a-half ago, we could only broadly lunge after armed with the self-reflective insights of philosophy and the nascent science of psychology. Over the ensuing years, however, there has emerged an entrancing dance between two disciplines, mathematics and neurobiology, that has allowed us to harness the magisterial modeling capabilities of math to the breathtaking complexity of the human mind in order to finally start rigorously answering such seemingly basic questions like, 'Just what am I *doing*, when I'm "focusing" on something?'

The history of computational neuroscience is filled with colourful characters, of mathematicians who dared to leave the safe fields of abstraction to wade into the squishy, goopy irregularities of biology, and of equally stalwart biologists willing to forego the beautiful chaotic messiness of their chosen field and see the value in idealised models. With one foot in each of two highly demanding intellectual fields, computational neuroscientists are a rare breed, but rarer still are those who can not only do the research which deepens our understanding of how billions of neurons add up to make a brain, but can effectively communicate those insights to those of us who don't happen to be multi-faceted geniuses.

Dr. Grace Lindsay is one such individual, though given her list of activities my working hypothesis is that she is in fact three such individuals. Her 2021 book, *Models of the Mind: How Physics, Engineering and Mathematics Have Shaped Our Understanding of the Brain* is a grand tour through the history of computational neuroscience, from its humble beginnings in information theory and neuron structure up to its modern manifestations harnessing supercomputers to run large scale convolutional neural networks that model important brain systems. It is a profound book which has already entered the pantheon of classic general reader neuroscience texts like LeDoux's *Synaptic Self* or Montague's *Why Choose This Book?* which don't shy away from close detailing of important experiments but also don't lean on unapproachable jargon to convey their nature or import.

That ability, to rigorously explain complicated ideas to a general audience, is one Lindsay honed over four years of co-hosting the podcast *Unsupervised Thinking*, which in forty-nine episodes from 2015 to 2019 dove into the deep history and promising future of neuroscience. Those episodes are all still available and worth listening to not only for the insights they reveal into the state of modern computational neuroscience, but as an object lesson in how scientific podcasts should be done. Neither a slickly-produced but wafer-thin science appetiser, nor a stodgy exercise in impenetrable academic posturing, *UT* is exactly what you want – three colleagues sitting around, talking honestly about the thing they've devoted their lives to, showing us not only how new ideas and procedures are formed, but the more fundamental issue of how to pose new questions and evaluate their potential fruitfulness. It's the sort of skill we don't tend to teach very well in our high school curricula, and just listening to Lindsay and *UT*'s other hosts working through how to best formulate the questions they would like to see answered is a great example of fundamental scientific thinking that should be required listening in science classes.

Good SciComm is one of the most important things a gifted scientific writer can do in these troubled times, but it's only one string in Lindsay's bow, for primarily she is a researcher who caught the computational neuroscience bug while at the University of Pittsburgh when she attended a lecture by Dr. Brent Dorion as an undergraduate. Dorion demonstrated how neurons can be modeled by mathematical functions, thereby revealing to his audience a new way to think about why the brain is able to do the things it does, in terms of the flow of information and the physical connections needed to optimise the equations that describe it.

After some time in Freiburg at the Bernstein Centre where she attempted to narrow down the particular area of neuroscience she wanted to model, she arrived at Columbia University, where she turned her focus on the mechanisms behind the phenomenon of 'attention,' which brings us squarely back to that red dot that we began with – what do our brains do when we try and 'focus' on the red dot, and how are those processes potentially de-railed by the arrival of new events?

It turns out that our brains are the scene of a rich and continuous tug-of-war between 'top-down' and 'bottom-up' processes – between the areas of the brain which want to direct what we should be paying attention to, and those which are predisposed to react to certain features of the environment and excitedly pass that information upwards regardless of what the official directives from the top might be. Those bottom-level neurons which react preferentially to particular features and events – to, for example, lines and motion – are fascinating in their own right and we'll get to talk more about them in the portrait of Jennifer Groh, but Lindsay's work on attention has been primarily concerned with how the brain is able to 'pay attention' to a particular object or feature, and how to build computational models which are able to represent how attention is organised and what benefits it produces.

That work centres upon evaluating the feature similarity gain model (FSGM) of S. Treue and J.C. Martinez-Trujillo, which holds that one way that your brain organises 'attention' is by giving preferential weights to those neurons which react more strongly to the features that characterise what the brain is trying to get us to pay attention to. If you're staring at a screen and need to press a button every time a green blip shows up, your brain would do well to make sure that any neurons which respond particularly strongly to green get amplified, while those that do not get subdued.

Lindsay put this theory to the test by building a convolutional neural network (or CNN – and if every time you read the phrase 'convolutional neural network' from here on out you do so in the voice of James Earl Jones, you're doing it right), a vast computer model that is often used to mimic the connectivity of mammalian visual systems. She gave it the task of identifying whether particular objects were located in blended or compound pictures. Without any extra 'attention'-type architecture built in, the CNN was still able to, more often than not, respond correctly to the images fed into it. However, after building in a preferential tuning system, and experimenting with different layers of implementation for that system, she was able to produce a neural network that responds to visual stimuli and categorises them with near-human levels of accuracy, teaching us thereby both more about how we pull off the neat chemical trick of focusing on particular features, and how we can build machines with better optical recognition functionality.

Lindsay's work continues at University College London, where she has recently taken Josh Merel's virtual rodent, developed to understand motor systems, and employed it in the study of different theories of how appropriate behaviours are developed through various learning methods, and how the transfer of previous knowledge and behaviour developed to confront one type of task aids an individual in the performance of new, similar tasks. It's exciting work from one of this generation's most promising minds, and most gifted communicators, whose words and ideas we can all look forward to guiding us to a better knowledge of our squishy, squishy brains and the exquisite mathematics underlying them in the years and decades to come.

Chapter Thirty-Six

Brief Portraits: The Modern Age

Jessica Benjamin (b. 1946):

By 1988, psychoanalysis was due for a reboot, and in no aspect more than its conceptions of gender. Even though a generation of women psychoanalysts had devoted themselves to the problem in the post-Freud years, some persistent Victorian encrustations still clung tenaciously to the hull of the Good Ship Psychoanalysis, awaiting the right tool to scrape them off without simultaneously leaving so many holes behind as to render the vessel unseaworthy. Jessica Benjamin's *The Bonds of Love: Psychoanalysis, Feminism, and the Problem of Domination* (1988) and *Shadow of the Other: Intersubjectivity and Gender in Psychoanalysis* (1997) attempted just this task, bringing into Freudian psychoanalysis subsequent developments in philosophy that highlighted the Lacanian problem of what we do when faced with The Other, and how people develop roles of the dominator and dominated out of failures to recognise the subjectivity of other individuals, which has effects that ripple through all aspects of society, from gender relations to large scale institutional structures. Recast in this light, a number of the Freudian conflicts that seemed archaic by the late twentieth century took on new life, with the deep conflicts between children and parents recast as struggles for individuality against domination.

Shelley Elizabeth Taylor (b. 1946):
Top of the Head Phenomena. Positive Illusion. Tend and Befriend.

The work of Shelley Taylor is full of catchy sounding ideas that contain within them deep truths about social cognition, i.e. the ways in which we interpret and process the actions and intentions of other people, which in turn inform how we perceive the rules of social conduct. Since the late 1970s, she has been testing how different social settings, and the particular location of subjects within those settings, affect individuals' interpretations of the events they view, and, over time, work changes in their actual mental makeup. From over-estimating the importance of a designated 'leader' in the causation of what is happening around them (the top of the head effect), to automatic mental grouping of individuals within a setting via easy and ready-made categories of gender or ethnicity, to altering perceptions about what an individual's motives might be based on how much that individual resembles the group you see them in, to the long term health benefits of having positive if totally unrealistic notions of yourself and your future (positive illusions), to the long term neurochemical and neurophysiological detriments that result from prolonged exposure to stressful social structures, Taylor has led the way in showing us how much the structure of our surrounding society penetrates our minds and pushes us towards particular unintended perceptions of other people, which largely unconsciously flavor how we approach reality and interpret the chains of causation and motivation within it.

Candace Pert (1946–2013):
We met Candace Pert above as the individual who, in 1976, used autoradiography to discover striosomes, some two years before Ann Graybiel did so using histochemical methods. That is a big thing, particularly as we continue to uncover more about the role that striosomes play in different neurological diseases, but it isn't even the biggest thing there is to know about Pert. That would be her discovery, three years before striosomes, of the first opioid receptor, while working in the laboratory of Solomon Snyder. Though Pert was listed as lead author on the paper announcing the discovery ('Opiate Receptor: Demonstration in Nervous Tissue' (1973)), when it came time to award the Albert Lasker Award for the breakthrough, it went to Snyder, without even a citational mention of Pert.

The discovery of the first opioid/opiate receptor launched an entire new multi billion dollar industry devoted to identifying receptors and designing drugs to target them. Pert would turn her expertise to the development of targeted drugs, notably an early HIV inhibitor, but was known increasingly for her deeper dives into the implications of her findings. She interested herself in the interconnection between the different bodily systems through neurochemical intermediaries, and began to conceive of emotion as, at its heart, a communication between the body's systems, allowing her to conclude in her famous formulation, 'Your body is your unconscious mind.' Because of the interconnectedness of thought, emotion, and body through the intermediary of neuropeptides, it no longer made sense from Pert's perspective to talk about the old Cartesian mind-body duality, but rather we should start talking about a unified entity, or 'bodymind,' in order to allow us to harness the modalities of the body to aid us in solving the problems of the mind. Certainly, in the wrong hands these ideas could nudge the psychological sciences into dangerously nebulous pop-spirituality waters, but they also contained within them profound truths about the interconnectivity of bodily processes and mental states that have worked real and long-lasting good for millions of individuals, who have benefitted either from the drugs that her insights into peptide-receptor relations made possible, or the therapies developed to bring the chemical 'wisdom' of the body to the aid of the battle-worn mind.

Hermona Soreq (b. 1947):
A molecular neuroscientist with the Hebrew University of Jerusalem, Hermona Soreq is a world authority on the neurotransmitter acetylcholine and the mechanisms of its regulation. Acetylcholine is important both as the neurotransmitter that allows motor neurons to tell muscles to contract, and also as a brain chemical that plays a role in alertness, memory, REM sleep, and learning. Soreq's group (called ... the Soreq Group) is at the forefront of pushing new molecular techniques to investigate the function of acetylcholine, including the roles of non-coding RNA such as microRNA and transfer

RNA fragments in controlling the expression of acetylcholine related genes. Under light stress, our brains increase acetylcholine production, but maintaining the right balance is tricky, and under intense stress we tend to err in the direction of preventing excessive buildups of acetylcholine by producing large amounts of the enzyme that breaks it down, the problem being that the effect lingers long after the stressful situation is over, causing people who experience regular stress to have longer-term lowered acetylcholine levels that can lead to corresponding neurological problems.

Linda Buck (b. 1947):
Nobel laureate Linda Buck is the individual who told us how noses work. For other senses, the simplicity of the communicating medium suggested early the mechanism of the sense – sound is about vibrations through a fluid medium – the ear has little drums inside of it connected to nerves. Seems simple enough. Light is carried by wave-particles with energies corresponding to their frequencies. Trickier, but essentially there's one variable that determines the difference between photons, which is suggestive of how we might be structured to distinguish colours and intensities. But smell? SMELL? It's a sense which has to give us meaningful information about the vast and profound world of molecules around us, legion in their geometries. How do we pull *that* off?

In 1991, Linda Buck and Richard Axel published 'A Novel Multigene Family May Encode Odorant Receptors: A Molecular Basis for Odor Recognition' in *Cell*, and tied up the whole mystery in a neat bow. They discovered a group of hundreds of genes, each of which codes for a unique olfactory protein that, when it engages with one type of molecule, can pass a signal on to the olfactory bulb, alerting us that that particular molecule is present in the air. When you inhale a perfume, you are inhaling a rich mixture of different molecular geometries, each of which nestles into its own tailor made (or I should say gene made) receptor in the nose, causing each of those receptors to communicate the encounter to the brain, which takes the aggregate information and lets us know what we are likely smelling. For unlocking the puzzle of olfaction, Axel and Buck shared the Nobel Prize in Physiology or Medicine in 2004, which is just one of a half dozen high-profile prizes Buck was awarded in the wake of her landmark paper.

Paula Tallal (b. 1947):
Once we learned about the remarkable dynamism of our nervous systems, with their capacity to form new connections and assume new roles in the facing of changing situations (a capacity we call neuroplasticity), one of the immediate questions we had was how we might use this capacity to help people. Tallal was on the forefront of employing neuroplasticity to help people suffering from language-learning impairments. In her 1996 paper, 'Language Comprehension in Language Learning Impaired Children Improved With Acoustically Modified Speech' she put to the test her theory that what held a number of children back in their language learning was an inability to keep up with the speed of sound change of the phonemes within words. While the brain is still trying to figure out whether the syllable 'pa' or 'ba' was just thrown at it, the word

is continuing to march forward, creating ever-growing hurdles to comprehension. Tallal developed new technology that created artificial speech which exaggerated the differences in the acoustic properties of different sounds, and a program of exercises designed to train children's minds to distinguish those sounds more and more quickly, until their brains had reprogrammed themselves to understand regularly spoken language at regular levels. She co-founded Scientific Learning Corporation in 1996, a company which produces the Fast ForWord linguistic cognitive skills programs that have Tallal's research as their motivating core. In 2012, Tallal was named Inventor of the Year by the New Jersey Inventors Hall of Fame (insert Jersey-disparaging joke here), and has since seen Fast ForWord employed by language professionals in over forty countries.

Mary B. Kennedy (b. 1947):
Why do we remember some things more easily than others? If memory is just a matter of neurons connecting to other neurons, it seems like we should be able to access everything, always. But clearly that doesn't happen, which is why I never spell the word occasional the same way twice. One of the key components of how our brains solidify certain neural pathways was discovered in 1966 in the form of Long Term Potentiation (LTP). Essentially, part of what happens when you 'learn' something is you make certain connections between neurons more efficient – receiving neurons grow more sensitive to signals from neurons they 'talk' to more often, and grow more structures to receive messages from them. This allows neuron A to fire neuron B more easily, and allows us to correspondingly more easily access whatever it is that neuron B is connected to. One of the key molecules that lets neuron B make itself more receptive to neuron A is CaMKII (calcium/calmodulin-dependent protein kinase II), which was characterised by Mary Kennedy in 1983, and which makes up a solid 1–2 per cent of the brain's protein. The units that receive glutamate, the most common neurotransmitter we use to excite neural activity, are called AMPA receptors. The more you have of them on a receiving neuron, and the more sensitive they are, the more likely you are to fire when neuron A throws glutamate molecules at you. CaMKII plays a role in both making AMPA receptors more sensitive, and in the process of dragging new reserve AMPA receptors to the surface, and so lies at the centre of the whole LTP process.

Kennedy also discovered, in 1992, the protein PSD-95 which, together with PSD-93, is a major component of the postsynaptic scaffolding that uses its three PDZ domains (regions of 80ish amino acids that are really good at holding receptor proteins) to anchor important synaptic receptors like our friend the AMPA receptor, as well as NMDA receptors and potassium channels. So, the next time you find yourself remembering that one actor from that one movie you like, give a silent word of thanks to Mary Kennedy, who sleuthed out some of the major bits of architecture that make all of that recall work.

Carla Shatz (b. 1947):

If you know just one catch-phrase from the realms of neuroscience, it's probably 'Cells that fire together, wire together,' and if you know two, the second is probably, 'Out of sync, lose your link,' mottos which both originated from the ground-breaking work of Carla Shatz. Her studies of vision in the 1980s led to discoveries which were highly resisted in her time, and which form the basic lingo of our neuroscientific awareness now. When she first began her research, the fine tuning of vision, its capacity to take signals from adjacent retinal cells and flawlessly deliver them to adjacent regions of the brain for analysis, was so tightly controlled that it was assumed to be a hard-baked feature from our genetic code. Shatz found that, on the contrary, the vision we emerge with from the womb is an undeveloped, haphazard affair, and that our superb visual organisation emerges over time by spontaneous firings of retinal cells which create 'retinal waves' in their neighbors. Over time, these spontaneously generated waves cause neighboring cells to wire together and coordinate, producing that highly localised specificity of vision we take for granted daily.

We saw, in the story of Mary Kennedy, a little bit about how, mechanically, jointly firing neurons can become wired together, but that leaves the story about just how neurons that fall out of sync lose their link. That seems to imply some sort of pruning mechanism, whereby two retinal neurons that have figured out they aren't really associated with each other sever their ties in a semi-permanent manner so that their signals don't get crossed up. Shatz discovered this mechanism in an unlikely place – the MHC1 molecule. Recognised as an important immune molecule, it was widely believed to be relatively unimportant for the brain, and when Shatz presented her evidence that MHC1 played a role in preventing connections from forming between two neurons, many simply refused to believe it, until she subsequently also discovered PirB, a protein that MHC1 works with to lock out neurons from communication with each other.

The neuroscientific community is now fully on board with Shatz's flurry of inspired work from the last three decades, and has presented her with numerous awards in recognition of its importance, including the Gruber Prize in 2015, the Kavli Prize in 2016 and the Harvey Prize in 2017.

Riitta Kyllikki Hari (b. 1948):

Finnish neuroscientist Riitta Hari has published over 300 articles, mostly focused on the use of magnetometers to detect minute electrical changes in the brain during different activities (a branch of neuroscience called magnetoencephalography, or MEG). Believing that some of the most interesting things about human brains happen when they are engaged with other human brains, she was among the pioneers developing 'two-person neuroscience', and in 2012 constructed an MEG setup that allowed the simultaneous recording of brain signals of two individuals engaged in different activities with each other, to gather insights about socially produced emotions, and the neurophysiology of social interaction.

Star Vega (1948–2004):
Besides being the winner of the Best Name In This, Or Any Other, Book competition, the Mexico-born Star Vega was, like Patricia Arredondo, a psychologist deeply engaged with improving access to mental health resources for the Latinx community. One of the pieces of information you often see about her is that she was a co-author of the 2002 paper 'Translating Research into Action: Reducing Disparities in Mental Health Care for Mexican Americans' which *was* an important paper about how to actually, in practice, improve mental health care access for Mexican Americans, but which was co-authored by *William* Vega, not *Star* Vega. Star Vega was the first Latina to serve as president of the California Psychological Association, and her 1983 dissertation lay at an important junction between labour policy, psychology, and culture. In that piece, she noted that the awarding of worker's compensation in the United States is largely based upon the MMPI, a standardised personality test, which is meant as a tool for weeding out 'fake' claims. That test, she showed, has built-in biases in its psychological assumptions which mean that Mexican-Americans habitually receive results that translate into fewer honored worker's comp claims when compared to Anglo-Saxon claimants. The connection of culturally-skewed practices in psychological testing to flawed outcomes in industries that base their policies upon that testing was an important one, and in 2002 she received the Heiser Award for Advocacy.

Svetlana Alexandrovna Dambinova (b. 1949):
With over 350 papers, 7 co-authored books, and a couple of dozen patents to her name, the Siberian born Svetlana Dambinova ranks as one of the most prolific modern neuroscientists specialising in neuroreceptor research. At around the same time that Mary Kennedy was doing research into the mechanisms that increase the number and efficiency of glutamate receptors, Dambinova was studying how those receptors play a role in different brain pathologies. In ischemic stroke, a lack of oxygen to the brain can wreak havoc as electrochemical gradients start breaking down, causing glutamate and calcium ions to flood across neuronal barriers, eventually stopping the ability of the neurons to carry electrical signals, and ultimately leading to cell death. One of the things that can happen in these episodes is a surge in auto-antibodies (antibodies that attack your own cells instead of foreign ones) that target our NMDA receptors (which is, like AMPA, a glutamate receptor) and damage them, producing NR2 peptides in the process that Dambinova realised could be used as reliable markers of whether a stroke event was an actual ischemic episode, or a stroke-like phenomenon calling for different treatment.

Dambinova has also studied the importance of neuroreceptors in addiction, the value of deep intracranial electrical stimulation, the use of AMPAR peptides (products of the breakdown of AMPA receptors) as markers of concussion-style head injuries, and the development of new chemical techniques for detecting injury or ischemic effects in the spinal cord.

Baroness Susan Greenfield (b. 1950):
As a parent with some sense of history, I have developed a certain outer layer of resistance to each new wave of claims that 'X is destroying the minds of children.' First it was comic books, then rock and roll. In my youth, it was video games, and in that of my own children YouTube and social media have, between them, been called out as the elements of twenty-first century life that are definitively rewiring their brains for the worst. The question then becomes, *is* the mind being changed under the influence of social media manipulation, or is this just the latest in a string of criticisms that the older generation always launches against those activities of the younger generation that it doesn't understand? A key player in this debate has been the figure of Susan Greenfield, an experimental psychologist known in scientific circles for her work on therapies for Alzheimer's disease and distribution mechanisms for neurotransmitter release, and in the larger world both as a member of the House of Lords and a popular writer who has brought neurophysiological evidence to bear on the question of how brains respond to sustained exposure to social media. She has cited the negative impact that social media usage has on the release of striatal dopamine, a phenomenon also at the centre of drug addiction, and a particular danger for an activity that teenagers reportedly spend 10.75 hours engaged in daily. In *Mind Change: How Digital Technologies Are Leaving Their Mark on our Brains* (2014), Greenfield made her case for the negative impact that technology is having on childrens' depth of learning, authenticity of interpersonal relationships, grounded sense of self, and ability to focus attention, citing some 200 studies in support of her concerns, and beginning a debate about the psychological status of social media addiction that will likely not see any resolution anytime soon, but keep checking your profile for status updates.

Justine Sergent (1950–1994):
Sergent was a promising up-and-coming neuroscientist, whose specialty was the use of PET scanning to map cerebral activity under various circumstances – the sort of work that many in her generation would go on to make substantial careers exploring. In 1992, she discovered the Fusiform Face Area, which is the part of the brain responsible for perceiving faces, and in the same year she published a study of the musician brain while performing. In July of that year, however, an anonymous letter accused her of ethical violations, which, when it did not result in Sergent's termination, caused other anonymous letters to be sent to the press. The *Montreal Gazette* published one of them in April of 1994, and within two days Sergent and her husband were found dead, from suicide.

Established in 1999, the Justine and Yves Sergent Fund honors their memory and the tragedy of their loss by providing funding each year to acknowledge and encourage the work of women cognitive neuroscientists.

Catherine Vidal (b. 1951):
Though Vidal has done important work in prion research (2005–2007), working memory studies (1994), and the effects of non-noxious stress (1982), she is most

generally known for her theories about the intersection of neuroplasticity and gender. By this point, we have seen the pendulum of gender theory swing back and forth several times between biologically deterministic and socially constructed explanations. Vidal, using the insights of neuroplasticity research, asserts that, while genetics provides a basic template of abilities, it is cultural and social expectations that are seized upon in a neuroplastic sense by our minds as they wire themselves towards the behaviours and abilities that are expected of them. Through this process, the brains of men and women can *become* measurably different in various attributes, though they did not necessarily start out that way. These ideas, published in a series of articles and books in the mid mid-2000s to early 2010s, have come in for distinctly heated criticism from some segments of the neuroscientific community, of course, and while further research is being done on Vidal's theories, the great gender theory pendulum swings on…

Susan Hockfield (b. 1951):
Let's say that you are interested in one particular type of cell in the body – how do you go about finding instances of it among the 30 trillion cells that make up a human being? Ideally, you could unobtrusively introduce some trackable substance into the body whose entire purpose in life was to find, and bind to, that type of cell, and only that type of cell. Fortunately, this is precisely what antibodies do. Unfortunately, creating pure samples of a single type of antibody was not an option open to researchers until the mid-1970s, when 'monoclonal' antibody technology was developed, allowing the production of custom tracing bodies that could well and truly investigate the cellular diversity of the body. One of the earliest adopters of monoclonal technology to investigate cell types in the brain was Susan Hockfield, whose 1982 paper 'Monoclonal Antibodies Distinguish Antigenically Discrete Neuronal Types in the Vertebrate Central Nervous System' kicked off a decade-long research program of searching neural systems for particular organisational patterns, cell types and surface proteins, which resulted in the discovery of the potential role that the Cat-301 antigen plays in facilitating connections between neurons during early postnatal development, and of the part that the extracellular matrix protein BEHAB/brevican plays in making gliomas such a particularly deadly form of tumor.

In 1998, Hockfield began the pivot towards administration which saw her, in 2004, become the first woman president of the Massachusetts Institute of Technology (MIT), a role she held until 2012.

Gillian Einstein (b. 1952):
Globally, women make up about two thirds of all Alzheimer's patients, a fact that positively screams for explanation. While some wave the disparity off as an effect of women's generally longer lifespans, others have seen it as the result of some underlying biological mechanisms worthy of research. Chief among these is Gillian Einstein, whose work in the 2010s and into the 2020s on the links between women's shifting biochemistry and dementia has showed connections between the decrease of estrogen brought on by menopause or surgical procedures like oophorectomies, and the onset of

dementia-like symptoms. In a recent paper, 'Neurobiology of Aging' (2022), Einstein reported results of a comparison of women who had undergone oophorectomy, half of whom took estrogen supplements, and the other half of whom did not. The latter group demonstrated both decreased hippocampal activity and significantly lower results in memory performance. In the last half decade, the use of estrogen as a treatment for some forms of dementia has been tested, with so far promising results.

Susan Fiske (b. 1952):
Princeton social psychologist Susan Fiske has, for the past four decades, led the way in the psychology of person perception as modified by social power dynamics. In particular, she has investigated how differences in status feed into differences in perception, which in turn motivate the hardening of stereotypes and the rigidification of relations between individuals of different status. In her stereotype content model (published in 2002 and reiterated in 2018), she identifies two dimensions as lying at the heart of how social groups are perceived: warmth and competence, the former category reflecting our evaluation of the goodness of their intentions, and the latter reflecting our evaluation of their ability to carry out their goals. We can perceive a group as cold but competent (lawyers, for example), as warm but incompetent (the mentally handicapped, the elderly), as cold and incompetent (as immigrants are often perceived), or as warm and competent (which is generally how we perceive the group we happen to be in, particularly if we are middle class). We use these categories to differentially modulate the degree of attention we give to the members of a given group, based on our perception of our own status, which is where ideas of 'envy' and 'scorn' come into play (2010), whereby incessant status comparison forms a central component of how we evaluate ourselves and interact with others, with members of a high status able to disregard those of a lower status as having no meaningful role to play in their access to the resources they crave, while those of a lower status must keenly attend to those above them as the possessors of the resources they need to survive. Power dynamics, resource possession, and interpretation of motivations and the power to carry them out all play a role in the rigidification of stereotypes, which, Fiske found, can generally only be attenuated psychologically through placing one's self in teamwork situations with members of different groups, active practice of empathy, or having ready access to counter-stereotype information.

Vijayalakshmi Ravindranath (b. 1953):
As founder of the National Brain Research Centre, Guragon, founding chair of the Indian Institute of Science's Centre for Neuroscience, *and* founding director of the Centre for Brain Research, Ravindranath is a driving force in modern Indian neuroscience, engaged in research on brain metabolism and its role in brain disease. She has both researched aging brains' increasing difficulties with breaking down newly produced proteins (which are believed to be contributing factors in a number of brain diseases) and conducted longitudinal studies aimed at recording the progression of

mental disease, and the lifestyle practices that promote or protect against disease progression, in order to design better intervention strategies for dementia patients.

Adele Diamond (b. 1952):
For years upon years, educators focused on IQ as the central measure of a child's basic intellectual capacities and likelihood to succeed in a future career. By the late 1970s and early 1980s, researchers and occupational theorists began to challenge the dominance of IQ as a predictor of success. What is intellectual capacity, they asked, without the guiding habits to bring that capacity to bear on a problem at hand and see the working of a solution through to its end? Thus was born the concept of Executive Functioning, an outgrowth of the attention studies performed by Richard Schiffrin, Walter Schneider, and Michael Posner, which laid out the core functions that determine an individual's ability to recognise a problem, create a strategy to solve that problem, and organise and oversee the implementation of that strategy. These functions include inhibition control (resisting the siren call of enticing activities), working memory (being able to hold relevant information in your head and manipulate it), cognitive inhibition (keeping your mind On Task), attention control (ignoring things in the environment that are not central to what you are trying to do), and flexibility (adjusting plans in the light of new information or insights).

The question then became, given that EFs are important for the future success of children in school and life, how best do we instill those functions from an early age? In answering that question, one is not long in running across the name of Adele Diamond, who is one of psychology's most cited researchers, with numerous articles having been cited thousands of times, so central is her work and thought to the question of how education can best serve the development of EFs. Diamond's research in the early 1980s while a grad student at Harvard had led her to the idea that the prefrontal cortex was the likeliest centre of executive functioning, a domain of the brain which at the time Patricia Goldman-Rakic was the reigning expert on. Diamond, realising how much she could learn from working at Goldman-Rakic's lab, secured a fellowship to study there, learning the deep linkages between the neuron architectures of the prefrontal cortex, the molecular systems at work there, and the mental capacities those structures and systems created.

In a 1994 paper, 'An Animal Model of Early-Treated PKU', Diamond combined neurochemistry, executive function testing, and her experience of the prefrontal cortex to demonstrate how phenylalanine, when present in even moderately elevated levels in the bloodstream, could produce substantial reductions of neurochemicals in the prefrontal cortex, with resulting impairments to cognitive function.

Throughout the 1990s and 2000s, Diamond carried out her exhaustive studies of the educational approaches that best serve executive functioning. In her papers 'The Early Development of Executive Functions' (2006), 'Preschool Program Improves Cognitive Control' (2007), and 'Interventions Shown to Aid Executive Function Development in Children 4 to 12 Years Old' (2011), Diamond laid out what EFs could best be developed by what techniques, and how much the development of one

EF did or did not impact others. She found in particular that physical activity (and particularly physical activity with a self-discipline component, like martial arts, or a mindfulness component, like yoga) significantly improved executive functioning, as did role playing activities (as in the *Tools of the Trade* curriculum), student-to-student teaching (as often happens in Montessori settings), computerised working memory training, and programs that teach children to identify their emotions before acting on them. Taken in combination, the result is children with better impulse control, more flexible planning skills, improved social coordination abilities, and more focused and retentive minds, and if the next thirty years turns out anything like the last thirty, boy, will they need them.

Maria Fitzgerald (b. 1953):
Let's talk about pain. Particularly, about why some people are more sensitive to it than others. We come equipped with sets of sensory neurons devoted to the detection of noxious stimuli called nociceptors. Since the late 1980s Maria Fitzgerald has been a key figure in using modern technology to investigate how nociception develops, and how that development goes on to affect how children experience pain. She found that our pain processing circuitry is still developing in early life, and that because of that longer development process, inflammation or noxious events experienced by a young child have a higher chance of impacting the growth of that circuitry. Early Life Pain (ELP) changes the connectivity of our pain pathways, making us more hypersensitive to pain later on, and perhaps more disposed towards chronic pain in adult life. As a result of these studies, Fitzgerald has taken a leading role in advocating for and developing new procedures for recognising and treating infant pain, and new protocols for infants in intensive care to limit the amount of invasive procedures which they are exposed to.

For her efforts on behalf of understanding and reducing children's pain, and thereby that of the adults they will become, Fitzgerald has been made an honorary member of the British Pain Society (which is *certainly not* an underground British version of Fight Club), and the International Association for the Study of Pain, and in 2016 was made a Fellow of the Royal Society.

Ann McKee (b. 1953):
Before Ann McKee, American professional sports (and particularly American football and boxing) was littered with young people whose brains had been permanently damaged by the game they played, and their teams'/managers' refusal to recognise the seriousness of the injuries they had sustained. Then, in 2009 McKee, after dissecting and analysing the brains of former players, and finding repeated evidence of CTE (chronic traumatic encephalopathy), brought her findings to the NFL, starting a national conversation that saw McKee labeled both as The Woman Trying to Destroy Football and The Woman Trying to Save Football From Itself. Studying some 190 donated brains from players across different sports, she found that 116 had some form of CTE, which can result in mood disorders, difficulty in thinking, and eventually dementia. McKee's results on the potentially life-threatening seriousness of repeated

concussion, met with initial fierce resistance by the NFL, have since fundamentally changed how head injuries are treated in high school and college sports, with much more stringent guidelines in place for when and if players are allowed to return to play after experiencing a concussion.

Ann Elizabeth Kelley (1954–2007):
When Ann Kelley died of metastatic cancer in 2007, the world lost one of its most brilliant investigators into the brain's motivational and reward systems. In the early 1990s, she found that the injection of chemicals that activated mu receptors in the nucleus accumbens produced significant and long-lasting changes in animals' eating patterns. Mu receptors are more popularly known as the targets for morphine in the brain, and Kelley later found (2001) that, when they are stimulated, animals have a hyper tendency to eat more sucrose, salt, and fat laden foods. She further found (2004) that deactivating regions of the amygdala undoes this binging effect, reenforcing a proposed role for the amygdala in determining the value of a stimulus that had been highlighted by her and Ned Kalin's earlier demonstration that amygdala damage inhibits the natural fear responses in primates to dangerous stimuli (2001).

Maria Primitiva Paz Root (b. 1955):
Maria Root is a psychologist and activist most known for her work in the psychology and lived experience of multi-racial individuals. In 1993, she authored the Bill of Rights for People of Mixed Heritage, laying out the issues that people from different ethnic backgrounds have in navigating society's incessant bureaucratic need for single categorisations. How do people who present as one race, but identify most with another part of their heritage, make their way through organisations and social groups that treat that less-visible identity with indifference or even contempt, and how does that process affect their outlook on themselves and the world? Root's advocacy has aimed to make the life course of these individuals less fraught, and her efforts have resulted in governmental forms that allow people to declare multiple ethnicities, and a gradually evolving cultural landscape that allows for and understands a conception of ethnic identity that is allowed to change over time as individuals come to different conclusions about the various lineages that make up their identity, and how much those lines speak to their present place in life.

Vindhya Undurti (b. 1955):
According to a 2018 Lancet Public Health report, Indian women make up some 18 per cent of the world's population, but account for some 36 per cent of its suicides, exemplifying the continuation into the modern era of a grim trend of unusually deep depression and hopelessness for Indian women that has been evident ever since this sort of data started being collected some three quarters of a century ago. Vindhya Undurti has been for decades now a leading force in investigating the tragic persistence of Indian women's elevated depression and suicide rates, and in pushing for the large scale societal changes necessary to begin reversing the trend. She has contributed

to publications investigating the prevalence of domestic abuse in India (2013), the phenomenon of dowry deaths (by which husbands use persistent abuse, sometimes to the point of murder, to extort more dowry gifts from their wives' families) (2016), and exploring how Indian women's quality of life is affected by different family-work dynamics (2016). These studies have allowed Undurti and her colleagues to assemble the beginnings of a feminist psychology in India that recognises and accounts for the cultural uniqueness of India's multiplicity of traditional structures and expectations.

Melly Oitzl (b. 1955):
Austrian neuroscientist Melly Oitzl's research aims to answer the question of why different people have such different responses to stress. Some thrive under stressful conditions, while others buckle under it. When under stress, we usually produce glucocorticoids like cortisol, which help to lower inflammation, and which enhance memory of the crucial components of the event which caused stress, and suppress the irrelevant components of it. They do this by stimulating two different types of receptors, MR and GR, which act in balance to kick off the stress reaction, protect the brain, and process and remember the conditions which caused the stress for future reference. Oitzl argued that imbalances between the functioning of MR and GR lead to chain reactions under stress that throw off our regular neurotransmitter systems, and that these imbalances might develop from epigenetic factors arising from the maternal chemical environment, resulting in long-term diminished abilities to handle certain types of stress.

Nancy Ip (b. 1955):
Codeveloper of the China Brain Project (a fifteen year project approved in 2016 to improve diagnosis and treatment of brain disease, and to elucidate the neural foundations of cognition, and their application to ethical artificial intelligence), current president of the Hong Kong University of Science and Technology, and leader of the team that in 2018 rolled out the first whole-genome study of genetic risk factors for Alzheimer's disease in the Chinese population, Nancy Ip is a powerhouse in the neuroscientific community, directing impossibly large projects while also carrying out her own research goals. In 1992, she and George Yancopoulos published their finding that bone marrow cells involved in the immune system and neuronal cells shared common receptor systems, suggesting closer than expected relations between the immune and nervous systems. Upon her return to China from the United States in 1993, she began studying neural plasticity, and soon reported important results about the role of the protein Cdk5 in central nervous system development, the formation of new synapses that lies at the heart of neural plasticity, and cellular communication.

Turning to Alzheimer's research in the mid-2000s, Ip discovered the potential therapeutic value of IL-33, an immune protein, in treating cognitive disease, and the role that the protein Eph Receptor A4 plays in both neural plasticity and neurodegenerative disease, suggesting new treatment pathways involving the inhibition of EphA4. For her research on plasticity, neural disease, and neuroscientific leadership, Ip has received numerous honors, including the Medal of Honor from the Hong Kong government, the title of Chevalier de l'Ordre National du Merite, the appointment as an Academician (2001) and Fellow (2019) of the Chinese Academy of Sciences and a Fellow of the World Academy of Sciences.

Isabelle Peretz (b. 1956):
An expert on the psychology of music, Peretz studies what parts of the brain are involved in music listening for both musicians and non-musicians, what parts develop plastically for the musician brain, whether we possess music-specific neural regions or simply re-specialise components of the brain for musical processing, and what different disorders in musical processing like congenital amusia (what the rest of the world calls tone-deafness) can tell us about how regions of the brain responsible for musical processing interface (or don't, as the case appears to be) with other parts of the brain responsible for seemingly related linguistic and speech competencies.

Nora Volkow (b. 1956):
Currently director of the National Institute of Health's National Institute on Drug Abuse, Nora Volkow has spent her life researching the neural underpinnings of drug addiction, and educating the public about addiction's status as a brain disease, rather than a moral failing. Prior to the work of Volkow and her colleagues, limbic subcortical regions were the primary focus of addiction research, but beginning in 2000, Volkow published results of PET scans on active cocaine users and cocaine users in remission, finding substantially decreased blood flow to the prefrontal cortex, particularly to the regions responsible for goal forming and plan formation. Her results explained one of drug addiction's most confounding aspects – the tendency of drug users to continue seeking drugs even when their use has long since ceased to be pleasurable. When our frontal cortex is working properly, we are able to properly form goals based on the reward value of an activity – when we are hungry, we decide to eat, but once full our brain realises that eating more food will not carry much reward for us, and we start forming non-food-based goals. With drug addiction, however, the prefrontal cortex's ability to form value-based plans is reduced, and we get locked into a system whereby motivation and value are disentangled, leaving us in a state of neurochemically-locked keen motivation to seek more drugs in spite of the fact that they no longer do anything positive for us.

Fun-Facts: In addition to being one of the world's leading authorities on the neurophysiology of drug addiction, Volkow is also the grand-daughter of Leon Trotsky! Less pub-night trivia-centric, her 2010 paper 'Neurocircuitry of Addiction' has been cited over 5,400 times in medical literature.

Nancy Kanwisher (b. 1958):
Over the last two decades, Nancy Kanwisher and her team at the McGovern Institute have used fMRI technology to investigate the functional structure of the brain, detailing dozens of specialised cortical regions that carry out particular functions, such

as analysing faces, places, and visually presented words, processing specific motor and sensory data, and even such hyper-specific tasks as thinking about the thoughts of other people. This is a 'domain specific' approach to neuroscience, which holds that cognition is the result of evolutionarily developed specialised brain regions, rather than the emergent result of a more diffuse and distributed architecture. Kanwisher's lab continues today to investigate questions of how the connections between specialised regions allow new functions to arise, what computations different regions are capable of performing, and what regions of the brain might have as-yet undiscovered specialist roles.

Ingrid Scheffer (b. 1958):
Ingrid Scheffer is a world authority on the genetic causes of epileptic disorders, having discovered several hundred genes that contribute to epilepsy over the last three decades including the first ever recorded. This work in documenting the genes that play a role in epilepsy led Scheffer to a more nuanced view of the different types of epilepsy that present similarly in patients, but have fundamentally different genetic underpinnings, causing her to push for new and finer definitions of epilepsy types, and for more genetic screening practices to allow doctors to suggest treatments that are targeted to the patient's unique form of epilepsy.

Brigitte Kieffer (b. 1958):
After Candace Pert discovered the first opioid receptor in the 1970s, the floodgates opened for the study of how opioid receptors interface with our reward systems and addictive behaviours, a task that has been carried forward not only by Ann Kelley, who we met above during her studies of mu reception as it relates to food consumption patterns, but by French molecular neurobiologist Brigitte Kieffer, who studies not only the mu receptors that morphine acts upon, but also the delta receptor, the gene for which she isolated in 1992, which plays an important role in pain modification. Her research has established the role that opioid systems play not only in pain and addiction, but also in emotional states, with mu receptors found in higher densities of individuals suffering from depressive conditions, while delta and kappa receptor activity seems to correspond to mood improvement. For her work on opioid system genetics and functions, Kieffer has been awarded the Lamonica Prize (2012) and the L'Oreal-UNESCO Women in Science Award (2014), and became a member of the French Academy of Sciences in 2013.

Susan Nolen-Hoeksema (1959–2013):
Do you know someone (and that someone might be you) who, whenever a problem comes up in their lives, spins their wheels endlessly brooding over what caused their problem, and what the consequences of that problem might be, to the almost complete exclusion of actively finding solutions to the problem? In psychology, that is called 'rumination,' a response to difficulty which Susan Nolen-Hoeksema was able to link over the course of her all-too-brief career with susceptibility to anxiety disorders, depression, and substance abuse. In her studies, she found that women were more prone

to ruminative behaviour, with corresponding effects on the prevalence of depression among women, and that ruminators tend to have more long-term negative views of the past and future, and increasing difficulties in successfully navigating interpersonal problems. She published the results of her research in both academic texts like her 1990 *Sex Differences in Depression*, and in popular books meant to reach the people most in need of them, like *Women Who Think Too Much: How to Break Free of Overthinking and Reclaim Your Life* (2003) and *The Power of Women: Harnessing Your Unique Strengths at Home, at Work, and in Your Community* (2010).

Leeanne Carey (b. 1959):
For years, developing treatment plans for stroke victims was largely a matter of trial and error, groping forwards in the dark on the basis of some educated guesses grounded in a small collection of known neurological facts. Over the past quarter of a century, Leeanne Carey has led the way in applying MRI technology to probe the efficacy of existing treatments, and to create new ones that combine the physical facts of what she has observed with recent insights into how neuroplasticity can work to rewire the brain to regain lost functionality. Her task-specific brain training programs have yielded significant results in improving subjects' sense of body position, and tactile functioning (1993) as well as in upper limb functioning and resumption of everyday tasks (2009).

Nancy Bonini (b. 1959):
There are reasons that researches love the *Drosophila* fruit fly – it has a short lifespan, not too many chromosomes to keep track of, produces a lot of offspring each generation, and attracts Nobel Prizes like Star Trek conventions attract vector calculus enthusiasts (to date, five Nobels have been awarded for *Drosophila* research). A model organism for questions of heredity and development, what doesn't impress itself upon the casual armchair scientist when observing *Drosophila* is its potential as a model for human neural degeneration. It took a group of scientists of unique insight to make that connection, thereby harnessing all of the useful properties of the fruity fly to the study of diseases that we used to have to employ mice or primates to investigate. In 1998, Nancy Bonini was part of a team that reported the recreation of a human neurodegenerative disease (spinocerebellar ataxia type 3) in *Drosophila*, demonstrating the potential of using invertebrates to study the cellular-level changes at the heart of human neural disease.

This discovery opened the gates to brand new ways of looking at the genetic, molecular, and cellular underpinnings of disease, allowing the impact of mutant type proteins, diverted genetic expression pathways, and microRNAs to be quickly and exhaustively determined, with results that opened up new therapeutic potentials in the treatment of Alzheimer's disease, Parkinson's disease, and generalised neural aging.

Pamela Sklar (1959–2017):
Sometimes, nothing holds science back so much as its own success. After work in the 1980s had established strong links between single genes and single diseases (such as

the single mutations that cause the defective transmembrane protein at the heart of cystic fibrosis, or the extra repeats in the HTT gene that cause the exceedingly long huntingtin proteins that initiate Huntington's disease), scientists became used to that one gene-one disease explanatory model. The problem was that, as the twentieth century shaded into the twenty-first, research around the one gene model was repeatedly coming up empty on a number of our most important neural diseases with hereditary components, particularly bipolar disorder and schizophrenia. Pamela Sklar was part of a generation that saw the success researchers were having with using new genetic technologies to locate the cause of disease not in a single gene, but in the accumulated buildup of dozens of genetic factors, and decided to import those methods into the study of neural diseases.

In 2009, Sklar released a report, 'Common Polygenic Variation Contributes to Risk of Schizophrenia and Bipolar Disorder' which compiled genetic data from 3,300 schizophrenic and 3,600 non-schizophrenic individuals, and argued for the cumulative effect of thousands of genetic alleles in explaining the incidence of schizophrenia. This was a massive result in the polygenic explanation of neural disease, and was followed in 2014 by an even larger analysis (some 37,000 schizophrenic and 113,000 controls) that uncovered 108 distinct schizophrenia-associated genetic locations (83 of which had previously been unreported in the literature).

Lisa Goodman (b. 1961):
Ever since her pioneering studies of the psychological impact of homelessness, domestic abuse, and rape in 1991, Lisa Goodman has earned herself a reputation as a psychologist unafraid to address the mental health impact of some of modern civilization's ugliest aspects and practices. She has documented the social factors which lead to higher likelihoods of homelessness among mothers as well as the psychological damage done to individuals while in a prolonged homeless state to create a haunting picture of the cycle whereby those most isolated from support structures become those whose mental health is most aggressively attacked by their newfound status, leading to further isolation, and further psychological impact. She has also studied how women experiencing male violence undergo long-term symptoms that echo post traumatic stress disorder, which impact women's help-seeking behaviours and self-conceptions in a way that calls for the development of new specialised training for those prospective counseling psychologists who operate at the intersection of poverty and domestic abuse.

Valina L. Dawson (b. 1961):
Apoptosis is the process by which cells destroy themselves as part of the regular and scheduled operation of our bodies. Out with the old and worn cells, in with the new. The problem with developing an efficient death mechanism, of course, comes when its processes are hijacked by diseases that do not have our best interests at heart. Valina and her husband Ted Dawson have, for thirty years now, been at the forefront of sleuthing out the molecules responsible for misleading apoptosis in our neural cells. They discovered the role that nitric oxide (NO) buildup plays in activating poly (ADP-

ribose) polymerase (or PARP), which is generally a useful enzyme that helps repair damage to DNA, but which, when present in too great a quantity, can consume all a cell's ATP, leading to necrosis. Further, the Dawsons discovered, PARP can produce PAR, a molecule that has the ability to convince mitochondria to release their stores of apoptosis inducing factor (AIP), which teams up with yet another molecule in the cell to wander over to the nucleus and cut up the genetic material found there. This complicated death mechanism the Dawsons gave the unequivocally cool name of Parthanatos, and it ranks as just one of the many complicated molecular pathways that the team have uncovered which lure healthy cells to their death, playing roles in neurodegenerative disorders such as Parkinson's disease.

Katrin Amunts (b. 1962):

One of humanity's grandest efforts at self-knowledge was launched in 2013 in the form of the human brain project, a ten year effort to simulate the human brain down to the molecular level, develop brain-based computing technologies, and create international neuroscience resources to collate new research with existing knowledge and distribute it. The scientific research director of that project (and the only woman currently serving on the HBP's governing body, The Directorate) is Katrin Amunts, who was born in East Germany and earned her PhD at Moscow's Institute of Brain Research in 1989. Since the early 2000s she has been engaged in developing the mathematical, informational, and technological tools necessary to accurately map the brain, culminating in her role as head of the Julich Brain Atlas project, a 3-D dynamic cytoarchitecture map that employed probabilistic methods to smoothly combine the individual differences in the twenty-three post-mortem brains that were scanned using MRI devices and sliced into sections 20 micrometres thick for staining and digital imaging.

At the time of this writing, the original ten year deadline for the HBP is fast approaching, with critics stating that it has fallen far short of its original lofty goals, and supporters pointing out that the impact of the global pandemic on research programs should be taken into account when trying to hold the HBP strictly to its ten year time table, and that the importance of its work in developing new computational research infrastructure is of itself enough to justify the continuation of the project. No matter how all of that falls out in the months to come, I have a feeling that, one year hence, Katrin Amunts will still be out there, doing whatever she can to let us roam at will the sprawling labyrinth of our mind.

Magdalena Götz (b. 1962):

When I was in high school, one of the mottos that was drilled into our heads in biology class was, 'Adults can't make new neurons.' This was meant, I think, to give us a proper appreciation that learning and memory is a matter of making new connections rather than growing new cells, and possibly to make us treat our heads with care, since any injuries to our brains could not be healed the way that broken bones or torn muscles could (this was also before neuroplasticity had percolated its way into American biology textbooks). Though individuals like Joseph Altman and Michael Kaplan began challenging that orthodoxy as early as the 1960s, such discoveries were not heeded by the scientific establishment until the 1980s. In 1981, Fernando Nottebohm, while studying the neural changes that allow canaries to learn new seasonal songs, found that neurogenesis does occur in an adult brain, and in 1992 Samuel Weiss and B.A. Reynolds published their results in persuading neural striatal tissue to become new neurons and astrocytes. It wasn't until the early 2000s, however, that a mechanism for new neural formation was discovered, by Magdalena Götz. She found that radial glial cells, previously known in the cerebral cortex for their role in acting as scaffolding to guide migrating neurons to their ultimate destinations, also function as neural stem cells, with the capacity to generate new neurons. In subsequent years, Götz experimented with guiding glial cells to become specific types of functional neurons, with an eye towards using them in therapy for repairing injured brain tissues. In the late 2000s, Götz found that not only radial glial cells, but astrocytes as well (which are far more abundant than glial cells in the central nervous system) have neuroregenerative capacities, which come to the fore when the brain is injured. Today her work focuses on developing procedures to guide the *in vivo* development of specifically determined new neuron types for the treatment of neural injuries and degenerative diseases.

Elizabeth Gould (b. 1962):
In between Fernando Nottebohm's canaries of 1981, and Magdalena Götz's radial glial cells in the 2000s, the torch of neurogenesis was carried by Elizabeth Gould, who in the late 1980s believed she found evidence of the phenomenon in the hippocampus of stressed rats. Doing a deep dive in the literature, she discovered the lineage of neuroscientists who claimed to have found various aspects of neurogenesis, saw their ultimate fate, being sidelined by the larger academic system, and made the courageous decision to proceed with her investigations nonetheless, following the evidence of her studies rather than the dictates of career pragmatism. During the 1990s, she compiled observations of rodent and primate brains, building for herself an iron-clad case for the existence of neurogenesis occurring in the living brain. She published her rat results in the early to mid-1900s, and her primate results in the late 1990s, demonstrating the creation of new neurons in both the hippocampal and olfactory bulb regions of the brain with such a weight of data that by 1999 even the arch-priest of neuronal stability, Pasko Rakic, released a paper conceding the existence of neurogenesis.

Gould's work in the 2000s has explored what role adult-formed neurons might play, what conditions give rise to their formation, and what factors determine their ultimate retention. She has documented their seeming role in the formation of trace memories, and the role that stress and environmental complexity play in the rate of neuron formation and survival, with results suggesting that, even though we regularly form new neurons, if we keep ourselves in environments that lack social or organisational complexity, we tend to lose them just as quickly.

Misha Mahowald (1963–1996):

Misha Mahowald was a promising neuromorphic engineer, who was awarded a PhD in 1992 for the creation of a 'Silicon Retina,' which was a tour de force combination of biology, neuroscience, and electrical engineering that used a series of electrical circuits to mimic the functionality of the human retina. The invention won her awards and patents, and seemed to indicate a bright future harnessing the lightning pace of the electronics industry to the development of new sight technologies. All was not well, however, with Mahowald, who drove herself relentlessly trying to cut through the technological barriers between her and her futuristic goals, experimenting with yoga, LSD, Catholicism, and animism to jumpstart her mind, and falling into a deep depression when none of these quick fixes seemed to do the trick. At Chistmas in 1996 she asked one of her close friends over to Geneva to celebrate with her, and when he arrived, she told him that she was being invaded by creatures nested at the base of her spine. Later that night, while he was sleeping, she climbed into the bathtub with a knife and attempted to kill herself by slicing through her spinal cord then, when that did not work, she left her apartment, waited for an oncoming train, and stepped in front of it, ending a troubled life rich in promise.

Marina Picciotto (b. 1963):

For decades, nicotine lay at the centre of an interesting conundrum known as Nesbitt's Paradox. Taken in small quantities, nicotine acts as a stimulant, but taken in large quantities, it acts as a sedative. How can the same substance have two such diametrically opposed effects on the nervous system, and what is the mechanism by which that switch in effect at higher concentrations takes place? These are questions which Marina Picciotto has been providing answers to for just over a quarter of a century now. Whereas we might want a simple solution on how nicotine engages with our brains, Picciotto's research has shown that we might have to wait on that a bit, that the receptors which respond to nicotine (called nAChRs, which I love as an acronym because it looks like a cat just suddenly stepped on the keyboard) can both be activated *and* desensitised by contact with nicotine, and that the subtle balance of those events has a keen role to play in the engagement of drugs with our neural reward systems, the downscaling of most types of aggression, and the release of dopamine and acetylcholine in our brains. By investigating the subunits out of which different nAChRs are constructed, she has been able to pull back the veil to an important degree on why some receptors are more prone to desensitisation at higher concentrations than others, and why some of them engage with other molecules as readily as they do with nicotine.

Jennifer Eberhardt (b. 1965):

Eberhardt's work as a social psychologist has gone far to uncovering both the extensiveness of unconscious racial bias, and the role that it plays in some of the most important institutional structures of modern society. Using fMRI data, she discovered that brains spend naturally more time and energy processing the faces of individuals that they recognise as belonging to their own race than they do those of other races (2001). Following up on that result, she found that police officers, shown an individual and then asked to identify them in a lineup composed of members of the same race, tended more often to misidentify Black individuals, most often mistaking them for individuals who had more stereotypically African features, than they did for white individuals. Further, when cued that a presented scenario had some sort of crime aspect to it, subjects devoted more attention to the Black individuals in the scenario than in non-crime cued situations (2004). Of course, law enforcement officers aren't the only people whose unconscious biases have a profound impact on the lives of racial minorities, and Eberhardt soon applied her technique to the classroom, showing with Jason Okonofua in 2015 that teachers had stronger negative reactions to Black students engaging in disruptive behaviour than students of their own race, resulting in harsher disciplinary measures for those students.

Eberhardt's results on unconscious bias have since been used as the basis for new training programs for educators, officers, and managers in realising what their implicit biases are, and using that realisation to adjust their behaviour towards more objective analyses of what they unconsciously and automatically believe they are witnessing.

Madakasira Vasantha (M.V.) Padma Srivastava (b. 1965):
Srivastava is a professor of neurology at the All India Institute of Medical Sciences, where she has developed the code red program for the treatment of ischemic stroke victims. Code red is based on hyperacute reperfusion techniques, which include the direct delivery of clot-destroying drugs to the site of blockages through catheters (intravenous thrombolysis) and the use of balloon-tipped catheters guided through blood vessels to physically break up blockages (mechanical thrombectomy). Code red is the first program of its kind to be enacted in an Indian public hospital.

Sarah Joanna Tabrizi (b. 1965):
Striking approximately 1 in 10,000 individuals, Huntington's disease, once diagnosed (usually somewhere between the ages of 30 and 40), generally terminates in death after a period of from 10 to 30 years as the mutated form of huntingtin (HTT) produced by an error in a person's genes attacks their nerves. Sarah Tabrizi has dedicated vast swaths of her life to understanding HD, from showing that mutated htt proteins in immune cells hinder their migration to sites of neural damage (2012), to organising long-term tracking studies of how HD progresses in individuals carrying the defective htt gene who have yet to express symptoms, to developing new treatments that target HD's genetic core, including Tominersen (originally IONIS-HTT), a treatment employing antisense oligonucleotides (single strands of nucleic acid that can be specifically coded to interfere with the mRNA that carries out instructions from our DNA) that showed promising reduction in mutant htt levels in early testing, but which its prospective manufacturer, Roche, ultimately pulled the plug on in early 2021. Fear not, however, for Tominersen was just one arrow in Tabrizi's anti-HD quiver, and behind it there are a further plethora of potential promising treatments whose efficacy she explored and which are in development, including phosphorylation modulation, 'zinc-finger' transcriptional repression, CRISPR-based solutions, and phosphodiesterase inhibition.

Catherine Simpson Woolley (b. 1965):

Before we move on to our main task of talking about the work of American neuroendocrinologist Catherine Woolley, I expect many of you are positively straining at the urge to ask the question, 'Is Catherine Woolley related to psychology pioneer Helen Thompson Woolley?' I reached out to Catherine Woolley and asked her that very question, and her response was that, as far as she knows, no, it's just a coincidence, which is a bit of a downer but there it is.

Okay, genealogy sidebar complete. Let's get into the science. In the early 1990s, Woolley was part of the group around Elizabeth Gould investigating the possibility of new neuron growth in the rat hippocampus. One of the early results of her work there was the paper 'Gonadal Steroids Regulate Dendritic Spine Density in Hippocampal Pyramidal Cells in Adulthood' which established that, when deprived of estrogen, pyramidal neurons (which are found throughout the brain, and play particularly important roles in the prefrontal cortex) start losing dendritic spine density. Usually, the dendrites of pyramidal neurons are covered in thousands of little protrusions called spines which allow them to gather signals from other neurons. The more spines you have, the more connected you are, the easier signals flow. Following up on her original research, in 1992 Woolley reported that the rise and fall of hormone levels characteristic of the rat estrous cycle sees corresponding fluctuations in hippocampal synapse density. This was another nail in the coffin of Rakic's static brain model, and an important early result for the emerging study of neuroplasticity.

The importance of estrogen in regulating synapse density in some regions of the brain suggested that further studies should be done on how hormones differentially affect individuals of different genders, and Woolley has taken up this challenge in the twenty-first century, investigating drugs that have gendered impacts on neuronal activity, and digging beneath the surface on similar-seeming neural responses to hormones across the genders to find that they are actually caused by subtly different underlying neural mechanisms. The overall message of this research has been to confirm the need for careful testing of new drug therapies on members of both genders, not only to determine obvious differences in drug impact, but potential problems arising from brains taking different routes to get to the same end result.

Jennifer Groh: (b. 1966)

We are, at our essence, brains stuffed inside of a bone and muscle travel vehicle, which is itself immersed in a world of electromagnetic radiation, atmospheric variations, and plain old physical objects which those brains must harness to interpret the world and steer themselves safely through. That is a tall order – how does a brain keep track of where our limbs are at any given moment? How does it use differences in timing between when sounds hit our ears to construct spatial guesses about where that sound originated from? How does it break down the absolute bombardment of electromagnetic data it is exposed to in order to give us a relatively seamless sense of foreground, background, and motion? These questions, of how we have a sense of our own internal space, and of the space around us, are central to the research project of

Jennifer Groh, whose 2014 book *Making Space: How the Brain Knows Where Things Are* is among my favorite general readership neuroscience books, and whose lab at the Duke Institute of Brain Sciences is devoted to determining how the brain encodes locations based on sound and visual information, and how on a cognitive level our different senses communicate and coordinate with each other.

Groh's educational path wended its way through Princeton and the University of Michigan before culminating in her PhD at the University of Pennsylvania under David Sparks, and postdoctoral work at Stanford University. Since 1997 she has been associated primarily with Dartmouth (1997–2006) and Duke University (2006–present). Groh arrived on the neuroscientific scene with an influential early paper detailing some exquisite work on how we calculate the velocity of objects in our visual field in order to track them with our eyes. We have sets of neurons that preferentially respond to different directions of motion – some fire vigorously when things move horizontally across our vision, for example, but are indifferent to vertical motion. When we see something moving out in nature, then, different sets of neurons will respond in different degrees to that motion, and will send all of that information to our middle temporal visual area (MT), which has the task of having to sift through it all and send commands to the motor neurons that move our eyes about how best to track the object. The question Groh investigated was, how does it do that? Does it just believe whichever directional neuron fired the most vociferously, and ignore the inputs from the others? Does it perhaps take all of the pieces of information and perform a vector sum on them? No and no, it turns out – instead, it creates a running average of the reports it gets, which produces the same estimate of direction as the vector summation approach, but with a different guess on the overall speed. That information can then be packaged and sent to the eyes to tell them either how fast they need to move, and in what direction, to track the object or, if the object is one we need to rapidly jerk our eyes towards so that its photons will strike our foveas (the part of our retinas where we have the greatest visual acuity thanks to the concentration of cones there), which is an operation that takes longer, our MT can use that average velocity with an estimate of how long it will take the eye to jump to the future location of the object to send a message which will allow us to move our eyes to precisely where the object *will be* by the time all of the processing, messaging, and eye motion have happened.

Today, Groh's lab continues its investigations of how we make sense of our place in the world, including a discovery that, when our eyes move, our eardrums do as well, providing another layer of explanation for how we combine and synthesis auditory and visual data, and new research on 'multiplexing' which illuminates how we are able to process simultaneous stimuli. In a 2018 paper, the members of Groh's lab documented how a single auditory neuron, presented with two different sound stimuli, was able to encode and pass on information about both by alternating its firing pattern between signals characteristic of each stimulus, allowing one neuron to enhance its processing capacity by carrying multiple pieces of information differentiated by firing rates.

Sophie Scott (b. 1966):

What are we doing, when we laugh? According to Sophie Scott, who has spent the 2010s researching and writing about laughter extensively, most of the time what we are doing is simply communicating. In social situations, we generally engage in voluntary laughter as a complicated way of assuring people of the strength of our mutual bonds, the solidity of our social clique, and the agreeable overlap of our shared conditions. We are saying, 'You are my friends, this is our space, I am enjoying our time together, and am interested in further exploring this interpersonal dynamic,' without having to actually say *any* of that. As utterances go, it's pretty efficient, even as the underlying neural structures of laughter can be quite complex indeed.

The next question, after identifying why we voluntarily laugh (involuntary laughter, where we find something so funny we break down in laughter we can't help, works quite differently), is where that capacity came from, which is a part of a larger question of how evolution moved us from the 'simple' vocalisations of non-human primates to the 'complex' language perception we enjoy today. In the 2000s, Scott studied this question, believing that such a seemingly qualitative leap in capacities could not have arisen all at once, but must be based in auditory structures in primates that bear resemblances to the systems we humans use. In an influential series of papers, Scott demonstrated that all primates have some degree of hierarchical structure and functional processing in their primary auditory complex which allow for parallel streams of information originating from a given set of sound events, allowing for the multiple layers of processing which could be the foundation for how we are able to process complicated languages the way that we do.

Fei Xu (b. 1969):

When I was young, I remember getting new Transformers toys, gleefully paying no attention to the instructions about how to transform, say, Soundwave, from a robot into a tape recorder, and having the thing fully figured out within a matter of a few minutes, a feat of learning and cognition which boggles my mind today, when a fully printed set of instructions, a YouTube video, and the help of a couple of other people in the room are the minimal requirements for me to pull off a similar feat for any one of the small army of more recently acquired Transformers that happily roam the acreage of my desk. How did I figure things out so quickly then, and where did that skill go?

Berkeley developmental psychologist and cognitive scientist Fei Xu has some compelling answers to that question arising from her three decades of research into the field of infant cognition and learning. Xu's findings have led her to develop a 'rational constructivist' approach to how infants learn that builds off of the ground-breaking, Piaget-challenging work done by Elizabeth Spelke (b. 1949), Rochel Gelman (b. 1942), Renée Baillargeon (b. 1954), and Leda Cosmides (b. 1957) in the 1980s. Firstly, infants do not come into the world in the 'great blooming, buzzing confusion' described by William James, but rather come pre-loaded thanks to their evolutionary heritage with a core set of innate concepts about number, agent, object, space, and causality. The trick now comes in harnessing that set of basic concepts for the creation of linguistic systems and learning mechanisms, which Xu argues happens when infants make the deep change from viewing objects in terms of fundamental spatio-temporal rules, and start latching onto words as a means of allowing them to think of differences between objects as differences in object types, as codified within the underlying rules of a language's symbolism and categories.

New learning accompanies these tectonic changes in the form of Bayesian inferences, which is a computational game we play as we enter a new experience. We have a set of internal hypotheses about how that experience will play out, to each of which we have assigned different likelihoods based on past experiences. We then undergo the experience, which serves to reinforce certain hypotheses we had and contradict others, resulting in a mechanical re-writing of our internal probability tables, thereby impacting how we are likely to understand future experiences of the same sort. Repeating this mechanistic process, we get better and better hypotheses for how our world works. In short, we learn. Xu's work in figuring out how to get infants to reveal the secrets of their concepts of number, object, and probability have demonstrated that we are neither a blank slate which the world writes on, nor a complete analytical machine from birth, but rather a cunning hybrid – a creature that knows a few important things from the start, and has some computationally effective ways of expanding those into knowledge of the complicated world around us.

Lisa Goodrich (b. 1969):
Whenever somebody wants to make a point about how remarkable human senses are, they tend to head straight for the eye and its ability to sift information from the barrage of photons bombarding us at every moment. And yes, eyes are great, but for my money I've always been more impressed by ears. Think about the average office – at any moment there are perhaps six conversations going on, in addition to the sounds of printers, phones, a few different songs emanating from earbuds turned up way too loud, and a smattering of exterior noises, all doing their own part to jiggle the atmospheric soup in which we live in characteristic ways. And then there's our ear, at the centre of it all, which has the seemingly ludicrous task of using the compounded jiggles of that soup to create meaningful spatial, linguistic, musical, and temporal information which forms a major part of how we navigate our world.

To do that requires exquisite timing that is grounded in a simply beautiful neural architecture. We've known the basics of that architecture for some time – sound causes cochlear fluids to vibrate, those vibrations are picked up by a series of hairs which are laid out along the cochlea in the order of what frequencies they best distinguish, and the vibration of those hairs trigger firing of spiral ganglion neurons, with the number of neurons a hair cell is connected to playing a large part in how well we can pick up soft sounds that are of a frequency that hair specialises in. All of the spiral ganglion neurons then gather together to form the eighth nerve, which leads to the cochlear nuclear complex in the brainstem, where differences in signal timing between the ears allow us to make an auditory map of the space around us.

What we've historically dragged our feet in figuring out, however, either because the inner ear is so relatively inaccessible compared to the eye or the nose, or because there's more glamor to be had in studying sight, is just how the auditory neural circuitry lays itself out, and why we lose auditory functionality so inevitably over time. This is the work of Lisa Goodrich, whose lab at Harvard has carried out research to establish how much of auditory circuitry development is pre-programmed and how much is subject to random activity organisation events (like we saw in Carla Shatz's eye research) or environmental factors. The otic vesicle gives birth to auditory neurons, which Goodrich found decide very early whether they will become cochlear neurons (the ones carrying hair vibration data to the brainstem) or vestibular neurons (the ones responsible for balance). Once they've made the decision, it's off to the races. Unlike much neural growth, which is characterised by slow, exploratory behaviour, once a neuron makes up its mind that it is a cochlear neuron, it makes a beeline for precisely where it needs to go, with that decision largely determined by when that neuron was born, thereby harnessing the temporal organisation of birth order to create a spatial order of connectivity that lines up with the order of frequency-specific hairs in the cochlea.

Goodrich's lab is also working on the phenomenon of hearing loss. When we experience loud noises, or simply when we age, we lose synapses between our hair cells and our spiral ganglion neurons, effects which a different system composed of olivocochlear neurons which run from the brainstem to the cochlea, are able to moderate to some degree. Goodrich is currently working on how to apply the knowledge she has acquired of auditory neuron biology to help protect against synapse loss and, potentially, even reverse the process.

Eleanor Maguire (b. 1970):

Though Eleanor Maguire's name will probably always be remembered for the famous 'Taxi Driver Study' (2000) which found that London taxi drivers, in the course of their grueling 4-year training to memorise the city's 25,000 streets, experience profound growth in the rear region of their hippocampi, she is by no means an academic one-hit wonder. One of the most decorated of modern neuroscientists, and a Fellow of the Royal Society since 2016, Maguire has established a decades-long reputation for originality in probing the limits and structures of our memories. When considering what amnesiacs could teach us, for example, instead of the usual round of exploring how their knowledge of the past had been lost, she had the idea of testing how well they are able to imagine the future. In a 2007 paper publishing her results, she reported that hippocampal amnesiacs had markedly greater difficulty than control subjects in producing coherent visions of the future, that they were able to produce bits and pieces of images, but were unable to link it all together in a cohesive setting. She attributed this to the importance of the hippocampus in giving memories their spatial context, an ability which is also harnessed in creating future or fictional mental scenes.

Maguire is also an authority on episodic memory – the sort of autobiographical memory we have of events in our daily life that is distinct from our more rote memory of, say, who the lead singer of Men Without Hats is. This has historically been a tricky area of study, with different neuroimaging studies often giving seemingly contradictory results on what parts of the brain come into play when we are remembering life events. Maguire found that, while both the right and left hippocampi are active when recalling recent autobiographical events, as you ask people to recall events further back in time, the right hippocampus progressively checks out of the process, with the dorsal amygdalas following suit, indicating an intriguing physical asymmetry in our recall of distant memories (2003). Interestingly, however, Maguire also found, when comparing how young and old people retrieve memories, that the young lean far more on the left hippocampus, while the old tend to use both the right and left (2003). In studies of how we store and recall life memories, Maguire has argued for the importance of a 'scene construction' model whereby the parts of the brain responsible for generating a coherent spatial context work together with stored informational content to reconstruct a scene, and that any damage to our ability to generate spatial contexts will have a corresponding impact on our ability to tie together the elements of a past life event into a coherent memory.

The lead singer of Men Without Hats is Ivan Doroschuk.

Sharon Thompson-Schill (b. 1970):

When Brenda Milner revitalised the frontal lobes as areas of functional interest in the 1970s, she but little knew the maelstrom of research activity she had unleashed, which today includes a tsunami of research employing fMRI techniques to distinguish the biological underpinnings of the frontal lobe's varied and complex cognitive processes. Sharon Thompson-Schill, born in Washington DC, and receiving her degrees from Davidson College (BA, 1991) and Stanford University (PhD, 1996), has contributed some 190 articles to the elucidation of the brain's cognitive structures, including a much-cited 1997 paper on the function of the interior frontal gyrus that is worth a talk, because the experiment is just fundamentally cool.

What she and her colleagues were trying to establish was that the IFG, contrary to previous theories, isn't so much involved in the retrieval of semantic knowledge (the knowledge of facts and objects we have accrued over time, as opposed to our knowledge of how to do things), as it is in selecting a particular feature to focus on in a sea of semantic alternatives. Her study involved, then, designing tasks which presented subjects with comparisons of different degrees of difficulty, and measuring how the IFG responded. For example, when asked to generate a verb in response to a cue word, some tasks were designed to be 'Low Selection' by using objects that have one verb overwhelmingly associated with them – Kite almost always produces Fly, Scissors almost always produces Cut. There's not much agonising choice between alternatives, so the IFG shouldn't be too active, even though this task uses semantic memory of verbs. The 'High Selection' category, however, featured nouns that have multiple possible associated verbs – Rope could elicit Pull or Drag or Hang or Tie, and the act of choosing which feature of Rope to highlight and select should, by Thompson-Schill's theory, make the IFG kick into high gear. It is an elegant experiment that created high and low conditions across multiple types of semantic tests, and which found in all of them IFG activity that varied directly with the difficulty of selection in the given task.

This paper is just one example of the clever neural sleuthing that has allowed Thompson-Schill to find more refined distinctions of old functional definitions, and to unify competing theories within larger syntheses (as in her 2005 paper on how Broca's area, and the left interior frontal gyrus generally, participates in syntactic processing of sentences), which has given biological heft to categories that previously existed only in the rarefied air of philosophy courses, and probed how the brain copes with novel situations, such as having to come up with new and unconventional ways to use common tools. She is currently the Christopher H. Browne Distinguished Professor of Philosophy at the University of Pennsylvania and co-director of the Centre for Cognitive Neuroscience.

Aude Billard (b. 1971):

Robot psychology. It seems like something that only an Isaac Asimov would have to really hunker down and think about, but as we get better and better at constructing autonomous machines, we have to start thinking about the brains we are giving them, and how those brains are going to interact with the worlds they find themselves in. In effect, we are looking down the other end of the telescope now, towards the end of our journey. Instead of moving from the brain we happen to have, and investigating the psychological structures that brain produces, we are now increasingly able to ask what psychological features we want a being to have, and then construct the artificial brains that will produce them. Aude Billard is a Swiss engineer whose work takes place on the bleeding edge of machine learning. She has studied robot social interaction paradigms with the goal of creating multi-robot systems that are able to take in dynamic information, communicate it among themselves, and adapt their behaviours according to the needs of the situation and their own knowledge specialties and past experiences. As if that isn't cool enough (and, let's face it, that's probably the coolest sentence I'll ever write), she also works in the area of robot learning, using insights on how humans learn to design processes whereby robots can piece together the components required to solve a new problem, and develop for themselves new skills to meet new needs. If you want to be just entranced for a minute and a half, head over to YouTube and look up the EPFL video 'Ultra-fast, the robot arm catches objects on the fly' which shows not only some really hypnotising videos of robot arms just snatching objects out of the air, but also features Billard talking about the future applications of the type of learning systems she and her team have developed.

Jocelyne Bloch (b. 1971):

Like Billard, Jocelyne Bloch is a scientist doing work that, just fifteen years ago, would have sounded like the wildest dreams of science fiction. For Bloch, working at the Wyss Centre and EPFL, that research involves restoring functionality lost by neural damage, including exciting work to restore mobility to people who underwent severe spinal cord injury. By using spinal cord imaging to create a model spinal cord using procedures similar to those employed by Katrin Amunts to develop the Brain Atlas, Bloch and her colleagues were able to create a baseline structure that could be selectively altered for each individual case. Often, in the case of a severing of the spinal cord, the neurons above and below the break are still capable of operation, but signals simply cannot clear the gap to the lower body. Bloch was able to restore lower limb movement in humans using implants that bridged the subjects' spinal injuries, permitting neural signals to flow through lower body neurons again, allowing at first artificial control, and eventually voluntary control of lower body muscles (2018). Currently, Bloch is part of the ABILITY project, which is a wearable brain-computer interface that detects brain activity, decodes it, and then translates that activity into desired action through coordination with prosthetics or mechanical devices, a potential boon to individuals suffering from extreme conditions such as ALS. And on top of *that* she is also part of the Epios project, which is an unobtrusive implant which would record long-term data of brain activity to aid in both the diagnosis of brain disorders and their subsequent treatment.

Daniela Schiller (b. 1972):
As a researcher at the Icahn School of Medicine at Mount Sinai since 2010, Daniella Schiller has devoted herself to the investigation of how we acquire emotional responses to previously neutral objects and, perhaps even more intriguingly, how we go about un-acquiring them. The world is a complicated and shifting place, and to make our way through it, we need to evaluate novel situations and emotionally weigh them in an appropriate manner, protect ourselves with fear reactions when necessary, and un-learn those fear reactions when they get in the way of living life normally. In a series of papers spread over the past decade, Schiller has demonstrated how memories, rather than fixed objects, are dynamic beings that, when recalled, are vulnerable to 'updating.' By introducing new information during this vulnerable period, memories can be altered over time, and even extinguished. This work explains the earlier findings of Elizabeth Loftus as to how the way that questions are posed can alter what people remember about events – question posing compels recall, and opens the memory to alteration. Schiller's interest, however, seems to be less in the legal implications of these discoveries, and more in the therapeutic applications. Some people are so dominated by a traumatic memory that they cannot function in a way that they find meaningful or worthwhile, and for them, the idea of memory extinction might be the only way forward into a desirable life. Schiller, among other interests she is pursuing at Icahn, is investigating ways that this might be possible, through either pharmacological or drug-free methods.

Petra Ritter (b. 1974):
As cofounder of The Virtual Brain, Petra Ritter is engaged in the task of creating a brain model that doctors and researchers can use to investigate the structure and connectivity underpinning different brain processes. In the 'virtual brain', the significant unit is the node, a collection of neurons whose joint behaviour is given by a collection

of differential equations adapted from effective single-neuron models. Dividing the brain up into ninety-six distinct regions, Ritter's system is able to model novel aspects of overall connectivity and explain how the different neural paths involved in mental activity produce the intricate and exact timing required to pull off the chemical computations that motivate that activity. The model has also turned up interesting information about the brain's resting state, i.e. what our brain is doing when we are 'doing nothing.' Instead of reverting to a blank state lit only by the needs of our automatic functioning, the resting brain instead keeps itself in a state of readiness, keeping systems running on the off chance that they'll be needed to respond quickly to new happenings. This aspect of resting mental vigilance is particularly well modeled by the virtual brain.

There are many applications for this technology, including helping surgeons improve the accuracy of their methods through modeling individual patients' mental condition and therefore more minutely differentiating what regions are responsible for disruptions of function, outlining the brain dynamics involved in brain diseases such as Alzheimer's or injuries such as those caused by strokes, and creating an efficient study mechanism for long term processes associated with aging.

Camilla Bellone (b. 1975):
Bellone is at the forefront of the drive to understand neuropsychiatric disorders at the molecular level, and develop techniques to treat them accordingly. Her main area of focus has been elucidating the molecule structures that contribute to autism spectrum disorders (ASD), to which end she has participated in the discovery of the role that a mutation in the scaffolding protein Shank3 common to ASD individuals plays in reducing dopamine neuron activity in mice, resulting in subsequent impaired social activity, an impairment that she and her colleagues were able to undo through optogenetic neuron stimulation. She has also investigated the role of NMDA receptors in ASD, which are highly common ion channels that play major roles in signal transmission at synapses and neuroplasticity. Different NMDA receptors are made up of different collections of subunits, which give it particular physical, chemical, and signaling properties. By studying the structural and functional details of the different sub-types, Bellone has contributed to the development of a chemical-level explanation for linkage between under/over-performing NMDA receptors and neuropsychiatric disorders.

Joy Harden Bradford (b. 1979):
As host of the podcast *Therapy for Black Girls*, psychologist Joy Harden Bradford is the latest link in a long chain of women psychologists seeking to promote cultural competence in psychology, and work towards better mental health practices for individuals who have historically been denied access to, and culturally been warned against, professional help for their unique psychological challenges. With millions of episode downloads and hundreds of thousands of followers, *Therapy* has been an unambiguous success in bringing to light the particular problems faced by Black girls and women, particularly

in the United States. And, while resources like *Therapy* have proven important for rolling back the social stigmas against Black women seeking outside advice for their mental health, Bradford points out that motivation is only part of the problem. In her experience, one way to help Black women bridge their wariness of the professional mental health system is to have Black women psychologists available to meet with and yet, historically, Black women have only made up something on the order of 5 per cent of psychologists. And then, even if one *does* want help, and *does* find a psychologist one is comfortable with, there are still socio-economic factors related to lack of insurance in the American Black community that mean even highly motivated women with unique opportunities still might not be able to access the care that they need. *Therapy*, then, in conjunction with the Facebook group Bradford has formed and the online index of psychologists she has assembled to put Black women in touch with someone who can help them, represents a mighty individual effort to overcome large-scale institutional and historical forces which goes to show how much can be done by one talented person, motivated by a desire to redress systemic wrongs, and possessed of the cultural savvy to harness technology to address people where they are, as they are, and with compassion and awareness, lead them to where they would like to be.

Hannah Critchlow (b. 1980):
Through books like *Joined Up Thinking* (2022), *The Science of Fate* (2019), and *Consciousness* (2018), as well as her numerous media appearances on television and podcasts, Hannah Critchlow is one of the most prominent members of the modern neuroscience #SciComm community. As a scientist, her research has included a molecular-level investigation of the mechanisms of anti-psychotic drugs, particularly with regard to schizophrenia. Her 2006 study analysed two popular drugs prescribed for schizophrenia, clozapine and haloperidol, finding that the first acted by significantly increasing dendritic spine densities (remember from Catherine Woolley's research that the more spines there are on dendrites, the easier it is for signals to flow through those neural regions), while haloperidol, which is also prescribed for delirium, psychosis, and bipolar disorder related mania, has the opposite effect, significantly *decreasing* spine density.

Rosemary Bagot (b. 1981):
Rosemary Bagot studies the phenomenon of stress resiliency, which underlies the daily observed fact that some people function seemingly just fine under stress, while others develop symptoms of significant and even deep depression. Prior to Bagot's work, it had been known that the nucleus accumbens (NAc) played an important role in the development of depression. In 2015, Bagot published results demonstrating that the chemical reactions in the NAc which are associated with depression are triggered by inputs from the ventral hippocampus (vHIP) and medial prefrontal cortex (mPFC). By employing optogenetic methods, which allow activity of certain neurons to be amplified using light, Bagot was able to show that, when dialing up the inputs from the vHIP, mice became increasingly susceptible to stress, and when dialing those inputs down, they became increasingly resilient to it. Stimulation of the inputs from the mPFC has

the opposite effect, with stimulation producing resiliency and diminution producing depression. This suggests a fine-tuneable control system for an individual's reaction to stress based on the relative strengths of the mPFC versus vHIP input signals. Bagot suggests a 'Three Hit' model for how our brain determines that balance, based on (1) genetic predispositions, (2) early development (as we saw above in Melly Oitzl's studies of stress resiliency), and (3) later life environment (with great disparities between the stress profile of our early years and that of our later years being particularly effective at diminishing our resiliency and pushing us towards depression).

Janina Scarlet (b. 1983):
Born in Ukraine, Janina Scarlet was only 3 years old when the Chernobyl Nuclear Disaster occurred. Her family lived less than 200 miles from the plant, and she was poisoned by the radioactive fallout. Scarlet was racked by seizures, nosebleeds, and migraines in early life, and when she was 12 her family decided to move to the United States, where she faced taunting by her peers. She fell into depression, but was rescued by, of all things, the 2000 Patrick Stewart-Ian McKellan *X-Men* film. As with so many loners and societally rejected individuals before and since, she identified with the X-Men, spurned by society but resilient in their mutual connections, and inspiring in their personal hero's journey against the limitations placed on them by their origins.

Later in life, after earning her graduate degree in psychology, she experienced a breakthrough moment talking with a marine suffering PTSD. Like many in the armed forces, he identified strongly with super-heroes, and in the course of conversation, Scarlet got him to realise that, though Superman had weaknesses (magic, kryptonite), that did not make him any less of a hero, which allowed the patient to see himself not in terms of his failures and weaknesses, but in terms of his ideals and strengths. This was the beginning of 'Superhero Therapy,' Scarlet's novel approach to getting patients to open up about their problems by discussing them in the contexts of the characters they identify with, and using the links that people have with their fictional idols to drive their own self-reevaluations and paths of goal formation, allowing patients to become the heroes of their own life story, often flawed, but never beyond hope.

Urtė Neniškytė (b. 1983):
How do you make a brain? Well, one thing you could do is create some sort of code whereby every neuron grows and connects to precisely the other neurons it is supposed to connect to, and everything is perfectly wired from the start. That is connectively efficient, but also requires a great deal of pre-coding, and isn't terribly resilient against Stuff Going Wrong. We do see it with regard to certain systems, like the development of the cochlear neurons discovered by Lisa Goodrich, but for the most part the brain uses a second strategy, which is to just have every neuron connect with a whole bunch of stuff, and then to prune away the connections that don't make sense as you grow and develop. This minimises coding, but at the expense of creating a plethora of connections only to destroy them again a few years later. How this process works on the molecular level is the domain of Lithuanian neuroscientist Urtė Neniškytė, who

in 2018 co-authored a paper detailing the mechanisms by which wandering microglia 'trogocytose' the axons at synapses requiring pruning. Previously, it had been believed that microglia devour axons through phagocytosis, which would have involved them engulfing the entire synaptic region and internally digesting it. Neniškytė and her colleagues found evidence instead for trogocytosis, which is more of a nibbling effect by which pieces of the presynaptic neuron sink into depressions in the microglia which are then bitten off and digested by lysosomes.

Neniškytė has also studied the chemistry behind microglia's other great role, that of devourer of dying neurons. During episodes of neural trauma, microglia typically rush to the scene and employ phagocytosis to remove any damaged material they find there. This is usually a good thing, and it was long assumed that microglia only ate the dead and dying, but in 2011 Neniškytė published results finding that, during inflammation, microglia at the scene sent out chemicals that induced perfectly healthy neurons to place chemical 'Please Eat Me' signs along their surfaces, which the microglia would then engulf and eliminate. By employing processes that prevented healthy cells from displaying the 'Please Eat Me' markers, Neniškytė and her colleagues were able to prevent 90 per cent of viable neuron loss during inflammation events, a potential boon in the prevention of inflammation-related neurodegeneration.

Debra W. Soh (b. 1990):

Paraphilia is one of psychology's historically most contentious terms. Previously known as 'sexual deviation' it has in earlier DSM classifications included homosexuality and transvestism, but today is more strictly defined as sexual arousal caused by non-typical items, such as animals, children, or inanimate objects. As a society, we are both intensely interested in paraphiliacs and variously creeped out by them, both of which responses tend to shove them over into the category of some Other Manner of Being, instead of as individuals who are often as troubled by how they are and what they respond to as we are, and who want nothing more than to understand themselves. Debra Soh's PhD dissertation was a significant step in realising this latter goal, by studying paraphilic males, and determining that neurological conditions lay at the root of their behaviour and preferences. In subsequent articles, she has argued for a re-think of how we, as a society, approach paraphiliacs, stressing the need to understand them and help them understand themselves, rather than categorising them as irredeemable monsters.

Concurrent with her studies of paraphilia, Soh became embroiled in the modern gender transition and gender fluidity debate via a 2015 editorial when she expressed reservations about childhood gender transition. This predictably resulted in a Twitter backlash from the trans activist community, and in response Soh left academia, feeling that she couldn't do objective work in an academic environment that only accepted results that confirmed certain politically determined positions. Instead, she turned to journalism, and in 2021 published *The End of Gender: Debunking the Myths About Sex and Identity in Our Society*, which has at its core a set of 'Nine Myths' that Soh believes need to be re-evaluated in the light of modern scientific research. It has proven as polarizing as you might have expected.

So... What Does That Mean?

A Glossary of Frequently Used But Kind of Weird Terms

Amygdala: We have two amygdalae in our brain, one in each temporal lobe. They are most associated with emotional learning, and particularly with fear conditioning, or how we learn to recognise and avoid situations associated with harm. They also appear to have a role in the consolidation of memories.

Behaviourism: Though it has nineteenth century antecedents, behaviourism as we tend to know it today has its roots in the work of John B. Watson in the 1920s and B.F. Skinner in the 1930s, which was in response to early psychology's tendency to make claims about the component pieces of purely internal or mental states. Behaviourism rejects this approach as subjectively tainted, saying that only external events can be meaningfully measured and interpreted, and as such should be the primary, perhaps even sole, focus of psychology, with an emphasis on how our experience of the world drives our behaviour within it.

Clinical Psychology: As it emerged in the 1890s through the work of Lightner Witmer, clinical psychology represented an 'applied' approach to psychology, one which engaged with communities to use knowledge of psychology to provide services that would better people's lives, instead of remaining purely in the realm of abstract theory and academic discussion. Though clinical psychologists do still participate in academic debates, their focus is on engagement with the public, and using their expertise to assess individuals' problems and suggest treatment techniques.

Cognitive Psychology: By the 1950s, behaviourism was on the ropes. People wanted to know how thought, memory, and creativity, long branded forbidden realms of inquiry by behaviourism, worked, and with technology having emerged in the 1940s that made some of those areas probable through experiment, it was time for a cognitive revolution. What are the different types of memory, and how do they work? How do we attend to something? Might the structure of computers give us useful insights into how brains process information? These juicy questions were returned to the table by CP, as psychologists pursued how the structures of the brain could explain the structure of cognition.

Developmental Psychology: Developmental psychology is interested in the formation of psychological structures over time - what aspects of a child's environment can be traced to later behavioural tendencies? How does aging affect our various cognitive abilities? What behaviours become more likely during adolescence, and why? Developmental psychologists in the early twentieth century often engaged in large longitudinal studies to measure multiple aspects of individuals' lives over the course of decades to produce baselines of ordinary development, and detect experiences that had magnified impacts on later behaviours, and devised novel experiments to see just when certain cognitive/behavioural milestones are reached, such as the ability to show empathy, the awareness of inanimate objects, or the ability to form complex sentences.

Eugenics: Aw jeez. Popular in the late nineteenth and early twentieth centuries, eugenics grew out of concerns that civilization was on the decline, and needed to be set on the right path again. Because science had learned that some aspects of individuals were hereditary, eugenicists believed that a better society could be created by maximising the number of desirable traits in the population, which would then be passed on, and by minimising the number of undesirable traits in the population, which would then go on the decline. Many psychologists (particularly in America) were also eugenicists, who held that testing could be used to identify promising and unpromising individuals from ever earlier ages, with some going so far as to advocate for the internment or sterilisation of the latter, for the 'good of society'.

Evolutionary Psychology (EP): Though boasting an origin going back to Darwin himself, EP really started going in the 1960s and 1970s as researchers began looking at social interactions in an evolutionary context – what reactions and social instincts would have tended to allow distant humanity to better pass on their genetic material, and how do those still manifest today? Instead of thinking of the brain as a large general organ that specialises through experience, EP psychologists see it as an amalgamation of particular circuits accumulated over time, each of which efficiently solved a problem in our collective past.

Experimental Psychology: Emerging in the late nineteenth century, experimental psychology tasked itself with moving psychology from a branch of philosophy into an experimental science, with tightly controlled variables and procedures for isolating the different phenomena of our psychological lives. Wilhelm Wundt, who as we have seen touched the lives of just about every figure of significance in the early decades of professional academic psychology, was the founding figure here, bringing the weight of mathematical analysis and statistical methods to bear on the investigation of mental phenomena, even applying it to the seemingly highly subjective practice of introspection, popular in early American psychology, whereby subjects attempted to atomise their internal states under different circumstances, and break them down into basic units of mental experience, an approach associated with the 'structuralist' school of psychology.

Frontal Lobe: The largest part of the cerebrum, and the one containing the prefrontal cortex, the frontal lobe for a long time in the early twentieth century was considered a poor cousin to the posterior of the brain, where extensive but way scientific poking had revealed that much of our sensory centres lie. By the mid-twentieth century, and thanks to a number of the figures we've read about in this book, it came into its own as a centre of movement control and, thanks to the prefrontal cortex, of the working memory and internal mental modeling facets that are the hallmarks of humans.

Gestalt Psychology: Unlike structuralism, which sought to reduce psychological phenomena to their component parts through tightly controlled experiments and analyses, and then rebuild humans psychologically from the ground up using those components, Gestalt psychologists argue that we interact with an object or situation as a whole in a way that is more than just the sum total of all the individual parts in it we recognise. There is some extra bit of something at work when we approach and solve problems in the world, or engage with individuals or systems, that will never be arrived at through atomistically investigating our psychological processes.

Hippocampus: A seahorse shaped structure located on either side of the brain, the hippocampus structure plays a major role in allowing short term memories to make their way into long-term storage. Individuals like the famous patient H.M., who through surgery lost significant chunks of his hippocampi, are unable to make new long-term memories and live in a constant and never-ending present. It also appears to play a role in the formation of mental maps of a region.

Humanistic Psychology: Humanistic psychology (HP) sought a new role for what a therapist is supposed to do for their patients. As opposed to psychoanalysis, whereby the therapist is the expert, compelling the patient to confront deep truths that have their origins in earliest childhood, in HP the therapist is there to listen non-judgmentally as a supportive presence while the patient comes to grips with themselves as a complete individual, and to use empathy and compassion to help that patient formulate and achieve their unique goals.

Long Term Potentiation: 'Cells that fire together, wire together.' Long term potentiation (LTP) is the process whereby two neurons, after repeatedly communicating with each other across a synapse, develop new structures that allow their future communications to be more efficient still. It is a key ability behind neural plasticity, and an important cellular-level explanation for why practice allows us to improve our abilities and recall.

Myelination: Myelin is the material that encases most of our neurons. The combination of myelinated lengths with interspersed gaps in the myelin are responsible for our ability to send signals quickly through the nerves in our body – a myelinated nerve is some 200 *times* faster in sending a signal than an unmyelinated one. Diseases that degrade myelin, then, such as multiple sclerosis, can have profound effects on our ability to coordinate movement.

Neurons and Glia: In your classic biology textbook, neurons are the main players in the nervous system, with one long axon leading from an octopus-like cell body, and ending in a number of branching termina that, by communicating over gaps called synapses with the arms (or dendrites) of neighboring cell bodies, are able to transmit signals throughout our body. But it's not all about the neurons – sure, there are tens of billions of them in your brain, but there also tens of billions of another cell type, the glia, which are responsible for all manner of maintenance tasks, including clearing neurotransmitters out of synapses, repairing neurons, and creating myelin, but are also under investigation for being potentially responsible for inflammation effects that lie at the root of some of our most debilitating brain diseases.

Neurotransmitter: These are chemicals that allow nerves to communicate with each other, or with other tissues, and include some of the most important chemicals for the regular functioning of our brains, such as dopamine, serotonin, epinephrine, acetylcholine, and GABA. By releasing different types of neurotransmitters from the ends of my axons, I can either encourage nearby neurons to fire, or discourage them from doing so, allowing a wonderful fine-tuning of mental responses and reactions to situations.

Nucleus Accumbens: Part of the striatum located in our forebrain, the nucleus accumbens is a key region for the weighing of reward and determining of motivation which lies at the centre of when we decide to do things. If the amygdala is all about fear training, the NA is all about positive reinforcement, determining what we need and pushing us to do the things that will secure those needs. As such, when things go wrong with the NA, it can lead to a feeling of compulsion to repeat actions in spite of diminishing returns, which is a major hallmark of addiction.

Occipital Lobe: The part of your cerebrum located at the very back, this is where visual processing happens. Damage to the occipital lobe can result in colour agnosia (experiencing the world as essentially a variety of greys even though your eyes are capturing the colour just fine), hallucinations, or significant losses to reading ability.

Operant Conditioning: A major technique of behaviourism, operant conditioning is the systematic use of rewards or punishments to train animals in the performance of certain behaviours. B.F. Skinner's famous 'Skinner boxes' are the classic example of an OC system, whereby mice or pigeons could be trained, using shocks or food rewards, to perform increasingly complex tasks.

Parietal Lobe: In between the frontal lobe and the occipital lobe lies the parietal lobe, which you can think of roughly as the top-ish, back-ish but not too-far-backish part of your cerebrum, and it is here where all of your body's touch-based sensory data goes for processing, with larger areas of the parietal lobe given over to the parts of our body with a finer sense of touch, such as our hands, lips, and genitals.

Plasticity: Neuroplasticity is our brain's ability to change itself to meet new needs. On the smaller scale, this can be as simple as using LTP to strengthen associations between neurons we need for the tasks we do a lot, but on the more dramatic side this can include the rewiring of the brain to compensate for severe damage, finding new ways and wirings to accomplish the tasks normally assigned to the damaged region. (Incidentally, this capacity is called compensatory masquerade, which is definitely going to be the title of my autobiography.)

Prefrontal Cortex: The largest part of the frontal lobe, the prefrontal cortex (PFC) hosts most of those activities that we think of as at the root of our nature as humans, including the ability to plan towards long term goals, to hold items for extended periods of time in working memory, to run hypothetical models of situations and evaluate them, as well as a host of other 'executive' functions. Damage to the PFC severely limits our ability to form coherent life plans, to exhibit socially appropriate behaviours, and to pull off those thousand little daily acts of semi-temporary memory that allow us to move meaningfully from moment to moment.

Psychiatry: Unlike psychologists, who tend to possess PhDs and who treat patients through a variety of talking, behaviour modification, roleplay, or group dynamic techniques (to name a very paltry few), a psychiatrist is an MD who has the ability to prescribe medications to treat mental conditions that have their roots in neurochemistry.

Psychoanalysis: An outgrowth of the theories of Sigmund Freud, psychoanalysis hypothesises that individuals, in going through the stages of childhood, develop neuroses that arise from tensions between the different parts of their unconscious mind and the restrictions placed upon them by the outside world, including particularly intense and sexually charged rivalries with parental figures. The therapist, then, is tasked with leading the patient to the realisation, against that patient's often vociferous resistance to the idea, that the root of their present mental conditions lies in those childhood traumas and frustrated urges, and that only by acknowledging and accepting their existence can progress be made.

Temporal Lobe: The last of our lobes, the temporal is named not because it has anything to do with time, but because it lies, roughly, under our temples, on the sides of the cerebrum. The hippocampi live here, with their role in consolidating experiences into long-term memories, and it is also here that auditory information is processed, and the semantic structure of language is interpreted. It *also* plays a role in object recognition, so yeah, *do* try and take good care of it.

Working Memory: The type of memory that allows us to keep something in mind for a little while without having it directly in front of us, but not necessarily forever. Every time you look at the security number on your credit card, then put the card down and type the number onto the screen from memory, or run back into a room to pick up the keys that you left on the piano, you are using working memory. These are things that don't need to be remembered for the rest of time, but that are really handy to know for a little while, and there are some positively beautiful structures in the prefrontal cortex that allow us to create and then dismiss these temporary memories.

Selected Bibliography

General Histories
Bookwala, Jamila and Nicky J. Newton, ed. *Reflections From Pioneering Women in Psychology.* (Cambridge, 2022).
Gilder, George. *The Silicon Eye: How A Silicon Valley Company Aims to Make All Current Computers, Cameras, and Cell Phones Obsolete.* (Atlas Books, 2005).
Haines, Catherine M.C. *International Women in Science: A Biographical Dictionary to 1950.* (ABC-Clio, 2001).
McGrayne, Sharon Bertsch. *Nobel Prize Women in Science: Their Struggles, and Momentous Discoveries.* (Joseph Henry Press, 1993).
O'Connell, Agnes N. and Nancy Felipe Russo, ed. *Models of Achievement: Reflections of Eminent Women in Psychology.* (Columbia University Press, 1983).
———. *Women in Psychology: A Bio-Bibliographic Sourcebook.* (Greenwood Press, 1990).
Ogilvie, Marilyn Bailey. *Women in Science: Antiquity Through the Nineteenth Century. A Biographical Dictionary with Annotated Bibliography.* (MIT, 1986).
Ogilvie, Marilyn and Joy Harvey, ed. *The Biographical Dictionary of Women in Science: Pioneering Lives from Ancient Times to the Mid-Twentieth Century.* 2 vols. (Routledge, 2000).
Rossiter, Margaret W. *Women Scientists in America: Volume I: Struggles and Strategies to 1940.* (Johns Hopkins University Press, 1982).
——— *Women Scientists in America: Volume 2: Before Affirmative Action: 1940-1972.* (Johns Hopkins University Press, 1998).
———. *Women Scientists in America: Volume 3: Forging a New World Since 1972.* (Johns Hopkins University Press, 2012).
Sayers, Janet. *Mothers of Psychoanalysis.* (WW Norton, 1991).
Scarborough, Elizabeth and Laurel Furumoto. *Untold Lives: The First Generation of American Women Psychologists.* (Columbia University Press, 1987).
Strohmeier, Renate. *Lexikon der Naturwissenschaftlerinnen und naturkundigen Frauen Europas.* (Verlag Harri Deutsch, 1998).
Volkmann-Raue, Sibylle and Helmut E. Luck. *Bedeutende Psychologinnen des 20. Jahrhunderts.* (VS Verlag, 2011).
Warren, Wini. *Black Women Scientists in the United States.* (Indiana University Press, 1999).
Yost, Edna. *American Women of Science.* (Frederick A. Stokes Company, 1943).

Books By Great Women Psychologists and Neuroscientists
Ainsworth, Mary. *Patterns of Attachment: A Psychological Study of the Strange Situation.* (Psychology Press, 1979).
Barrett, Deirdre. *Supernormal Stimuli: How Primal Urges Overran Their Evolutionary Purpose.* (Norton, 2010).
Bechtereva, N.P. *The Neurophysiological Aspects of Human Mental Activity.* (Oxford University Press, 1978).
Bellugi, Ursula and Roger Brown, ed. *The Acquisition of Language.* (Chicago, 1964).

Bellugi, Ursula and Edward Klima. *The Signs of Language*. (Harvard University Press, 1979).
Bibring, Grete. *Grete Bibring: A Culinary Biography*. (Boston Psychoanalytic Society, 2015).
Brazier, Mary A.B. *The Electrical Activity of the Nervous System, 2nd. Ed.* (Macmillan, 1958).
——. *A History of Neurophysiology in the 17th & 18th Centuries*. (Raven Press, 1984).
——. *A History of Neurophysiology in the 19th Century*. (Raven Press, 1988).
Browne, Rose Butler and James W. English. *Love my Children: The Education of a Teacher*. (David C. Cook, 1969).
Chin, Jean Lau. *Learning from My Mother's Voice: Family Legend and the Chinese American Experience*. (Teachers College Press, 2005).
Corkin, Suzanne. *Permanent Present Tense: The Unforgettable Life of the Amnesic Patient, H.M.* (Basic Books, 2013).
Cosmides, Leda, Jerome H. Barkow, and John Tooby. *The Adapted Mind: Evolutionary Psychology and the Generation of Culture*. (Oxford University Press, 1992).
Critchlow, Hannah. *The Science of Fate. The New Science of Who We Are – and How to Shape our Best Future*. (Hodder, 2019).
Crosby, Elizabeth Caroline, C.U. Ariens Kappers, and G. Carl Huber. *The Comparative Anatomy of the Nervous System of Vertebrates, Including Man*. 2 Vols. (Hafner Publishing, 1936).
Diamond, Marian Cleeves. *Enriching Heredity: The Impact of the Environment on the Anatomy of the Brain*. (The Free Press, 1988).
Fernald, Grace M. *Remedial Techniques in Basic School Subjects*. (McGraw-Hill Book Company, 1943).
Firlik, Katrina. *Another Day in the Frontal Lobe. A Brain Surgeon Exposes Life on the Inside*. (Random House, 2006).
Goodnow, Jacqueline, Jerome S. Bruner, and George A. Austin. *A Study of Thinking*. (Routledge, 1956).
Graybiel, Ann M. and S. Grillner. *Microcircuits: The Interface between Neurons and Global Brain Function*. (MIT Press, 2006).
Greenfield, Susan. *A Day in the Life of the Brain: The Neuroscience of Consciousness from Dawn Till Dusk*. (Penguin, 2017).
——. *Mind Change: How Digital Technologies are Leaving their Mark on our Brains*. (Random House, 2014).
Groh, Jennifer M. *Making Space: How the Brain Knows Where Things Are*. (Belknap Press, 2014).
Haraguchi, Tsuruko, transl. by Yoko Kamei. *In America: 1907-1912*. (2006).
Heidbreder, Edna. *Seven Psychologies*. (Appleton-Century-Crofts, 1933).
Hertz, Rachel. *The Scent of Desire: Discovering Our Enigmatic Sense of Smell*. (William Morrow, 2007).
Hollingworth, Leta S. *The Psychology of the Adolescent*. (D. Appleton & Co, 1928).
Jahoda, Marie. *Ich habe die Welt nicht veraendert: Lebenserinnerungen einer Pionierin der Sozialforschung*. (Beltz, 2002).
Jameson, Dorothea and Leo M. Hurvich. *The Perception of Brightness and Darkness*. (Allyn and Bacon, 1966).
Klein, Melanie. *Love, Guilt and Reparation and Other Works 1921-1945*. (Free Press, 1975).
LaFrance, Marianne and Clara Mayo. *Moving Bodies: Nonverbal Communication in Social Relationships*. (Brooks/Cole, 1978).
Levi-Montalcini, Rita. *In Praise of Imperfection: My Life and Work*. (Basic Books, 1988).
Lindsay, Grace. *Models of the Mind: How Physics, Engineering and Mathematics Have Shaped Our Understanding of the Brain*. (Bloomsbury Sigma, 2021).
Lineha, Marsha M. *Building a Life Worth Living: A Memoir*. (Random House, 2020).

Loftus, Elizabeth and Ketcham, Katherin. *Witness for the Defense: The Accused, the Eyewitness, and the Expert Who Puts Memory on Trial.* (St. Martin's Press, 1991).
Maccoby, Eleanor Emmons and Jacklin, Carol Nagy. *The Psychology of Sex Differences.* (Stanford, 1974).
———. *A Memoir: 1917-2017.* (2017).
Masters, William and Johnson, Virginia. *Human Sexual Response.* (Little Brown and Company, 1966).
Montessori, Maria. *The Montessori Method.* (Frederick A. Stokes, 1912).
Newcombe, Freda. *Missile Wounds of the Brain: A Study of Psychological Deficits.* (Oxford University Press 1969).
Pert, Candace B. *Molecules of Emotion: Why You Feel the Way You Feel.* (Scribner, 1997).
Rhine, Louisa E. *Psi: What Is It?* (Harper and Row, 1975).
Ridenour, Rita. *Mental Health in the United States: A Fifty Year History.* (Harvard, 1961).
Soh, Debra. *The End of Gender: Debunking the Myths About Sex and Identity in our Society.* (Threshold editions, 2020).
Thorndike, Edward L. and Cobb, Margaret. *The Psychology of Algebra.* (Macmillan, 1923).
Washburn, Margaret Floy. *Movement and Mental Imagery.* (Houghton Mifflin, 1916).
Weisstein, Naomi. *Naomi Weisstein: Brain Scientist. Rock Band Leader. Feminist Rebel. Her Collected Essays.* (Off the Common, 2020).
Wolf, Maryanne. *Proust and the Squid: The Story and Science of the Reading Brain.* (Harper Perennial, 2007).

Books About Great Women Psychologists
Brothers, Barbara Jo. *Well-Being Writ Large: The Essential Work of Virginia Satir.* (Beyond Words, 2019).
Dittrich, Luke. *Patient H.M. A Story of Memory, Madness, and Family Secrets.* (Vintage, 2017).
Fardeau, Michel. *Passion Neurologie: Jules Et Augusta Déjerine.* (Odile Jacob Sciences, 2017).
Grosskurth, Phyllis. *Melanie Klein: Her World and Her Work.* (Harvard University Press, 1987).
Grossman, Klaus E., Bretherton, Inge, Waters, Everett and Grossmann, Karin eds. *Maternal Sensitivity: Mary Ainsworth's Enduring Influence on Attachment Theory, Research, and Clinical Applications.* (Routledge, 2015).
Hinshelwood, R.D. and Fortuna, Tomasz. *Melanie Klein: The Basics.* (Routledge, 2018).
Klein, Ann G. *A Forgotten Voice: A Biography of Leta Stetter Hollingworth.* (Great Potential Press, 2002).
Kofler-Bettschart, Birgit. *Cecile Vogt: Pionierin der Hirnforschung.* (Ueberreuter, 2022).
Kramer, Rita. *Maria Montessori: A Biography.* (Addison-Wesley, 1976).
Lancaster, Jane. *Making Time: Lillian Moller Gilbreth – Life Beyond 'Cheaper by the Dozen'.* (Northeastern University, 2004).
Launer, John. *Sex vs. Survival: The Life and Ideas of Sabina Spielrein.* (Overlook, 2014).
Maier, Thomas. *Masters of Sex: The Life & Times of William Masters and Virginia Johnson.* (Basic Books, 2009).
Medvedeva, C.V. *Natalya Bextereva: Kakoi Mui Yeyo Znalee.* (Cova, 2009).
Quinn, Susan. *A Mind of Her Own: The Life of Karen Horney.* (Summit Books, 1987).
Richebacher, Sabine. *Sabina Spielrein: Eine Fast Grausame Liebe zur Wissenschaft.* (Dorlemann Verlag, 2005).
Roazen, Paul. *Helene Deutsch: A Psychoanalyst's Life.* (Anchor Press, 1985).
Robinson, Paul A. *The Modernization of Sex: Havelock Ellis, Alfred Kinsey, William Masters & Virginia Johnson.* (Cornell University Press, 1976).

Robinson, Virginia. *Jessie Taft: Therapist and Social Worker - A Professional Biography.* (University of Pennsylvania, 1962).
Valentine, Elizabeth. *Beatrice Edgell: Pioneer Woman Psychologist.* (Nova Science, 2006).
Young-Bruehl, Elisabeth. *Anna Freud: A Biography.* (Yale, 1988).
Yost, Edna. *Frank and Lillian Gilbreth: Partners for Life.* (Rutgers University, 1949).

Index

Abel, Theodora, 166–7
Adler, Alexandra, 176–7
Adler, Alfred, 176–7
Ainsworth, Mary, 91–4
Albe-Fessard, Denise, 199–200
Allen, Doris, 209
Allen, Ingrid, 215–6
Altinok, Aysiana, 212
Ames, Frances, 202
Amunts, Katrin, 278
Anastasi, Anne, 193–4
Aphasia, 183–4
Arlitt, Ada, 146
Arnold, Magda, 92, 181–2
Arredondo, Patricia, 230
Arthur, Mary, 135
Arvanitaki, Angelique, 177
Attachment Theory, 93–4
Attention, 218

Babcock, Harriet, 128–9
Bagot, Rosemary, 291–2
Baker, Emma, 30
Ball, Josephine, 164
Baumgarten-Tramer, Franziska, 135–6
Bayley, Nancy, 167–8
Bayley Scales (BSID), 168
Beauvallet, Marcelle, 181
Beck, Diana, 211
Behaviorism, 20, 165, 294
Bekhterev, Vladimir, 137
Bekhtereva, Natalia, 205
Bellone, Camilla, 290
Bellugi, Ursula, 231–4
Bem, Sandra, 226
Benjamin, Jessica, 261
Berkeley Growth Study, 168
Berkeley, University of (Cal), 32, 124, 152, 164, 284
Better Baby Contests, 36

Bibring, Great, 168–9
Billard, Aude, 288
Bills, Marion, 147–8, 185
Binet-Simon Test, 43
Blanchard, Phyllis, 156–7
Blatz, William, 92
Bloch, Jocelyne, 288–9
Block, Jeanne, 204
Bonini, Nancy, 276
Boysen, Gudrun, 222
Bradford, Joy, 290–1
Brazier, Mary, 187–8
Bronner, Augusta, 133–4, 142
Brousseau, Kate, 35
Brown v. Board of Education, 95–6
Browne, Rose, 162
Buck, Linda, 263
Bühler, Charlotte, 153–4, 210
Burr, Emily, 133

Cacioppo, Stephanie, 255–7
Calkins, Mary Whiton, xi, 10–14, 37
Canady, Alexa, 212
Carey, Leeanne, 276
Casagrande, Vivien, 225
Cattell, James, 19, 154
Cattell, Psyche, 154
Charcot, Jean-Martin, xi, 6
Chin, Jean Lau, 227
Clark, Mamie, 95–8, 197, 201
Classical Conditioning, 141–2
Clinical Psychology, 44, 294
Cobb, Margaret, 136
Computational Neuroscience, 224, 258–60
Corkin, Suzanne, 111, 235–7
Cosmides, Leda, 123–7, 284
Crissey, Marie, 196–7
Critchlow, Hannah, 291
Crosby, Elizabeth, 145–6

Dambinova, Svetlana, 266
Davis, Katharine, 33–4
Dawson, Valina, 277–8
Déjerine-Klumpke, Augusta, xi, 5–9, 22
Dembo, Tamara, 177–8
Denmark, Florence, 215
Deutsch, Helene, 58–63
Dewey, John, 45, 149
Diamond, Adele, 270–1
Diamond, Marian, 123–5, 127
Downey, June, 46–7
Drummond, Margaret, 39
Dunbar, Helen, 178–9

Eberhardt, Jennifer, 280–1
Eckstein, Emma, 36–7
Edgell, Beatrice, 39–40
Edwards, Blanche, 6–7
Einstein, Gillian, 268–9
Eisenhardt, Louise, 211
Electroencephalograms, 187
Eng, Helga, 47–8
Eugenics, 34, 44, 79, 196, 295
Evolution, 42–3
Evolutionary Psychology, 125–7, 295
Executive Functions, 270–1

Family Therapy, 100–102
Fernald, Grace, 132
Ferrero-Lombroso, Gina, 40
Fielding, Una, 144
Field Theory, 172–3
Fiske, Susan, 269
Fitzgerald, Maria, 271
Foa, Edna, 220
Franklin, Marjorie, 141
Franz, Marie-Louise von, 198
Freud, Anna, 56, 71, 77, 84–7
Freud, Sigmund, 12, 36, 54, 56, 59, 64–5, 71, 84–6, 134, 137, 155, 194
Frey, Łucja, 146–7
Fromm, Erika, 194
Frostig, Marianne, 189

Galton, Francis, 154, 196
Gamble, Eleanor, 37–8
Gaw, Esther, 132–3
Gestalt Psychology, 172, 178, 296
Gibson, Eleanor, 195–6

Gibson, William C., 115
Gilbreth, Lillian, 31, 49–52
Goldman-Rakic, Patricia, 242–5
Goldsmith, Marie, 42
Goodenough, Florence, 139–41, 149
Goodman, Lisa, 277
Goodnow, Jacqueline, 205–207
Goodrich, Lisa, 285–6, 292
Götz, Magdalena, 278
Gouin-Décarie, Thérèse, 204
Gould, Elizabeth, 279
Gracheva, Yekaterina, 37
Grafstein, Bernice, 212–13
Gray, Susan Walton, 198
Graybiel, Ann, 225
Greenfield, Susan, 267
Grid Cells, 248
Griffith, Monty, 104
Groh, Jennifer, 282–3
Group Therapy, 173
Grzegorzewska, Maria, 144–5

Hall, G. Stanley, 32, 156
Hampstead Clinic, 86–7
Hanfmann, Eugenia, 188–9
Haraguchi, Tsuruko, 73–6, 160
Hari, Riitta, 265
Harvard University, 10–13, 48, 155, 162, 185, 286
Hawk, Sara, 138
Hebb, Donald, 109, 224
Heidbreder, Edna, x, 148–9
Heinlein, Julia, 157
Helmholtz, Hermann von, 2–3
Hering, Ewald, 2–3, 202
Hilgard, Josephine, 189–90
Hippocampus, 111, 296
H.M. (Henry Molaison), 110–12, 235–7
Hockfield, Susan, 268
Hollingworth, Leta, 45, 77–80, 140, 166
Horney, Karen, 64–8
Howard, Ruth, 83, 170
Howes, Ethel, 41
Humanistic Psychology, 153, 210, 296
Hurlock, Elizabeth, 164

Ionescu, Sofia, 211–2
Ip, Nancy, 273–4
Isaacs, Susan, 137

Jahoda, Marie, 191–2
James, William, xi, 11–12, 20, 181
Jameson, Dorothea, 202
Johns Hopkins, 4, 138, 164
Johnson, Virginia, 118–22
Johnstone, Eve, 229
Jones, Mary Cover, 159–60, 165
Jung, Carl, 69–70, 72, 155, 198

Kanwisher, Nancy, 274–5
Kelley, Ann, 272
Kennedy, Mary, 264–5
Kent, Grace, 48
Kieffer, Brigitte, 275
Kimura, Doreen, 216–17
Kinsey, Alfred, 118
Klein, Melanie, 14, 53–7, 60, 71, 85–6
Koch, Helen, 157–8
Koffka, Kurt, 178
Köhler, Wolfgang, 172
Kopell, Nancy, 224
Kōra, Tomi, 160–1
Krasnogorski, Nikolai, 142
Kreek, Mary, 219–20
Kübler-Ross, Elisabeth, 208–209

Ladd-Franklin, Christine, xi–xii, 1–4, 19, 30
Lampl-de Groot, Jeanne, 158
Lapicque, Marcelle, 43
Levi-Montalcini, Rita, 88–90
Levine, Lena, 182
Lévy, Gabrielle, 139
Levy, Jerre, 221–2
Lewin, Kurt, 172–3, 178
Lewis, Margaret, 174
Linehan, Marsha, 226
Lindsay, Grace, 258–60
Little Albert Experiment, 165
Little Peter Experiment, 159–60
Loftus, Elizabeth, 227–8
Long Term Memory, 110–12
Lost In The Mall Experiment, 228
Lubinska, liliana, 186–7

Maccoby, Eleanor, 103–107
Maguire, Eleanor, 286–7
Mahowald, Misha, 279–80
Mahler, Margaret, 162–3

Marie, Pierre, 7, 23
Martin, Lillien, xii, 28–
Maslowe, Alfred, 210
Masters, William, 119–22
Mateer, Florence, 141–2
Maturationism, 169
Mauss, Iris, 252–14
Mayo, Clara, 213–14
McAdoo, Harriette, 223
McBride, Katharine, 183–4
McDougall, William, 149, 151, 157
McGeer, Edith, 114–17
McGraw, Myrtle, 169–70
McKee, Ann, 271–2
McLachlan, Elspeth, 223–4
Merrill, Maud, 142–4, 150
Miles, Catharine, 149–50
Milner, Brenda, 108–13, 235
Mirror Neurons, 249–50
Mitchell, Mildred, 183
Montessori, Maria, xii, 15–18, 77
Moore, Kate, 131–2
Moser, May Britt, 246–8
Münsterberg, Hugo, 13, 41, 48
Murphy, Lois, 179–80
Murray, Margaret, 174–5

Nash, Dorothy, 211
Neff, Mary, 35–6
Neniškytė, Urtė, 292–3
Neugarten, Bernice, 198–9
Neuroplasticity, 264
Neurotransmitters, 26–7, 115–16, 264, 266, 297
Newcombe, Freda, 207–208
Nobel Prize, 89, 142, 177, 246, 263
Nolen-Hoeksema, Susan, 275–6
Norsworthy, Naomi, 129–30

Oakland Growth Study, 160
Oitzl, Melly, 273
O'Shea, Harriet, 158

Parapsychology/PSI, 151
Parrish, Celestia, 29
Pavlov, Ivan, 138, 141
Payton, Carolyn, 207
Peak, Helen, 170–1

Penfield, Wilder, 109
Peretz, Isabelle, 274
Pert, Candace, 225, 262
Picciotto, Marina, 280
Platt, Julia, 31–2, 35
Plessy v. Ferguson, 81–3, 95
Positive Mental Health, 192
Prosser, Inez, 81–3
Pruette, Lorine, 161–2, 180
Psychoanalysis, 36–7, 54–72, 134–5, 158, 162–3, 261, 298
Psychosomatic Medicine, 178

Radke-Yarrow, Marian, 201
Rand, Marie, 138–9
Rapin, Isabelle, 209–10
Ravindranath, Vijayalakshmi, 269–70
Rayner, Rosalie, 164–5
Reed College, 103–104
Rees, Sandra, 224–5
Restorff, Hedwig von, 191
Rhine, Louisa, 151, 157
Richards, Ellen Swallow, 33
Rickers-Ovsiankina, Maria, 165–6
Ridenour, Nina, 184–5
Rioch, Margaret, 192
Ritter, Petra, 289–90
Roboz-Einstein, Elizabeth, 185–6
Roe, Anne, 185
Rogers, Carl, 153, 210
Rogers, Natalie, 210
Root, Maria, 272
Rosenthal, Tatiana, 137–8
Rosenstein, Alice, 211
Royce, Josiah, 11
Rubinstein, Susanna, xi, 28

Saffran, Eleanor, 221
Sanford, Edmund, 12
Santhanakrishna, Thanjavur, 212
Satir, Virginia, 99–102
Scarlet, Janina, 292
Scheffer, Ingrid, 275
Schiller, Daniela, 289
Scott, Sophie, 284
Security Theory, 92
Sergent, Justine, 267
Seward, Georgene, 180–1

Shatz, Carla, 265
Sherif, Carolyn Wood, 202–203
Shinn, Milicent, xii, 32–3
Singer, Tania, 249–51
Sklar, Pamela, 276–7
Smith College, 41–2
Smith, Margaret, 30–1
Soh, Debra, 293
Somerfeld-Ziskin, Esther, 175
Soreq, Hermona, 262–3
Sperry, Roger, 221
Spielrein, Sabina, 69–72
Spock, Benjamin, 165
Srivastava, M.V., 281
Stanford-Binet Test, 78, 143, 157
Stanford University, xii, 28, 125, 149–50, 152, 252
Stanton, Hazel, 150–1
Stern, Clara, 130–1
Stokoe, William, 231
Stolz, Lois, 151–3
Strang, Ruth, 158–9
Strange Situation, 93
Structuralism, 29, 45, 172
Syková, Eva, 228–9

Tabrizi, Sarah, 281
Taft, Jessie, 134–5
Tallal, Paula, 263–4
Tangri, Sandra, 220–1
Tanner, Amy, 38–9
TAT testing, 182
Taylor, Shelley, 261–2
Terman, Lewis, 78, 143, 150
Tewari, Sujata, 222
Thompson, Clara, 155–6
Thompson-Schill, Sharon, 287–8
Thorndike, Edward, 74, 136, 160
Thurstone, Thelma, 163
Titchener, E.B., 19, 29, 37, 45
Tolman, Ruth, 15
Town, Clara, 43–4
Treisman, Anne, 218–9
True, Reiko, 217–8, 227
Tum-Suden, Caroline, 171
Turner, Alberta, 194–5
Twitchell-Allen, Doris, 173–4
Tyler, Leona, 190–1

Undurti, Vindhya, 272–3
University of Leipzig, xi

Vassar College, 28, 33, 165, 166
Vega, Star, 266
Vernon, Magdalen, 175–6
Vidal, Catherine, 267–8
Vogt, Cécile, 22–7
Vogt, Marthe, 22–7
Volkow, Nora, 274
Vulpian, Alfred, xi, 6

Warrington, Elizabeth, 213
Washburn, Margaret, 19–21, 166
Watson, John, 20, 165
Weisstein, Naomi, 238–41

Wellesley College, 38
Witmer, Lightner, 44
Wellman, Beth, 159
Whitehurst, Keturah, 197–8
Woolley, Catherine, 282
Woolley, Helen, 45, 152
Working Memory, 242–3, 298–9
Wright, Beatrice, 200
Wundt, Wilhelm, xi, 11, 19, 29, 40, 44, 47, 74, 147, 154, 172
Wyatt, Gail, 229–30

Xu, Fei, 284–5

Zeigarnik, Bluma, 166, 171–3
Zylberlast-Zand, Nathalie, 137